NNFM Editor Addresses

MEGAFLOW - Numerical Flow Simulation for Aircraft Design

Results of the second phase of the German CFD initiative MEGAFLOW, presented during its closing symposium at DLR, Braunschweig, Germany, December 10 and 11, 2002

Norbert Kroll
Jens K. Fassbender (Editors)

Springer

Professor Dr. Norbert Kroll
Dr.-Ing. Jens K. Fassbender
Deutsches Zentrum für Luft- und Raumfahrt e.V. (DLR)
in der Helmholtz-Gemeinschaft
German Aerospace Center
Member of the Helmholtz Association
Institute of Aerodynamics and Flow Technology
Lilienthalplatz 7
38108 Braunschweig
Germany

ISSN 1612-2909
ISBN 3-540-24383-6 Springer Berlin Heidelberg NewYork
ISBN 978-3-540-24383-0 Springer Berlin Heidelberg New York

Library of Congress Control Number: 2004117724

Springer is a part of Springer Science+Business Media
springeronline.com

Printed in Germany

Typesetting: Digital data supplied by editors
Cover design: deblik Berlin
Printed on acid free paper 89/3141/M - 5 4 3 2 1 0

Preface

The aerospace industry is increasingly relying on advanced numerical simulation tools in the early aircraft design phase. Today, under the pressure of economic and ecological requirements, not a single new aircraft development can be done without the intensive support of computational fluid dynamics (CFD). Nevertheless, there is still a great need for improvement of numerical methods, as standards for simulation accuracy and efficiency are constantly rising in industrial applications. Moreover, it is crucial to reduce the response time for complex simulations, although the complexity of relevant geometries and underlying physical flow models are constantly increasing.

In order to meet the requirements of German aircraft industry, the national project MEGAFLOW was initiated some years ago under the leadership of DLR. The main goal was to focus and direct development activities carried out in industry, DLR and universities towards industrial needs. The close collaboration between the partners led to the development and validation of a common aerodynamic simulation system providing both a structured and an unstructured prediction capability for complex applications. The software is constantly updated to meet industrial requirements.

In the first phase of the project (1996-1998) the main emphasis was on the improvement and enhancement of the block-structured grid generator MegaCads and the Navier-Stokes solver FLOWer. In the second phase (1999-2002) activities were focused on the development of the unstructured/hybrid Navier-Stokes solver TAU. Due to a comprehensive and cooperative validation effort and quality controlled software development processes, both flow solvers have reached a high level of maturity and reliability. The MEGAFLOW software is used in the German aeronautic industry and in research organizations for a wide range of applications. At universities the software is used for improvements of physical modeling and investigations of specific flow problems. Due to the use of a common software base, the process of transferring latest research and technology results into production codes has been considerably accelerated.

This volume entitled "MEGAFLOW — Numerical Flow Simulation for Aircraft Design" contains results presented during the closing symposium of the project which took place at DLR in Braunschweig, Germany, on December 10th and 11th 2002. Contributions are from DLR, aircraft industry and several universities. The selected papers focus on the activities of the second phase of the project. They give a good overview of the algorithmic features and physical modeling capabilities of the MEGAFLOW software. The prediction capabilities of the software are demonstrated by several validation test cases and large scale applications for aircraft design.

During the course of the MEGAFLOW project an efficient and open minded German network with partners from universities, research organizations, the aircraft industry and small enterprises has been created. This network has proved of great value for the establishment of numerical simulation as a well recognized and essential tool in the aircraft design process. Based on this network the numerical capabilities for aerodynamic shape design and multidisciplinary optimization will be further developed and improved within the follow-on project MEGADESIGN (2003-2007).

Thanks are due to all partners who have contributed in the context of the MEGAFLOW project in an open and collaborative manner. The knowledge and engagement of each individual contributed to the success and world wide appreciation of the MEGAFLOW project and software.

Furthermore, the funding of partial activities through the German Government in the framework of the air transport research program is gratefully acknowledged.

The editors are grateful to Prof. Dr. E. H. Hirschel as the general editor of the Springer series "Notes on Numerical Fluid Mechanics and Multidisciplinary Design" and also the staff of the Springer Verlag for the opportunity to publish the technical results of the MEGAFLOW project in this series.

Braunschweig,
Oktober 2004

Norbert Kroll
Jens K. Fassbender

Contents

Part VII Exploitation of MEGAFLOW Software

grid generation process is available. These sources are of the following types: geometric (simple 3-D shapes) and CAD based (panels and curves of the geometry).

1.2.2 Surface Mesh Generation

The first step after the geometry data preparation from the CAD package has been processed is to mesh the portions of the surface with triangles or quadrilaterals discretise arbitrary. The unstructured triangular elements provide the most flexibility to model arbitrarily complex geometries. An automatic approach is used to generate the unstructured surface mesh. The generation is robust and provides the user complete control over point placement with user specified clustering via sources, an automatic clustering based on geometric features (e.g. special geometric features, curvature or proximity of different surfaces), and an automatic quality control with certain smoothing steps.

1.2.3 Prismatic and Hexahedral Grid Generation

The next step after the surface mesh generation for Navier-Stokes grids is to march nodes away from the body surface to create prismatic or hexahedral elements. If no viscous features should be resolved, the tetrahedra generator (see next section) starts immediate from the surface mesh.

The structured nature of the prisms or hexahedral in the normal to surface direction exploits the advantages of traditional structured grid approaches. In particular, this grid offers good orthogonality and clustering capabilities. The algebraic marching procedure uses an automatic detection of gaps and cavities for a reduction of the marching step to avoid grid overlap. During the generation of prismatic and hexahedral elements various grid validity checks (e.g. aspect ratio, twist) are accomplished and if necessary an automatic improvement by an iterative process is used.

The *Centaur* grid generator also provides the added feature of chopping prismatic layers in certain regions and create transition elements such as pyramids and tetrahedra. This is useful to to treat non-manifold configurations, to handle zero-visibility surface topologies and improve overall grid quality. Finally the user can adjust the generation parameters locally via sources (e.g. initial wall spacing, number of layers, stretching and adjacent cell size variations).

1.2.4 Tetrahedral Grid Generation

The remainder of the computational domain is filled with tetrahedral elements because of the ease with which these elements fill arbitrary volumes in a single block. Tetrahedra also possess good grid adaptation qualities, which can be used to better resolve the features of interest. The tetrahedra are generated via an advancing front method beginning with triangular or quadrilateral faces of outer prismatic or hexahedral-layers and matching the local prismatic or hexahedral thickness. If necessary an automatic local regeneration is used to improves robustness and quality of the tetrahedral grid generation. The user has control over the maximum tetrahedron size, the stretching and via sources an user specified clustering is available. After the tetrahedral grid generation an automatic grid quality enhanced through post processing measures (e.g. volume ratios, slivers/skewed elements) is implemented.

1.2.5 Overall Hybrid Grid

The prismatic and tetrahedral grids are combined into one hybrid grid which can then be used for simulations. As described previously, hybrid grids offer several advantages over traditional grid generation approaches. For example, the grid requires no special solution interpolation or grid interfacing techniques. The semi-structured prismatic layers exploit the advantages of structured grids and the unstructured tetrahedra allow for meshing complex domains in a single block. Besides the obvious advantages of hybrid meshes, the *Centaur* system offers a broad collection of features, that partly described in section 1.4.

1.2.6 Hybrid Mesh Adaptation

An integral part of the generator is a dual mesh adaptation capability which employs both local refinement of the hybrid mesh, and local redistribution of the prisms close to wall surfaces. The adaptation is guided by flow features sensors based on local flow parameters spatial variations and by the user who can define regions to be refined.

1.3 Cooperation CentaurSoft with DLR and Airbus-Deutschland

Since three years a cooperation is established between CentaurSoft, DLR and Airbus-Deutschland. The goal is to enhance hybrid mesh generation to handle realistic aircraft simulations with the MEGAFLOW applications at DLR and Airbus. CentaurSoft collaborates with DLR and Airbus on defining important enhancements and receives feedback on effectiveness of the enhancements to the hybrid mesh capability. Further on CentaurSoft supports the users at DLR and Airbus using *Centaur* grid generator.

The major areas of the cooperation are an improved grid quality (e.g. hexahedral, fewer chopping), an improved user interaction for mesh generation, more automatic assistance with CAD issues for industrial aircraft geometries or the handling of wakes. Beyond this the generation of "MegaMeshes" for complex aircraft configurations is a major task, which can be divided into subtasks like the reduction of the number of elements (e.g. via mesh adaptation, improved sources for more control of local grid resolution or unstructured hexahedra), a general speedup, parallel mesh generation and regarding the computing environment a RAM memory and disc storage reduction to allow to handle such "MegaMeshes".

In the following examples of using *Centaur* grids at DLR and Airbus-Deutschland and developments following from the cooperation between CentaurSoft, DLR and Airbus are presented.

1.4 Examples and Developments using *Centaur* at DLR and Airbus-Deutschland

1.4.1 2D Hybrid Unstructured Mesh Generation

Beside the 3D "MegaMeshes" there is still a necessity of 2D hybrid meshes for daily research or industrial applications. The grid generator *Centaur* allows the creation of two dimensional hybrid meshes consisting of structured quadrilaterals in the boundary layer and unstructured triangles remainder of domain. It uses the same user interface and input file formats and supports the most of the 3D features like periodic meshes, wake meshes (see in the following sections) and many kinds of sources to control the grid resolution.

In Fig. 1.2a a so called gourney flap (a device at the trailing edge of the flap) was investigated at Airbus to improve the overall lift of a high lift configuration with simple devices. The grid has $76 \cdot 10^3$ points and was build in the ProHMS research project. In Fig. 1.2b a hybrid unstructured grid around a high lift configuration with spoiler is shown. This grid with $66 \cdot 10^3$ points was used to investigate the loads on an Airbus-A340 in landing condition.

1.4.2 Anisotropic Surface Elements

A flow field is often dominated by a gradient in one direction, for example a wing with dominating gradients of the flow in flow direction, whereas the gradients in spanwise direction are small and barely influence the flow behavior. This effect can be exploited to reduce the number of points in a grid by stretching the cells in direction of low gradients. Using block-structured grids this anisotropic stretching can reach a factor of 20 with a not negligible points saving. This value cannot be reached using hybrid unstructured meshes because of problems between at the interface of the prismatic/hexahedral and tetrahedral elements. Here a typical value of the anisotropy is about $2-3$, which leads to a typical point saving of 50% for a complex high lift configuration.

For example in Fig. 1.3a the surface grid for the M6 wing is shown with isotropic cells, whereas in Fig. 1.3b the same grid with anisotropic stretching in spanwise direction is depicted. The resolution in chordwise direction is the same for both grids, in spanwise direction the anisotropic grid has a lower resolution leading to the same aerodynamic results. Because the prismatic/hexahedra grid is build using a marching algorithm, the stretching is continued normal to the surface. In the tetrahedra the anisotropy disappears.

The direction of the anisotropy is controlled by the user by anisotropic line or panel sources. For complex configurations the anisotropic panel sources are mostly used because they allow an alignment of mesh elements with the local parametric surface directions and can be used for aligning elements in the middle of large panels without affecting neighboring panels (e.g. anisotropic wing and isotropic pylon and engine).

1.4.3 Hexahedral Meshes

Another technique to reduce the number of points and elements is the usage of quadrilaterals instead of triangles on the surface respective hexahedral instead of prismatic elements in the boundary layer. In the *Centaur* grid generator two kinds of quadrilaterals can be build: structured and unstructured.

The first type is restricted on panels with four sides, see Fig. 1.4a on the leading edge. They are aligned with the two directions of the panel and with it improving the solution quality if the panel is aligned with the flow direction. For complex configurations this cannot be assumed for all panels of the geometry and there are often panels with more or less then four sides.

In this cases the unstructured quadrilaterals can be used on user specified panels. In Fig. 1.4b for example the DLR ALVAST high lift configuration [3] is shown with a complete quadrilateral surface grid, which leads to hexahedral elements in the boundary layer a tetrahedra in the remaining domain.

1.4.4 Grid Quality Improvement

A key feature to obtain realistic flow simulations is besides the flow solver a high quality grid. This means that all features of the geometry are resolved in the surface grid (location and size of automatic/manual sources), ideally no chopping in hybrid unstructured grids and general good quality elements (e.g. low sliver, edge length ratio, face dihedral, skewness or aspect ratio).

The first goal is mainly reached by user control and likewise by automatic CAD depending sources. The second goal, reduced chopping, was one of the work packages in the last three years between CentaurSoft, DLR and Airbus. For example for a complex cruise configuration its last prismatic layer is shown in Fig. 1.5a. The areas, where chopping occurs are highlighted by brighter spots and iso-lines. The same configuration meshed with a current version of *Centaur* is shown in Fig. 1.5b. In this case no chopping occurs and so the improvements of the grid generator leads to an improved grid quality.

Concerning to the third goal the elements of the grid are improved in the different grid generator stages. Especially the prismatic/hexahedral elements

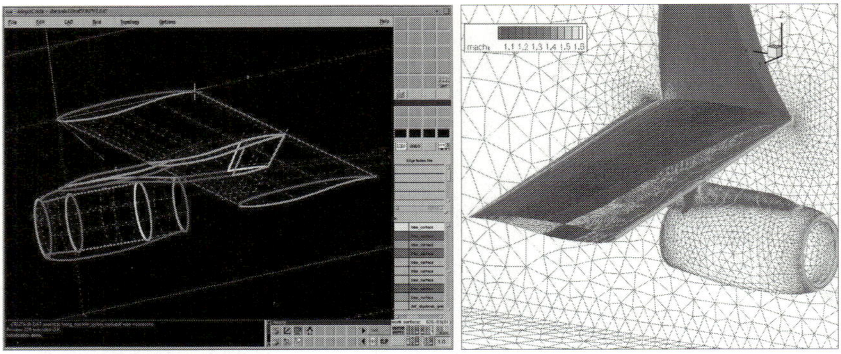

(a) (b)

Fig. 1.7. Connection between MegaCads and *Centaur* on example of Hyperflex wind tunnel model: a) MegaCads b) Centaur/DLR TAU.

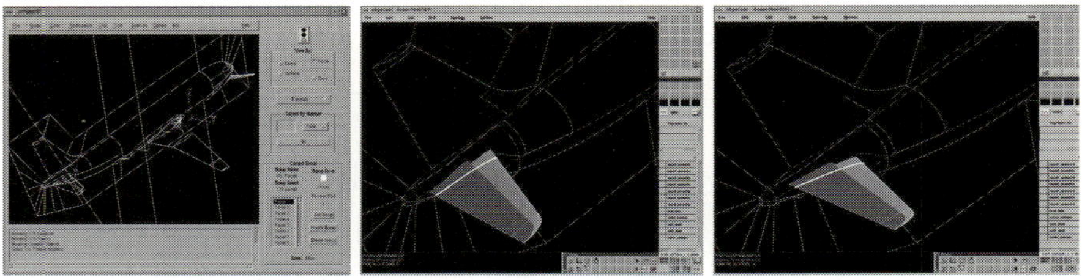

(a) (b) (c)

Fig. 1.8. 3FF geometry of DLR: a) geometry, b) initial elevator setting in MegaCads and c) changed elevator setting.

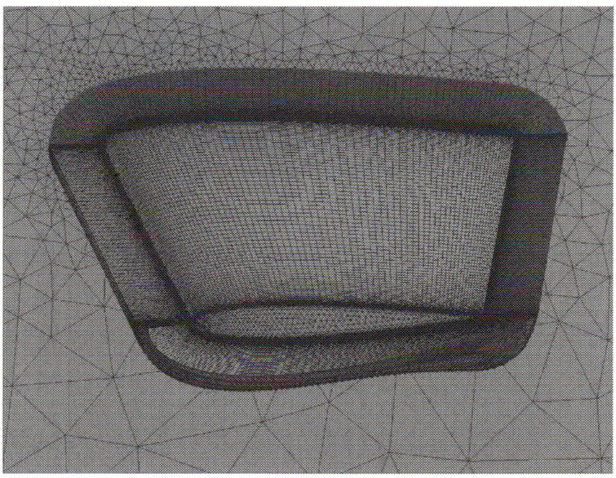

Fig. 1.9. Semi-structured Navier-Stokes grid for M6-wing.

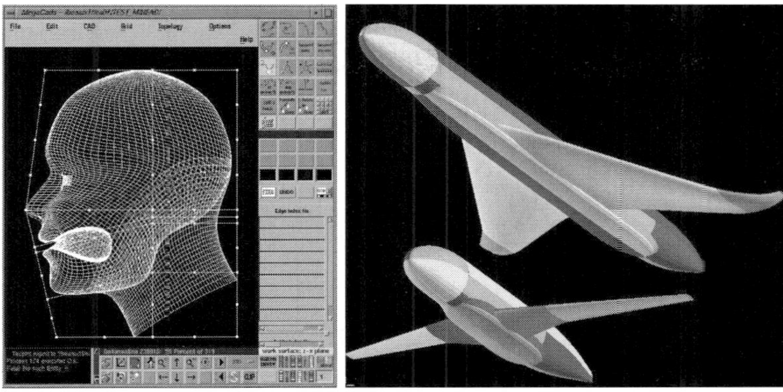

(a) (b)

Fig. 1.10. Free-form deformation a) implemented in MegaCads and b) modification of an aircraft geometry.

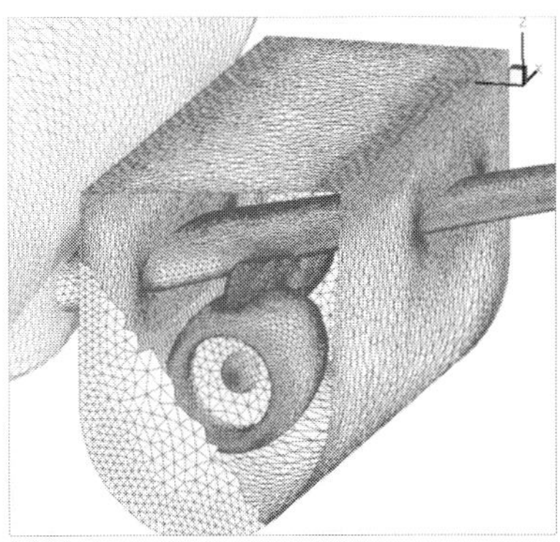

Fig. 1.11. Modular grid generation for complex aircraft configuration.

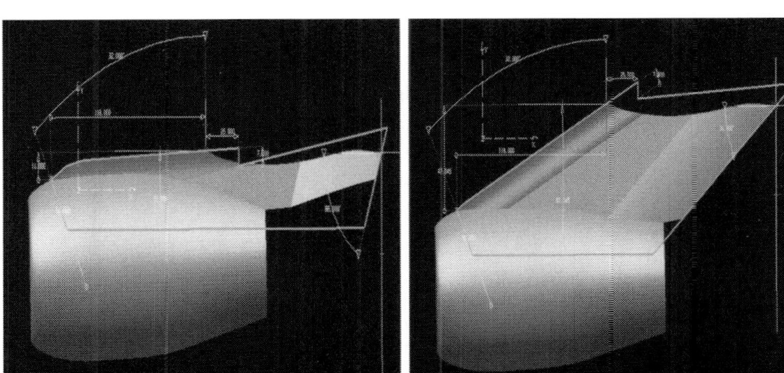

(a) (b)

Fig. 1.12. Pylon shape modification in CAD system for modular grid generation.

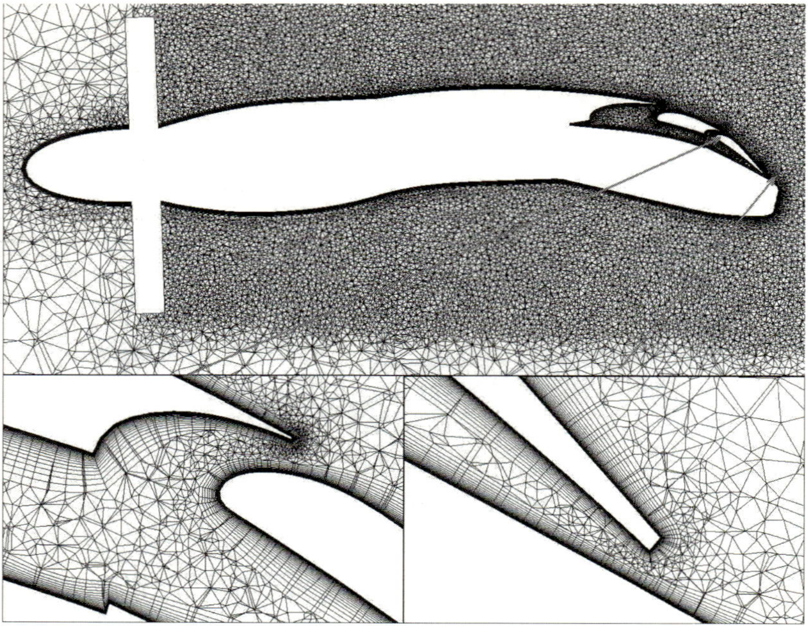

Fig. 1.17. Cut through nacelle and high lift system at A400m.

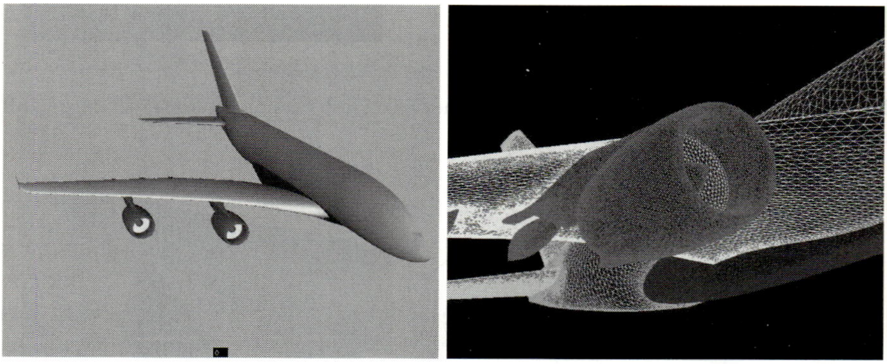

(a) (b)

Fig. 1.18. Airbus A380 civil aircraft: a) clean configuration and b) detail view on surface grid.

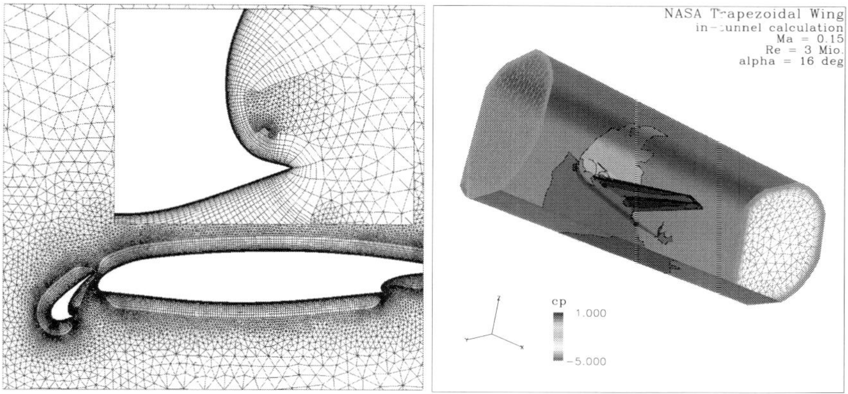

(a) (b)

Fig. 1.19. 2D 3 element configuration b) generic high lift configuration in wind tunnel.

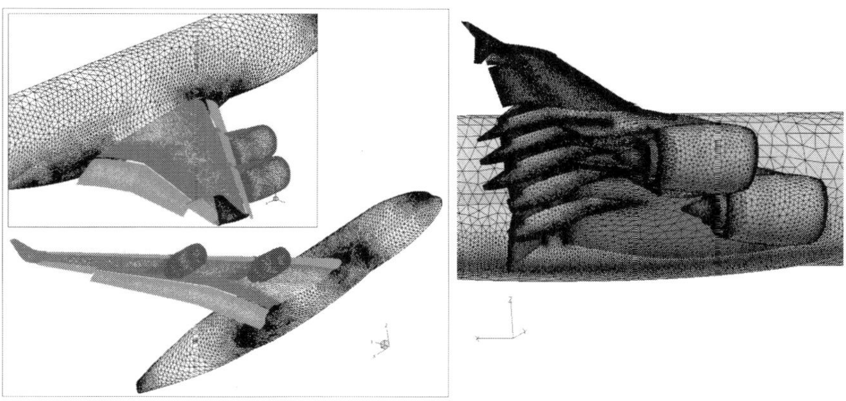

(a) (b)

Fig. 1.20. Surface grid for complex high lift configurations.

2

Block Structured Navier-Stokes Solver FLOWer

Jochen Raddatz and Jens K. Fassbender

DLR, Institute of Aerodynamics and Flow Technology, Lilienthalplatz 7, D-38108 Braunschweig

Summary. This paper comprises the developments of the block structured Navier-Stokes solver FLOWer during the DLR project MEGAFLOW II. At first the status of the FLOWer code including its numerical and physical capabilities at the end of MEGAFLOW II is presented. After introducing the objectives of the FLOWer development for MEGAFLOW II the major developments and corresponding results are discussed.

2.1 Current Status

2.1.1 General Overview

The FLOWer-Code solves the three dimensional compressible Reynolds averaged Navier-Stokes (RANS) equations in integral form in the subsonic, transonic and supersonic flow regime. Turbulence is modeled by either algebraic or transport equation models. The code is well adapted to the simulation of exterior flow fields around aircraft-like configurations. Furthermore, FLOWer includes an inverse design option, allowing the inverse design of airfoils and wings based on prescribed pressure distributions. A special variant of the FLOWer code is able to solve the adjoint equations [1]. This variant is used for aerodynamic shape optimization to calculate the gradients independently of the number of design variables [2].

The code is developed by a joined team from DLR, universities, Airbus Germany and EADS-M. The software management, including integration work and quality assurance, is performed at DLR.

Besides the MEGAFLOW projects other projects contributed to the development of the FLOWer code as well. A major work package has been the extension of the applicability of FLOWer to helicopter simulations within the CHANCE project [3].

4. Extension of the block interfaces to the treatment of hanging grid nodes.
   The ability of a code to use grids with non-matching cut simplifies the grid generation considerably. Additionally, it allows the automatic generation of adapted background grids for Chimera computations with a low number of grid points. Details will be described in chapter 2.3.3.
5. Improvement of Chimera technique for use of overset grids.
   This work package has been a joined effort together with the CHANCE project [3]. For more details see chapter 2.3.4.
6. Implementation of advanced turbulence models and improvement of existing models.
   A variety of turbulence models is now implemented in FLOWer, ranging from simple algebraic eddy viscosity models over one- and two-equation models to the nonlinear EARSM of Wallin [13]. For improved prediction of vortical flows enhancements are achieved combining the two-equation models with Kok's modification [12].
   Besides the modeling accuracy for turbulent flows, the numerical robustness of the respective transport equation turbulence models for complex applications has been a major issue. Results are presented in [8].
7. Improvement of the transition prescription technique.
   A general transition prescription technique has been implemented, which is e.g. able to set the transition lines of slat, main wing and all flaps of any high lift device as required [14].
8. Coupling of FLOWer with an e^{N} database for transition prediction.
   Besides the accurate modeling of turbulence, the correct prediction of the transition point or line is one of the major issues for the improvement of prediction accuracy. The procedure implemented in FLOWer during MEGAFLOW II is based on an e^{N} database method and works for 2-D cases and 3-D wing computations. Further details concerning transition prescription and transition prediction in FLOWer are given in [14].

2.3 Developments and Results

2.3.1 Upgrade of Robustness and Efficiency

The improvements of FLOWer concerning robustness and efficiency will be demonstrated by the ALVAST start configuration, which was the last FLOWer milestone of the MEGAFLOW I project. Fig. 2.1 shows the mesh, pressure distribution and convergence history of two computations. The computation with FLOWer 115 has been performed at the end of the MEGAFLOW I project. At that time, the parallel mode and the multigrid acceleration technique did not work for all test cases. Therefore, this computation could only run in sequential mode and without multigrid acceleration on a NEC / SX4 vector computer. On this computer FLOWer performs 40 singlegrid iterations

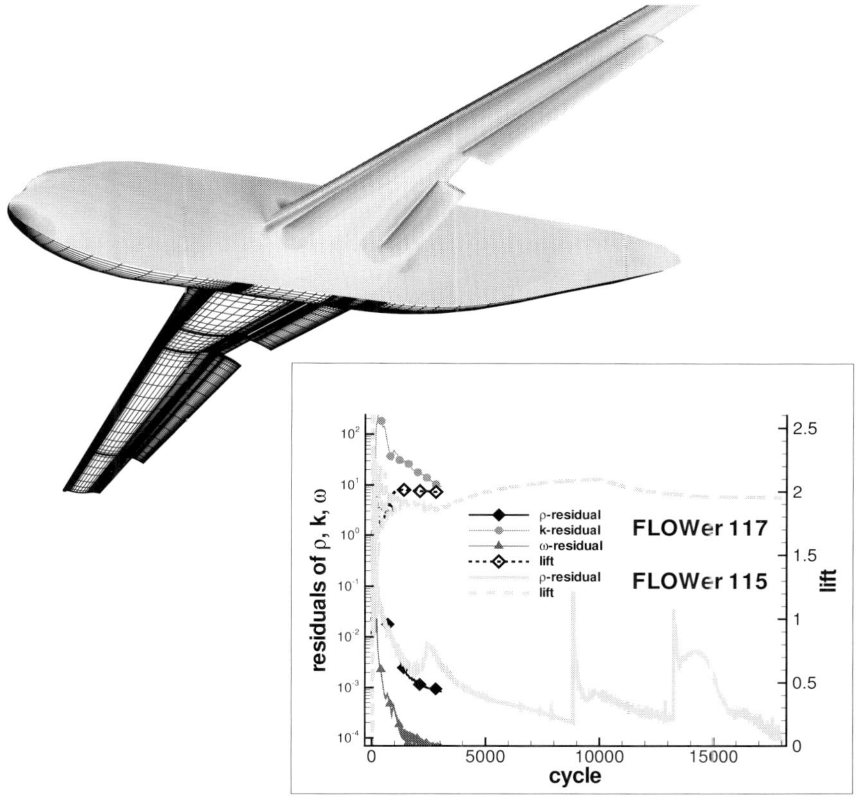

Fig. 2.1. Example for improvements of robustness and efficiency — ALVAST start configuration (8.2 million cells, 48 blocks; Ma = 0.22, Re $= 2.0 * 10^6$, $\alpha = 12.03°$)

per hour for this test case. Overall computing time was about 450 hours, resulting in a turn around time of 3 weeks including adaptation of the control parameters. Performing the same computation on a NEC / SX5 the cpu-times reduce to the half, that means 225 hours total computing time and about 10 days turn around time.

Nowadays, using the current FLOWer release 117, multigrid and preconditioning are available and the number of cycles can be reduced from 18000 to 2000. Additionally, the required time per cycle has been decreased. FLOWer 117 computes 90 multigrid cycles per hour on the NEC / SX5 in sequential mode, which adds up to a overall computing time of 22 hours for the ALVAST case. If the computation runs in parallel mode on four CPU's, a wall clock time of only 7 hours is required.

In general FLOWer 117 needs no special adaptation of the control parameters. This allows better comparison of similar cases, as identical control parameters could be used. Additionally, the turn around time for a computation may be reduced considerably as standard parameter sets may be used for the most applications.

2.3.2 Implicit Scheme for Turbulence Equations

In Fig. 2.2 the convergence behaviour of turbulent flow computations for the RAE 2822 airfoil over a range of Reynolds numbers is shown. The computational grids have been equipped with Reynolds number adapted boundary resolutions. The convergence of the RANS equations is represented by the density residual. For Reynolds numbers up to $20 \cdot 10^6$ convergence is achieved but not to machine accuracy. Approaching even larger Reynolds numbers no convergence is obtained at all.

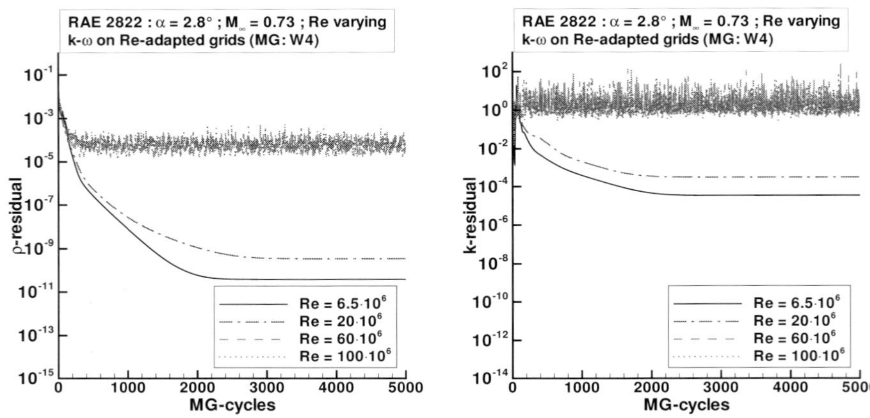

Fig. 2.2. RAE 2822: Convergence of density residual (left) and k residual (right) for varying Reynolds number; multigrid acceleration used for all equations (RANS + k-ω)

Since the k-ω turbulence model of Wilcox [9] has been used for these computations the k residual exemplarily stands for the convergence of the turbulence equations. It shows in principle the same behaviour as the density residual. Turbulence equations have been treated by means of the same time integration approach as the RANS equations: A 5-stage Runge-Kutta method is sped up by local time stepping, implicit residual smoothing (IRS) and multigrid. Neither the time step nor the coefficients used for IRS were adjusted for a more special treatment of the turbulence equations. Also the multigrid method applied is the same for RANS and turbulence equations. Only a point implicit treatment of the turbulence source terms is incorporated in the Runge-Kutta method.

A systematic and comprehensive investigation on the numerical treatment of turbulence equations utilizing computations and Fourier analyses has been performed. The objective of this investigation has been to find out reasons for robustness problems encountered when approaching flight Reynolds numbers [15]. The multigrid treatment of (productive) turbulence source terms has been identified to counteract the improvements in damping of low-frequency error modes usually gained by multigrid. Thus a singlegrid treatment for the turbulence equations has been deduced to be most robust in combination with an unchanged multigrid treatment for the RANS equations. Corresponding results are shown in Fig. 2.3.

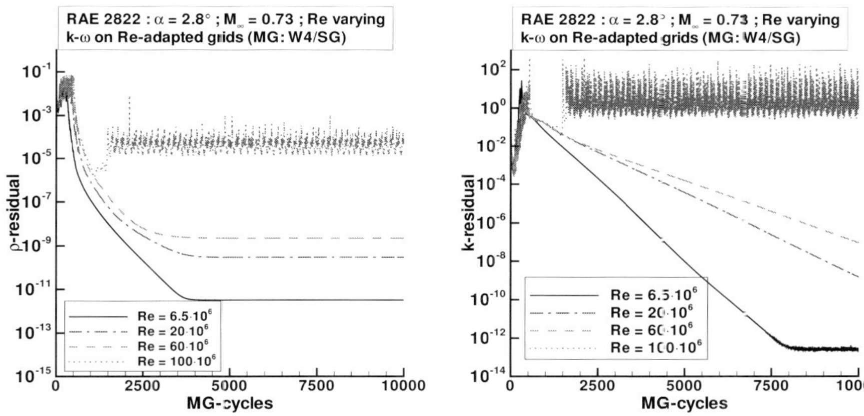

Fig. 2.3. RAE 2822: Convergence of density residual (left) and k residual (right) for varying Reynolds number; multigrid acceleration used only for RANS equations, k-ω treated only on finest grid

The convergence of the density residual is slowed down by a factor of about less than 2 if the multigrid treatment for the turbulence equations is foregone. Still the density residual does not reach machine accuracy. Despite this the robustness of the complete method is improved as the computation at $\mathrm{Re} = 60 \cdot 10^6$ converges only based on a singlegrid treatment of the turbulence equations. Moreover, the turbulence equations, here also represented by the k residual, converge to machine accuracy in case of omitting their multigrid treatment.

In order to increase robustness and to speed up the time integration of the turbulence equations a fully implicit time integration method has been applied to the turbulence equations. The resulting left hand side matrix is approximatively factorized with a diagonal dominant alternating direction implicit (DDADI) method [16]. The DDADI method differs from the standard ADI approach by employing identical main diagonals for all coordinate directions i. e. the main diagonals are independent of the coordinate direction. Since the

k-ω equations are coupled via the source term only, the source term derivative matrix which contributes to the left hand side matrix is diagonalized in the same way as it is used within the point implicit treatment of the source terms. By this decoupling of the turbulence equations scalar tridiagonal matrices have to solved for each turbulence equation separately. This approach – multigrid treated RANS equations loosely coupled with turbulence equations which are implicitly treated on a singlegrid basis – yields superior convergence behaviour for the density residual and for the turbulence residuals as shown in Fig. 2.4.

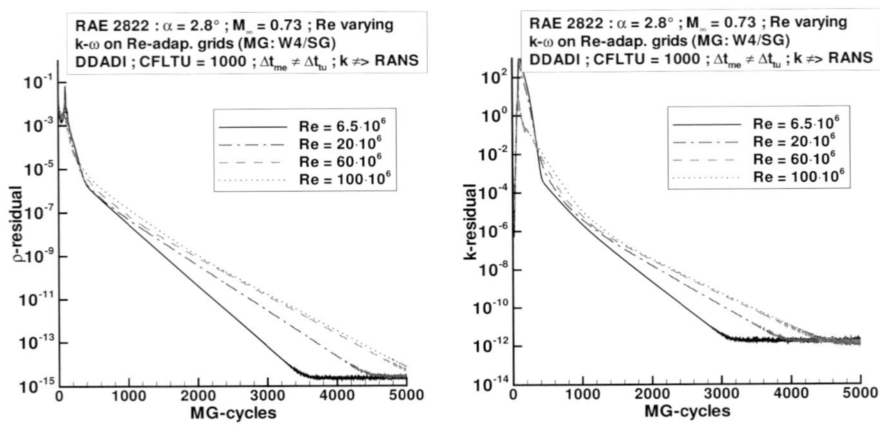

Fig. 2.4. RAE 2822: Convergence of density residual (left) and k residual (right) for varying Reynolds number; multigrid acceleration used only for RANS equations, k-ω treated only on finest grid employing implicit time integration scheme

These computations include several minor improvements to the numerical treatment of the turbulence equations. These are presented in detail in [15]. The resulting robustness enables convergence to machine accuracy within about 5000 iterations for all residual at all Reynolds numbers investigated.

The gain in robustness based on the implicit treatment of the turbulence equations may allow additional speed-up of the convergence of the density residual. The computations shown in Fig. 2.1 and discussed in chapter 2.3.1 differ in the number of multigrid levels used for the computation of the RANS equations: With FLOWer 115 only a singlegrid computation has been possible due to robustness problems otherwise. The DDADI scheme for the turbulence equations implemented in FLOWer 117 enabled the use of all multigrid levels of the computational grid for the treatment of the RANS equations.

2.3.3 Extension of Flexibility and Application Range by use of Block Interfaces with Hanging Nodes

A simplification in the grid generation process was achieved with the implementation of patched grid interfaces with hanging nodes. An example is shown in Fig. 2.5. In order to calculate a cooling fan or propeller it is sufficient for

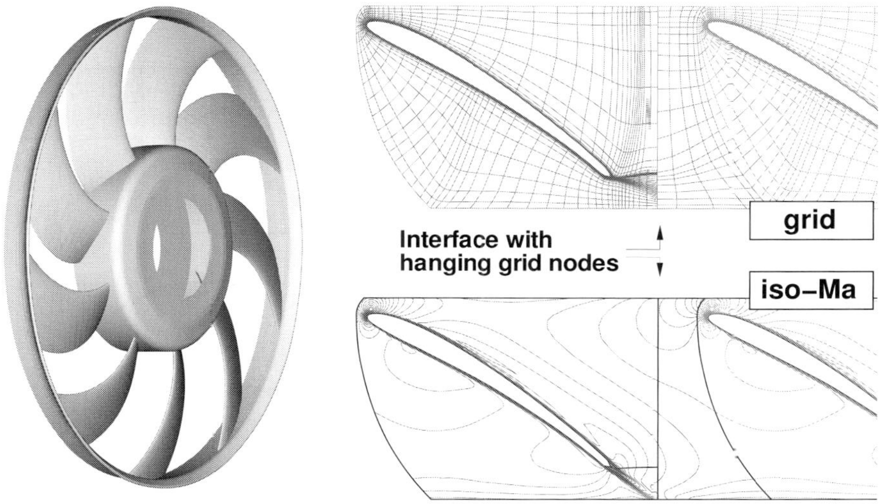

Fig. 2.5. Hanging nodes at block interfaces - cooling fan application

many cases to compute the flow around one blade. The effect of the other blades is taken into account by periodicity boundary conditions. Major difficulty of this boundary condition has been that the block boundaries belonging together, i. e. the boundary in front of the fan blade and the boundary behind the blade, had to have the same shape and an identical point distribution. If meshes of a C-type are used the second requirement is nearly impossible to fulfill. With FLOWer, this problem can easily be solved by use of the patched grid interfaces. The interfaces are currently not flux conservative but this will be changed in near future.

Another application of patched grid interfaces is the use for local grid refinement. A 2-D example is shown in Fig. 2.6. In the present stage of development the local grid refinement of structured grids is performed by an external tool. It works fully automatically without any user input and generates multiblock grids with hanging nodes on the new block to block interfaces. The sensors currently implemented for the detection of regions to be refined work on gradients or vorticity. In Fig. 2.7 a more advanced application of local grid refinement is given. Here the refinement is used to improve the rep-

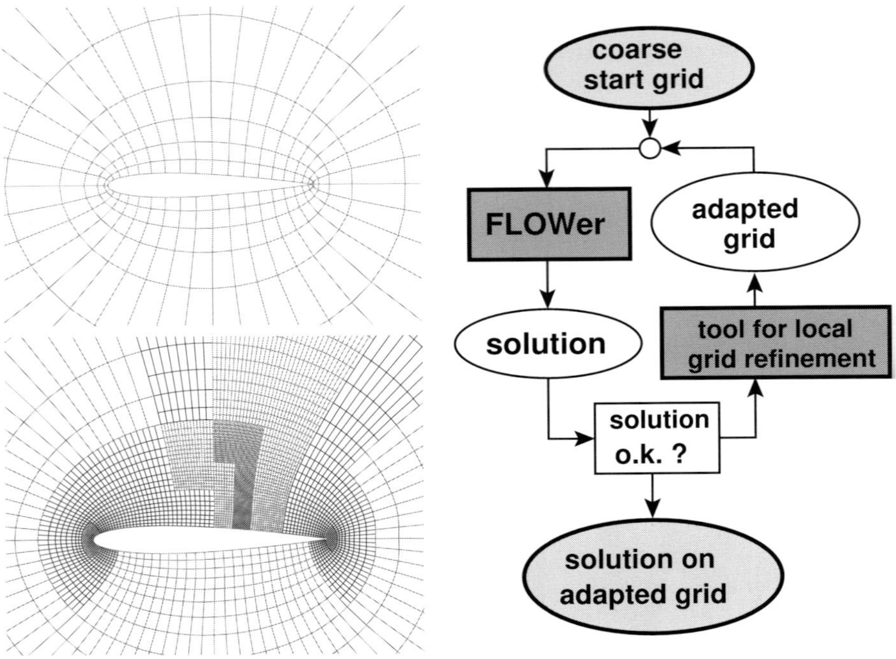

Fig. 2.6. Local grid refinement in FLOWer

resentation of the tip vortices of a helicopter rotor under hover conditions.

2.3.4 Extension of Flexibility and Application Range by use of Chimera technique

A considerable extension of the application range was achieved with the implementation of the Chimera technique [17] into the FLOWer code. This approach allows the grid blocks to overlap each other. The communication from mesh to mesh is realized through interpolation in the overlapped area. Hence, searching algorithms have been implemented to locate the donor cells. In the case when a mesh overlaps a body which lies inside another mesh, hole cutting procedures have to be used in order to exclude the invalid points from the computation. Roughly speaking, the Chimera technique consists of an overlapping boundary condition and a masking condition for the points that lie inside a body.

FLOWer has the full Chimera functionality [18]. Flow data are transferred by trilinear interpolation. The search for cells, required for interpolation, is performed by an Alternating Digital Tree (ADT) search method [19]. In order to mark points being inside a solid body, a simple auxiliary grid which encloses

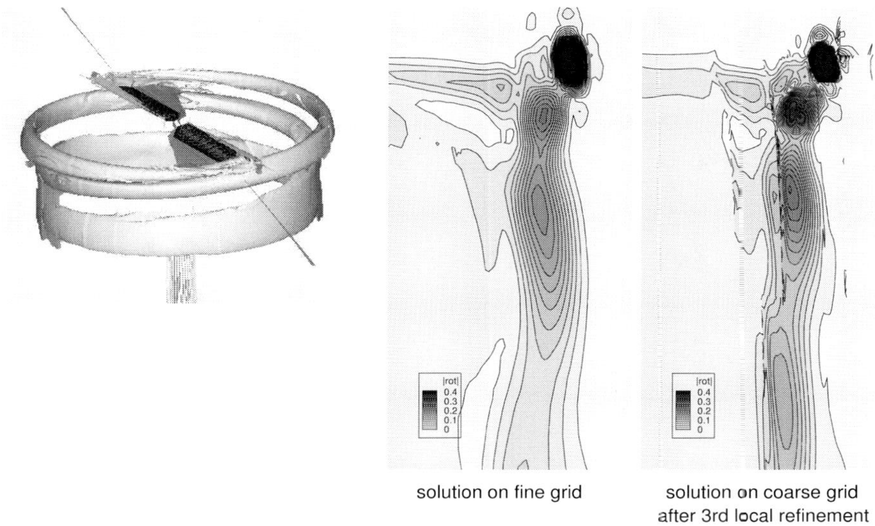

Fig. 2.7. Improvement of vortex capturing using local grid refinement: vorticity distributions from an Euler calculation of a 2-bladed model rotor in hover (fine grid: 2.35 million cells; refined coarse grid: 1.85 million cells)

the solid body must be provided by the user. All points of the grid inside the auxiliary grid are excluded from the flow calculation. Parallelization of the Chimera method is currently underway.

The Chimera approach allows an efficient treatment of bodies in relative motion, e.g. the computation of the flow around a helicopter in forward flight, as shown in Fig. 2.8. An example concerning flexibility enhancement in FLOWer due to the Chimera approach is presented in Fig. 2.9. The task for this application was the investigation of the impact of a spoiler on a given aircraft wing. In order to avoid the generation of a complete new mesh, only a component grid around the spoiler has been made and added to the existing wing-body mesh using the Chimera technique. Additionally, this approach opens an easy and fast possibility to investigate the best position and the optimal shape of the spoiler device. On the other hand the use of overset grids can be used to simplify the whole grid generation of complex configurations considerably. The basic idea is to decompose the configuration into simple components. Only near-field grids have to be generated for each component without having to take care about the other components, as presented in Fig. 2.10. The sole requirement is that the grids around the components do overlap each other. All component grids are placed inside a simple background grid which covers the whole computational domain. If some grid points are placed inside a solid body of a neighboring component grid, they are marked and excluded from the flow calculation.

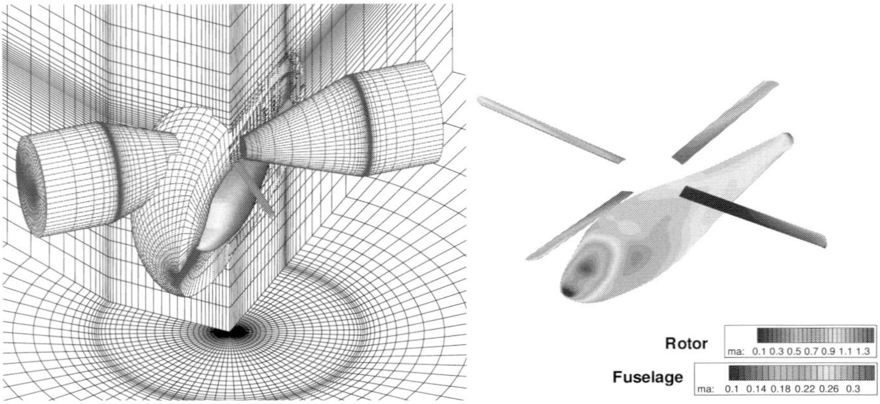

Fig. 2.8. Chimera grid and surface Mach number distribution of a helicopter in forward flight

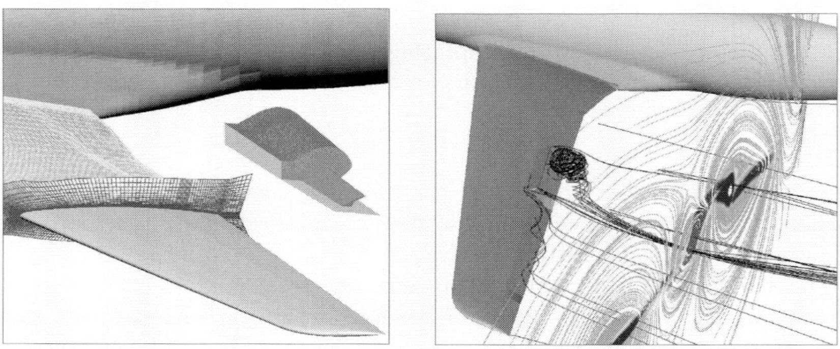

Fig. 2.9. Increased flexibility in applications due to the Chimera approach (example: adding a spoiler device to an existing wing-body mesh)

In order to further simplify the grid generation procedure, a fully automatic cartesian grid generator has been developed. The grid generator places fine grids around the component grids and puts successively coarsened grids around the fine grids. Patched grid interfaces with hanging nodes are used at the interfaces between the grid blocks of the cartesian mesh. In the vicinity of the configuration, the cartesian grid generator creates anisotropic (non cubic) cells, which are adapted to the size of the cells in the component grids. This ensures accuracy in the overlap region. Fig. 2.11 shows the cartesian background grid for the EUROLIFT wing-body configuration. The background grid consists of 1.56 mio cells, located in 300 blocks. Its works fully automatic and requires about 200 seconds on a SGI workstation.

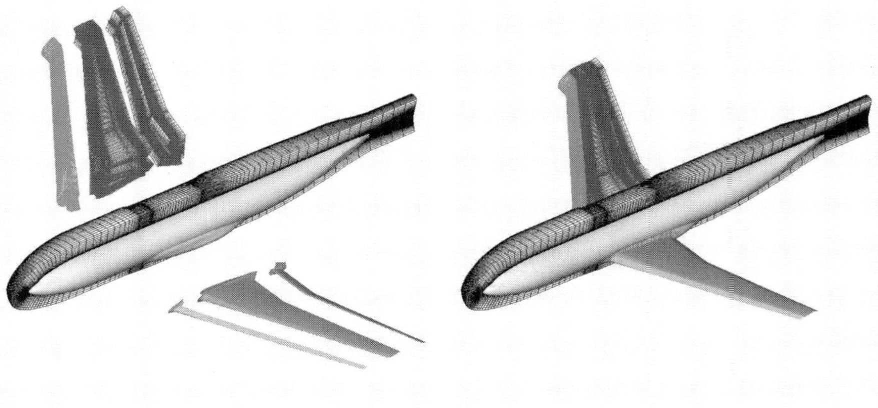

Fig. 2.10. Simplification of grid generation based on Chimera approach (example: EUROLIFT configuration with 33 blocks and 2.3 mio cells)

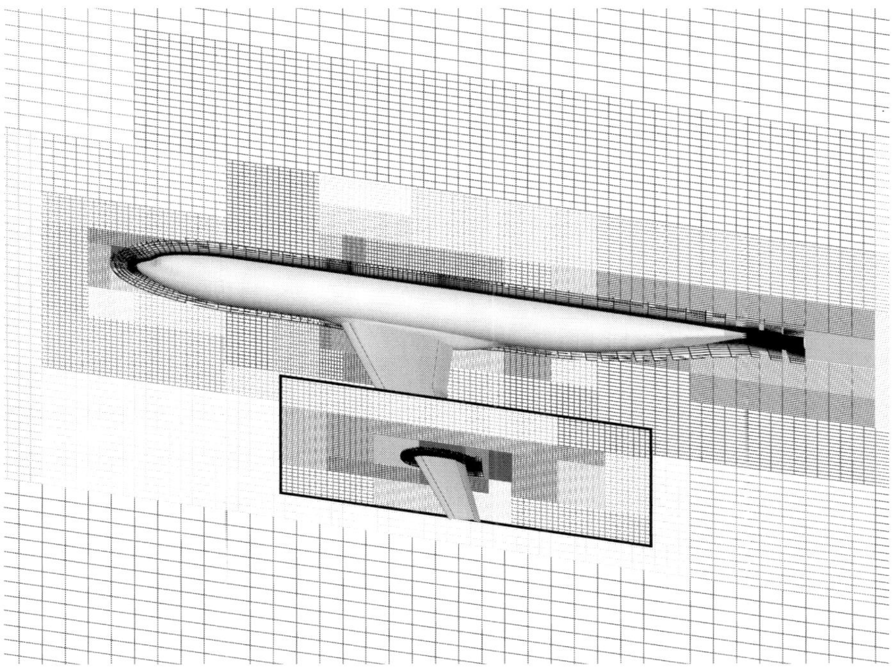

Fig. 2.11. Automatically generated and adapted cartesian background grid

The development work of the Chimera approach in FLOWer in the last years was a joined effort of the projects CHANCE and MEGAFLOW II.

CHANCE [3] is a common French-German project of DLR, ONERA and helicopter industry, which is dedicated to the adaptation and validation of CFD codes for helicopter aerodynamics.

The following development work in the Chimera field has been performed in FLOWer:

- Simplification of user input.
- Introduction of an Alternating Digital Tree search algorithm (ADT) to improve performance and robustness.
- Development of an accurate interpolation technique for boundary layers. The impact of this improvement is demonstrated in Fig. 2.12. The figure presents the bottom view of an helicopter fuselage including the wind tunnel strut, which is introduced into the fuselage grid as separate block structure using the Chimera approach. If the "standard interpolation" is used the computed streamlines in the boundary layer are interfered due to the Chimera boundary. Using the improved interpolation technique the streamlines are computed correctly, even within the region of overlapping grids.

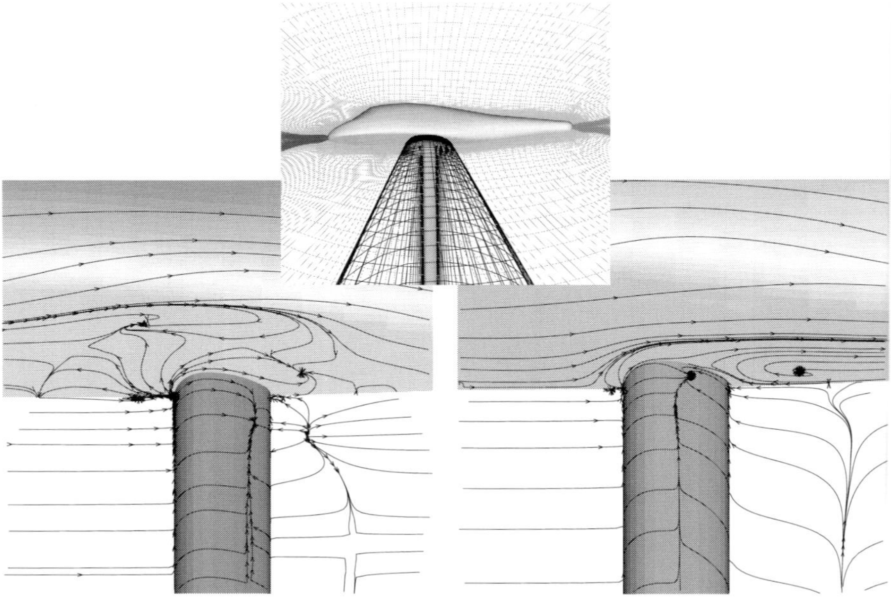

standard interpolation improved interpolation

Fig. 2.12. Interpolation inside boundary layers

- Introduction of a method for the correct calculation of global aerodynamic coefficients (C_L, C_D, etc.) on surfaces with grid overlap, as shown in Fig. 2.13. To get the correct coefficients the overlapping zone is removed for the force calculation.

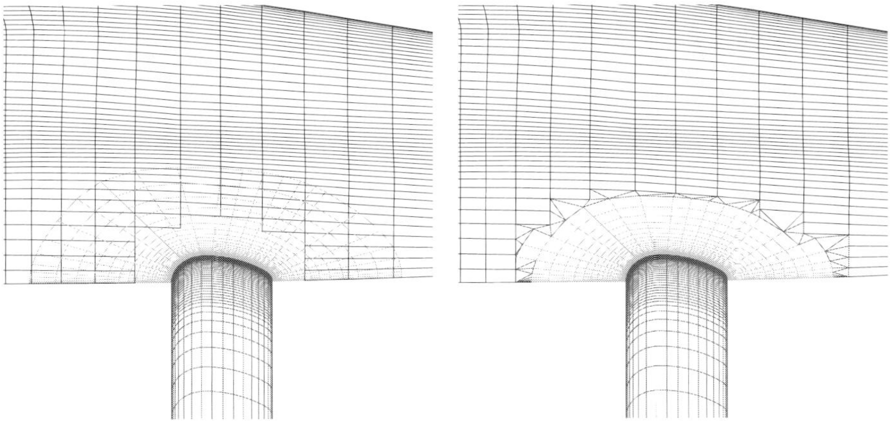

grid with surface overlap unique surface

Fig. 2.13. Remeshing for calculation of global aerodynamic coefficients

- Parallelization of the Chimera approach. The use of overlapping grids in parallel mode is already available but needs further improvements to increase the parallel speed-up.

2.3.5 Unsteady Applications

Standard method for time accurate applications in FLOWer is the "Dual Time Stepping Method". For details see [20] and [21]. It is combined with a moving grid approach, which allows all translational and rotational degrees of freedom. The use of overset grids and flexible grids is possible. Improvements in the last years consider the extension of the dual time stepping scheme to 3rd order time discretization and the introduction of an automatic time step adaptation technique [22].

Using unconditionally stable implicit methods for time accurate computations, like the dual time stepping scheme, the time step can be solely determined by the flow physics. It is not limited by stability requirements as in explicit methods. The problem is however, to determine what is the time step value in accordance with the physics. Based on the algorithm presented in [22] a method has been developed which automatically determines the physical time step in dependence of the truncation error of the time derivative. An example is given in Fig. 2.14. The figure shows evolution of the truncation error and the time step over the time for a calculation of an oscillating airfoil.

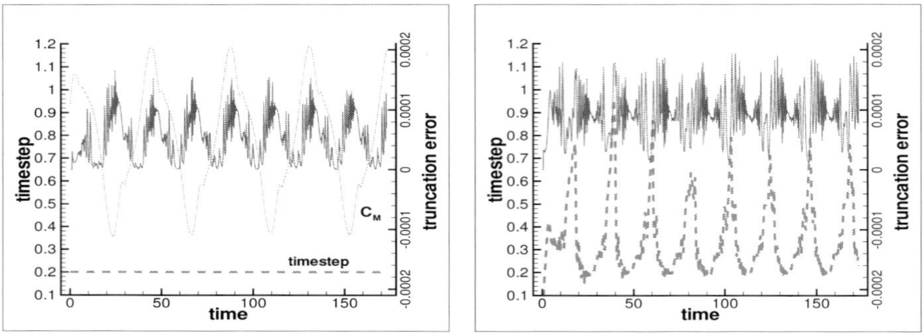

Fig. 2.14. Evolution of time step and truncation error for the computation of an oscillating NACA 0012 airfoil, in case of a constant time step (left side) and for an adapted time step size (right side)

Using an adapted time step size results in a truncation error comparable to the truncation error achieved in case of a constant time step size. On the other hand the adapted time step is sometimes up to 5 times larger than the constant time step necessary for a sufficient resolution of the oscillation. Thus it is expected to reduce the number of time steps to be computed by adapting the time step size. Moreover, the main achievement of this technique to adapt the physical time step for time accurate applications is that the user no longer needs to figure out a suitable time step size for the resolution of all relevant but possibly unknown physical phenomena in his application.

2.4 Summary

Within MEGAFLOW II emphasis related to the development of the FLOWer code has been laid on the improvement of its robustness and efficiency, on the extension of its flexibility and its application range and on the modeling of turbulent flows. For more details on turbulence modeling and transition prescription and prediction techniques it is referred here to the related sections in this volume. A major part of the improved robustness and efficiency is gained by the development of a DDADI scheme for the treatment of turbulence equations. The introduction of block interfaces with hanging nodes extended the flexibility of FLOWer considerably. The application range of FLOWer has been enhanced by further development of the Chimera approach. The use of FLOWer for time accurate computations has been eased by the implementation of an automatically adapted physical time step size.

References

1. N. Gauger and J. Brezillon: "The Continuous Adjoint Approach in Aerodynamic Shape Optimization", this volume.
2. J. Brezillon: "Application of the Adjoint Technique with the Optimization Framework Synaps", this volume.
3. J. Sidès and K. Pahlke: "Progress Towards the CFD Computation of the Complete Helicopter: Recent Results Obtained by Research Centers in the Framework of the Franco-German CHANCE Project", CEAS Aerospace Aerodynamics Research Conference, 10-12 June 2002, Cambridge, UK.
4. N. Kroll, R. Radespiel and C.C. Rossow: "Accurate and Efficient Flow Solvers for 3D Applications on Structured Meshes", AGARD R-807, 4.1-4.59, 1995.
5. P. Aumann, H. Barnewitz, H. Schwarten, K. Becker, R. Heinrich, B. Roll, M. Galle, N. Kroll, Th. Gerhold, D. Schwamborn and M. Franke: "MEGAFLOW: Parallel Complete Aircraft CFD", Parallel Computing, Vol. 27, pp. 415-440, 2001.
6. R. Heinrich, R. Ahrem, G. Günther, H.P. Kersken, W. Krüger and J. Neumann: "Aeroelastic Computation Using the AMANDA Simulation Environment", Proceedings of CEAS Conference on Mulitdisciplinary Aircraft Design and Optimization, 25-26 June 2001, Cologne, Germany.
7. K. Pahlke and B. v.d.Wall: "Progress in Weak Fluid-Structure-Coupling for Multibladed Rotors in High-Speed Forward Flight", 28th European Rotorcraft Forum, Paper 67, Bristol, UK, 2002.
8. B. Eisfeld: "Turbulence Models in FLOWer", this volume.
9. D.C. Wilcox: "Reassessment of the Scale-Determination Equation for Advanced Turbulence Models", AIAA J., Vol. 26 (11), pp. 1299-1310, 1988.
10. T. Rung, H. Lübcke, M. Franke, L. Xue, F. Thiele and S. Fu: "Assessment of Explicit Algebraic Stress Models in Transonic Flows", Proceedings of the 4th Symposium on Engineering Turbulence Modeling and Measurements, France, pp. 659-668, 1999.
11. M. Rakowitz, M. Sutcliffe, B. Eisfeld, D. Schwamborn, H. Bleecke and J. Fassbender: "Structured and Unstructured Computations on the DLR-F4 Wing-Body Configuration", AIAA Paper 2002-0837, 2002.
12. J.C. Kok and F.J. Brandsma: "Turbulence Model Based Vortical Flow Computations for a Sharp Edged Delta Wing in Transonic Flow Using the Full Navier-Stokes Equations", NLR-CR-2000-342, 2000.
13. S. Wallin and A.V. Johansson: "An Explicit Algebraic Reynolds Stress Model for Incompressible and Compressible Turbulent Flows", J. Fluid Mech., Vol. 403, pp. 89-132, 2000.
14. A. Krumbein: "Transition Modeling in FLOWer – Transition Prescription and Prediction", this volume.
15. J.K. Fassbender: "Improved Robustness for Numerical Simulation of Turbulent Flows around Civil Transport Aircraft at Flight Reynolds Numbers", Doctoral thesis, DLR research report DLR–FB 2003–09 (ISSN 1434-8454).
16. R.W. MacCormack: "A New Implicit Algorithm for Fluid Flow", AIAA Paper 97-2100, 1997.
17. J. A. Benek, J. L. Steger and F. C. Dougherty: "A Flexible Grid Embedding Technique with Application to the Euler Equations", AIAA Paper 83-1944, 1983.
18. T. Schwarz: "Development of a Wall Treatment for Navier-Stokes Computations using the Overset-Grid Technique", 26th European Rotorcraft Forum, The Hague, The Netherlands, 26-29 Sep. 2000.

19. J. Bonet and J. Peraire: "An Alternating Digital Tree (ADT) Algorithm for 3D Geometric Searching an Intersection Problems", International Journal for Numerical Problems in Engineering, Vol. 31, pp. 1-17, 1991.
20. A. Jameson: "Time Dependent Calculation Using Multigrid with Applications to Unsteady Flow past Airfoils and Wings", AIAA Paper 91-1596, 1991.
21. R. Heinrich and H. Bleecke: "Simulation of Unsteady Three Dimensional Viscous Flows Using a Dual Time Stepping Method", Notes on Numerical Fluid Mechanics, Vol. 60, pp. 15 - 23, Vieweg Verlag, 1997.
22. P. Rogiest, M. Delanaye and J.A. Essers: "Implicit Computations of Unsteady Separated Flows with a Quadratic Reconstruction Scheme", AIAA Paper 95-1734, 1997.

3

Transition Modeling in FLOWer — Transition Prescription and Prediction

A. Krumbein

DLR, Institute of Aerodynamics and Flow Technology, Lilienthalplatz 7, D-38108 Braunschweig

Summary. This paper summarizes the developments of transition prescription and transition prediction techniques which were implemented into the DLR Reynolds-averaged Navier-Stokes (RANS) solver FLOWer in the framework of the DLR projects MEGAFLOW and MEGAFLOW II and the German research project MEGAFLOW. The very basic transition handling functionalities which FLOWer provided before the projects started were generalized in order to prescribe arbitrary transition lines on very complex aircraft geometries with different components, such as wings, fuselages or nacelles. A number of transition prediction methods were incorporated into the code and an infrastructure was built up in order to handle the underlying transition prediction strategy which results in an iteration process within the solution process of the RANS equations. Finally, physical models for the modeling of transitional flow were implemented and tested.

3.1 Introduction

The modeling of laminar-turbulent transition in Reynolds-averaged Navier-Stokes (RANS) solvers is a necessary requirement for the computation of flows over airfoils and wings in the aerospace industry, as it is not possible to obtain quantitatively correct results if the laminar-turbulent transition is not taken into account.

A laminar-turbulent transition modeling consists of three major parts, the transition prescription, the transition prediction and the physical modeling of transitional flow. The transition prediction determines the transition locations on the surface of the configuration. The transition prescription applies the determined transition locations in the flow solver and thus brings the information from the transition prediction into the solution process of the RANS equations. The physical modeling of transitional flow is a means to improve the quality of the computational results on the one hand and to stabilize the iteration process of the transition locations on the other hand.

In this paper, results obtained with the RANS flow solver FLOWer, [1], of the German Aerospace Center (DLR) are presented. FLOWer is a 3-

dimensional, compressible RANS code for steady or unsteady flow problems and uses structured body-fitted multi-block meshes. The code is based on a finite volume method and a cell-vertex or a cell-centered spatial discretization scheme and uses an explicit Runge-Kutta time integration scheme with multi-grid acceleration. The influence of turbulence is taken into account by eddy viscosity turbulence models according to the Boussinesq approximation. All computational results shown in this paper were obtained with the cell-vertex discretization scheme.

The development of the transition prescription technique and the implementation of a coupling structure, which connects the flow solver to transition prediction methods has been partly realized in the German research project MEGAFLOW, [2].

3.2 Transition Prescription

The transition prescription technique in FLOWer can be applied in a general way to arbitrary, 3-dimensional geometric configurations and uses known transition locations, which have been determined previously, e.g. by experiments or separate calculations. A transition prescription technique is a necessary condition for transition prediction.

In general, a complex configuration consists of several components, each having an individual set of transition points. The transition points usually define a transition line on a part of the surface of the configuration, e.g. a wing or the fuselage of an aircraft. A single component of the configuration can have more than one transition line, e.g. a nacelle of a jet engine (outer surface of nacelle and inlet part) or a flap downstream of an engine with its jet blowing on the flap. The transition lines are curves in space bounded to the surface of the configuration. The shape of the surface of the configuration is arbitrary and, generally, the configuration itself is 3-dimensional.

Basically, the transition prescription technique splits the computational domain into laminar and turbulent flow regions. Its functioning is based on three steps, first the division of the surfaces of the configuration into laminar and turbulent regions, then the division of the field apart from the surfaces into laminar and turbulent regions and finally the different treatment of computational grid points in the laminar and turbulent regions during the solution process of the RANS equations.

3.2.1 Transition Setting on Surfaces

The boundary between a laminar and a turbulent flow region on the surface of the configuration is defined by a prescribed transition line usually in form of an oriented polygonal line defined by a definite, well-ordered sequence of transition points given in surface coordinates. The transition line must be mapped into the surface grid of the configuration, so that all surface grid points located upstream of the transition line can be detected as 'laminar' and all surface grid points located downstream as 'turbulent'.

The mapping procedure is described in detail in [3, 4]. It is applied one after another to every component of the configuration for which a transition line is prescribed. After every prescribed transition line has been mapped into the surface grid the surface is split into laminar and turbulent patches.

In Fig. 3.1 the nose of a helicopter fuselage is shown together with the prescribed transition line (pale) and its mapping (dark) into the surface grid. Both, the prescribed transition line and the surface contour are characterized by strong curvatures. In Fig. 3.2 a wing-body-pylon- nacelle configuration is shown with prescribed transition lines on the nose of the fuselage, on upper and lower side of the wing and on the outer and inner sides of the nacelle. The two examples illustrate that very demanding transition lines on complex configurations can be represented with excellent accuracy.

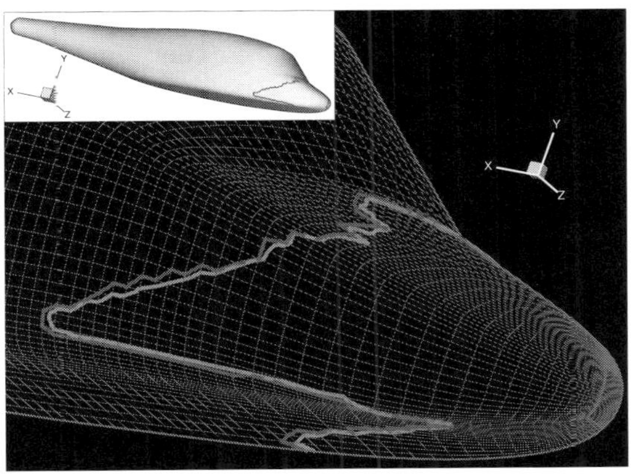

Fig. 3.1. Helicopter fuselage with prescribed transition line (pale) and its mapping (dark) into the surface grid

3.2.2 Transition Setting in the Field

To correctly simulate a laminar boundary layer a RANS solver needs 'laminar' grid points also in the flow field itself. Consequently, in the direct vicinity to laminar wall segments laminar spatial regions are generated adjacent to the laminar surface patches. The extent of the laminar zones in wall normal direction is not known a priori. The exact shape of a laminar zone is the shape of the laminar boundary layer associated with the current transition line, bounded in wall normal direction by the local laminar boundary layer thickness.

Fig. 3.2. Wing-body-pylon-nacelle configuration with prescribed transition lines on the fuselage, on the wing and on the nacelle

In the transition setting algorithm the exact shape of a laminar zone is approximated by a projection of the surface contour of a laminar surface patch into the flow field adjacent to the patch, so that a single variable is sufficient to generate a laminar zone, Fig. 3.3.

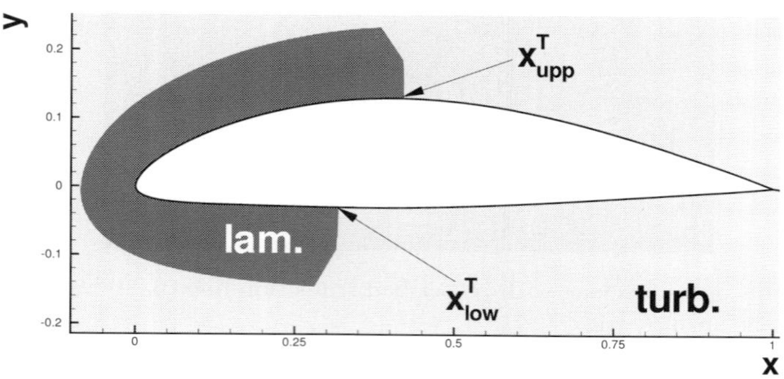

Fig. 3.3. Single-element airfoil with transition locations and laminar zone

This variable is the distance between a surface grid point and the outer edge of the laminar zone to be generated and is an input parameter for the flow solver.

The transition setting algorithm which is described in detail in [3, 4] is applied one after another to every component of the configuration for which a transition line is prescribed. After every component has been treated by this algorithm, every laminar surface patch has a laminar flow region adjacent to the patch, Fig. 3.4 and Fig. 3.5.

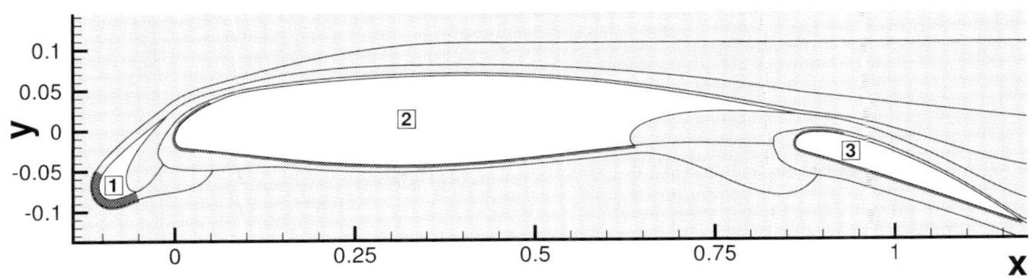

Fig. 3.4. 3-element airfoil with laminar zones of all elements

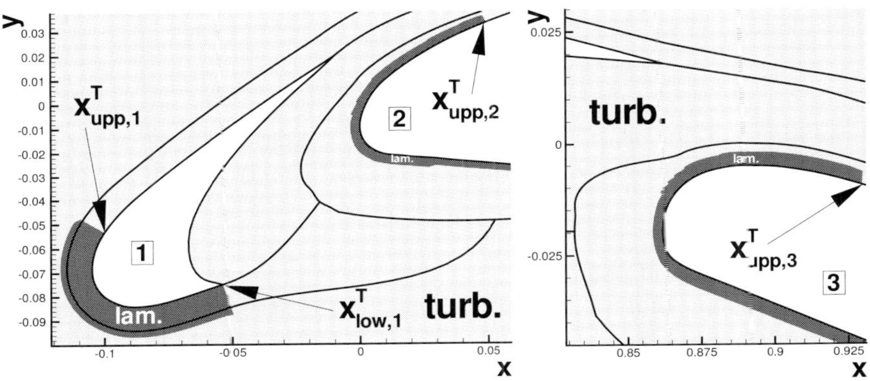

Fig. 3.5. Transition locations and laminar zones of the slat and the main airfoil (left) and of the flap (right)

The algorithm is applied after a 'fully turbulent' initialization of the complete flow field, so that its application leads to a 'laminarization' of the eddy viscosity in the case of algebraic turbulence models or to a 'laminarization' of source terms of the turbulence equations in the case of transport equation turbulence models respectively, chapter 3.2.3.

The transition setting algorithm works on structured and unstructured grids. In the case of structured grids, its application is independent of the topology of the computational grid, so that laminar flow regions are not bounded to the blocks, in which a single component of the configuration is embedded.

3.2.3 Different Treatment of Computational Grid Points

A typical way used in RANS solvers when transition prescription is performed is the different treatment of the grid points of the computational grid depending on their location either in a laminar or a turbulent flow region. The information, if a grid point is located in a laminar or a turbulent region, the laminar-turbulent status of the point, is stored via a laminar-turbulent flag, *ltflag*, as additional information at the grid point during the complete run of the flow solver.

Typically, in RANS solvers using eddy viscosity turbulence models applying the Boussinesq approximation transition onset is suppressed at grid points in laminar regions within the 'turbulent' remainder of the computational grid surrounding the laminar regions. Suppressing transition onset is realized by a manipulation of certain turbulence quantities. For algebraic turbulence models, the eddy viscosity μ_t is set to zero in laminar regions, $\mu_{t,\mathrm{lam}} = 0$, using the laminar-turbulent flag *ltflag*. Setting $ltflag = 1$ for a 'turbulent' grid point and $ltflag = 0$ for a 'laminar' one, the actual value of the eddy viscosity used in the flow solver μ_t code is applied as

$$\mu_t^{\mathrm{code}}(P) = ltflag(P)\mu_t(P) \tag{3.1}$$

P denoting the current grid point. For transport equation turbulence models, specific source terms of the model, S_ϕ , are controlled in such a way, [4], that

$$S_\phi(P)_{\mathrm{lam}} \leq 0 \tag{3.2}$$

holds at laminar grid points.

Applying this kind of grid partitioning and the described manipulation of turbulence quantities the transition process degenerates to a sudden tip over from laminar to turbulent at a single point (point transition).

3.2.4 Results

The following figures illustrate that an improved resolution of physical phenomena and better numerical results can be obtained when the laminar-turbulent transition is taken into account.

In Fig. 3.6 a comparison of the surface friction distribution of a helicopter fuselage, [5], of a fully turbulent computation and a computation with transition for the Mach number $M = 0.236$, the Reynolds number $Re = 30.1 \cdot 10^6$ and the angle of attack $\alpha = 0.016°$ using the Wilcox k-ω turbulence model, [6], is shown. The drag coefficient which results from the fully turbulent computation is $c_{d,ft} = 0.0133$, the one from the computation with 6 transition is $c_{d,tr} = 0.0128$ which corresponds to a difference of $\Delta c_d = 3.76\%$.

The same comparison is shown in Fig. 3.7 for a wing-body-pylon-nacelle configuration, [7], with $M = 0.8$, $Re = 9.97 \cdot 10^6$ and $\alpha = 2.2°$ using the Wilcox k-ω turbulence model. Here, the lift coefficient is $c_{l,ft} = 0.4977$ in the fully turbulent case and $c_{l,tr} = 0.5098$ in the case with transition so that the error with regard to the experimental value $c_{l,exp} = 0.515$ drops from 3.5% to 1%.

Fig. 3.8 compares the pressure and skin friction distributions of a 3-element airfoil, [8, 9], of a fully turbulent computation, a computation with transition

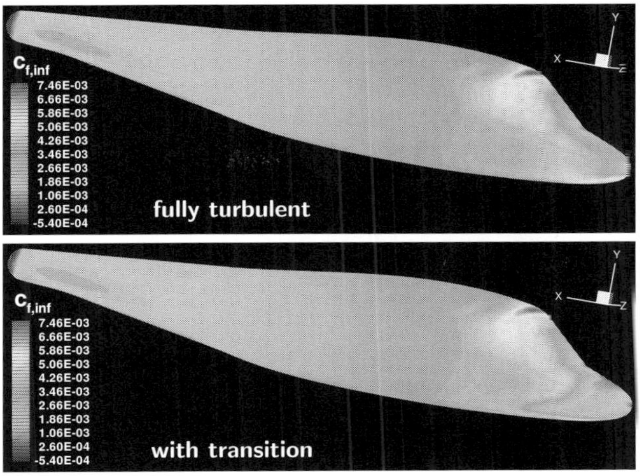

Fig. 3.6. Computed surface friction distribution of a helicopter fuselage, fully turbulent and with transition, $M = 0.236$, $Re = 30.1 \cdot 10^6$, $\alpha = 0.016°$, k-ω model

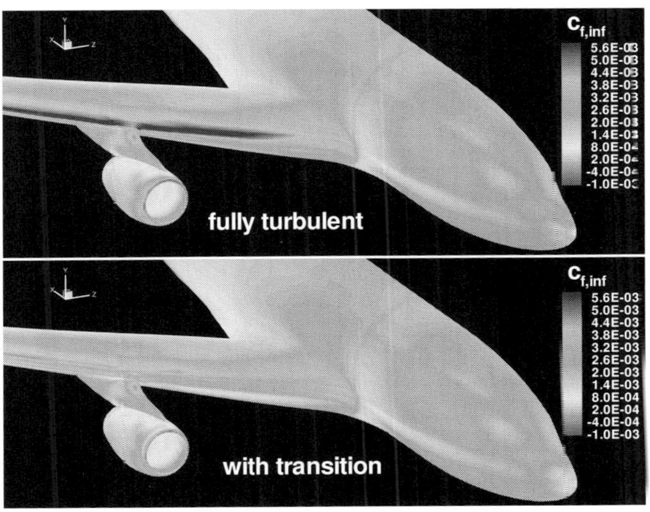

Fig. 3.7. Computed surface friction distribution of a wing-body-pylon-nacelle configuration, fully turbulent and with transition, $M = 0.8$, $Re = 9.97 \cdot 10^6$, $\alpha = 2.2°$, k-ω model fully turbulent with transition

and of experiments for $M = 0.2$, $Re = 3.52 \cdot 10^6$ and $\alpha = 20.18°$ using the Wilcox k-ω turbulence model. Only in the computation with transition the suction peak on the slat is reached. A comparison of the force coefficients is given in Table 3.1. The values in brackets are the deviations from the experimental values of the force coefficients.

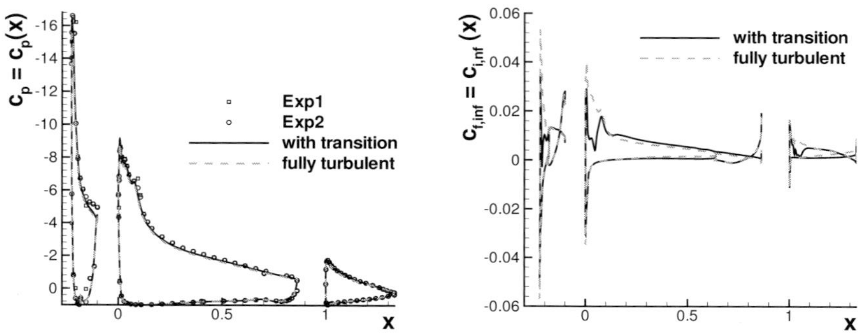

Fig. 3.8. Computed (fully turbulent and with transition, k-ω model) and experimental pressure and surface friction distribution of a 3-element airfoil, $M = 0.2$, $Re = 3.52 \cdot 10^6$, $\alpha = 20.18°$

Table 3.1. Force coefficients of a 3-element airfoil — experiment, fully turbulent computation, computation with transition (in brackets: deviations from the experiment)

	experiment	fully turbulent	with transition
lift c_l	4.071	3.822 (6%)	3.952 (3%)
drag c_d	0.0705	0.081 (13%)	0.071 (1%)

3.3 Transition Prediction

By 'pure' transition prescription the transition locations are imposed, [3, 4]. Transition prediction is the next step of a transition modeling. Numerous transition prediction methods are available, ranging in cost and complexity from simple empirical transition criteria via local linear, non-local linear and non-local non-linear stability methods to direct numerical simulations.

For the design process of wings in industry, there exists the demand for a RANS-based CFD tool that is able to automatically and autonomously handle flows with laminar-turbulent transition. The first steps towards the setup of such a tool were made e.g. in [10], where a RANS solver and an e^{N}-method, [11, 12], based on linear stability theory and the parallel flow assumption were applied, or in [13], where a RANS solver, a laminar boundary layer method, [14], and an e^{N}-method were coupled. There, the boundary layer method was used to produce highly accurate laminar, viscous layer data to experiment fully turbulent with transition 8 be analyzed by a linear stability code. Hence the very expensive grid adaptation necessary to produce accurate viscous layer data directly from the Navier-Stokes grid was avoided. The use of an e^{N}-database method, [15], results in a coupled program system that is able to automatically handle transition prediction. Alternative approaches using a transition closure model or a transition/turbulence model directly incorporated into the RANS solver are documented in [16, 17, 18].

In the MEGAFLOW projects, the FLOWer code was used together with the laminar boundary layer method in [14], the e^{N}-database method for Tollmien-Schlichting waves in [15] and the e^{N}-database method for crossflow instabilities in [19]. The laminar boundary layer method and the e^{N}-database methods form a so called 'transition prediction module' that is coupled to the RANS solver and that interacts with the solver during the computation, [4] and [20]. Presently, the transition prediction module of FLOWer can be applied to 2-dimensional multi-element configurations and to 3-dimensional single-element wings.

3.3.1 Transition Prediction Strategy

During the solution process of the RANS equations, the transition prediction module is called after a certain number of iteration cycles, k_{cyc}, of the RANS iteration process. With the call of the module the solution process is interrupted and the module analyzes the laminar boundary layers of previously specified components of the configuration, e.g. of an 2-dimensional airfoil or a wing section. The determined transition locations, $x_j^{\mathrm{T}}(cycle = k_{cyc})$ with $j = 1, \ldots, n_{loc}$, n_{loc} being the number of transition points, are communicated to the RANS solver, which performs transition prescription applying the transition setting algorithm, [4] and [20], and continues the solution process of the RANS equations. In so doing, the determination of the transition locations becomes an iteration process itself. The structure of the approach is outlined graphically in Fig. 3.9.

At every call of the module the surface pressure, $c_p(cycle = k_{cyc})$, along an airfoil or a wing section computed by the RANS solver is used as input to the boundary layer calculation. The viscous data calculated by the boundary layer method are then subsequently analyzed by the database methods.

The application of a boundary layer method for the computation of all viscous data necessary for the transition prediction method ensures the high

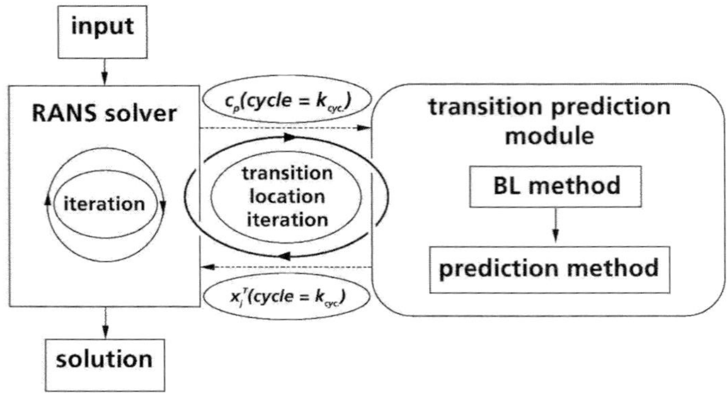

Fig. 3.9. Coupling structure of the RANS solver and the transition prediction module

accuracy of the viscous data required by the e^N-methods for the analysis of the laminar boundary layers. Thus, as shown in [13], the large number of grid points near the wall for a high resolution of the boundary layers, the adaptation of the Navier-Stokes grid in the laminar and turbulent boundary layer regions and the generation of a new adapted grid for the RANS solver after every step of the transition location iteration are avoided and the computational time can be massively reduced.

3.3.2 Transition Prediction Algorithm

(a) The RANS solver is started as if a computation with prescribed transition locations should be performed with transition locations set far downstream on the upper and lower sides of the airfoil or wing section, e.g. at the trailing edge. The RANS solver now computes a fully laminar flow over the airfoil.
(b) During the solution process of the RANS equations the laminar flow is checked for laminar separation. If laminar separation is detected, the separation point is used as approximation of the transition location and the computation is continued.
(c) The RANS equations are iterated until the lift coefficient c_l which can be represented as a function of the iteration cycles, $c_l = c_l(cycles)$, has become constant with respect to the iteration cycles.
(d) The transition prediction module is called.
(e) The determined transition locations $x_j^{\mathrm{T}}(cycle = k_{cyc})$ are underrelaxed, i.e. as new transition locations $\tilde{x}_j^{\mathrm{T}}(cycle = k_{cyc})$ coordinates located downstream of the coordinates $x_j^{\mathrm{T}}(cycle = k_{cyc})$ are used,

$$\tilde{x}^{\mathrm{T}}(k_{cyc}) = C_j^{\mathrm{T}}(k_{cyc}) x_j^{\mathrm{T}}(k_{cyc}) \quad \text{with} \quad j = 1, \ldots, n_{loc} , \qquad (3.3)$$

with $C_j^{\mathrm{T}}(k_{cyc}) > 1$. Only after the last step of the transition location iteration $C_j^{\mathrm{T}}(k_{cyc}) = 1$ is applied. The underrelaxation of the determined transition locations prevents the case that at an unconverged stage during the transition location iteration transition coordinates are determined too far upstream which might not be shifted downstream again.

(f) As convergence criterion $\Delta\tilde{x}_j^{\mathrm{T},l} < \varepsilon$ with $\Delta\tilde{x}_j^{\mathrm{T},l} = |\tilde{x}_j^{\mathrm{T}}(k_{cyc}^l) - \tilde{x}_j^{\mathrm{T}}(k_{cyc}^{l-1})|$ is applied, l being the current iteration step. In the case that the criterion is satisfied, the iteration for x_j^{T} is finished, else the algorithm loops back to station b).

3.3.3 Results

In the following figures, results for a subsonic and a transonic 2-dimensional single-element airfoil and for a 3-dimensional single-element wing are presented.

The transition prediction algorithm has been applied to the NLF(1)-0416 natural laminar airfoil, [21], with $M = 0.1$ and $Re = 4 \cdot 10^6$. In the computations the Baldwin-Lomax, [22], and the Wilcox k-ω turbulence models were used. The limiting N-factor was $N_{x_{tr}} = 11$ according to [23]. In Fig. 3.10 (left) the convergence of the transition locations and the corresponding influence on the force coefficients is illustrated for a single angle of attack $\alpha = -6.2°$ and the Baldwin-Lomax model, [4] and [20].

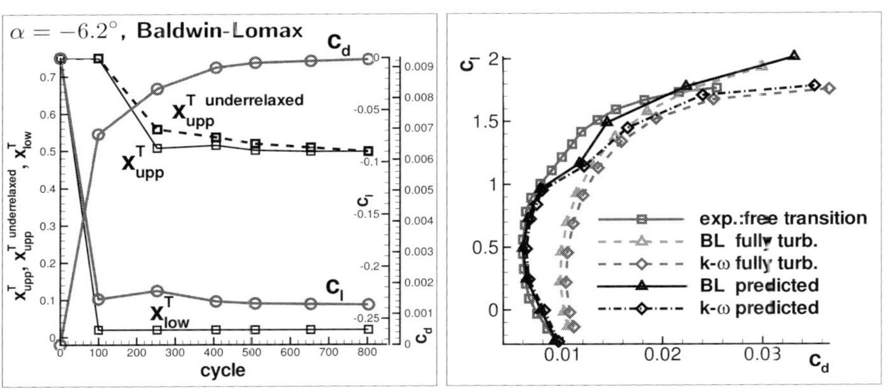

Fig. 3.10. NLF(1)-0416-airfoil — Convergence history of the iteration process of the transition locations (left) and $c_l = c_l(c_d)$-polars — experiment, fully turbulent computations, computations with transition (right), $M = 0.1$, $Re = 4 \cdot 10^6$ Baldwin-Lomax and Wilcox k-ω models, $N_{x_{tr}} = 11$

Fig. 3.10 (right) shows the $c_l = c_l(c_d)$-polars of the experiment with free transition, [21], of fully turbulent computations and of computations with transition using the transition prediction algorithm and the e^N-database for Tollmien-Schlichting waves, [15]. For both turbulence models a strong improvement of the computational results is achieved. For all points, there is a clear tendency towards the experimental data. Why the drag deviates at high values of α ($\alpha \geq 6°$) is still an open question.

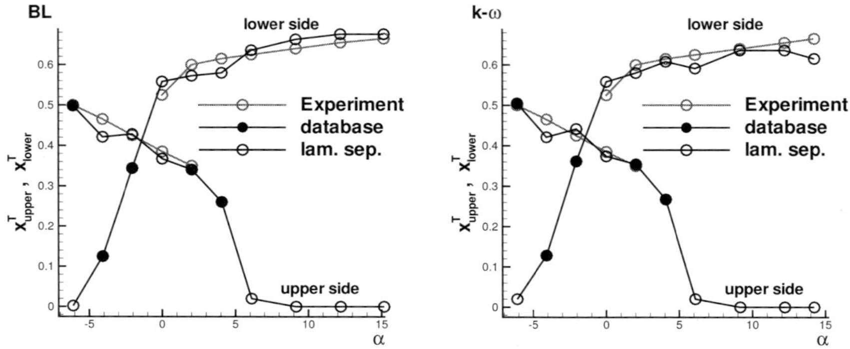

Fig. 3.11. Experimental and computational transition locations of the NLF(1)-0416-airfoil, $M = 0.1$, $Re = 4 \cdot 10^6$, Baldwin-Lomax model (left), Wilcox k-ω model (right), $N_{x_{tr}} = 11$

Fig. 3.11 compares the experimentally determined transition locations, as given in [21], with the transition locations determined by the transition prediction algorithm using the database. Indicated by filled, black symbols are the transition locations determined by the database. The hollow, black symbols mark the transition locations approximated by the x-coordinates of laminar separation, whose locations were detected by the laminar boundary layer method, [14]. The accuracy of the predicted transition locations is excellent.

Fig. 3.12 compares the experimental pressure distribution and the transition locations of the CAST 10 airfoil, [24], at $M = 0.73$, $Re = 3.9 \cdot 10^6$ and $\alpha = -0.25°$ and the computational results at $M = 0.71$, $Re = 3.9 \cdot 10^6$ and $\alpha = -0.3°$, [25], using the Wilcox k-ω turbulence model and applying the same procedure described above. The limiting N-factor was $N_{x_{tr}} = 7.84$ according to [25]. The pressure distributions were computed for fixed transition locations whose values had been detected in the experiment and for predicted transition locations.

The transition location on the upper side of the airfoil is the result of a Tollmien-Schlichting instability and was predicted with excellent accuracy. On the lower side, the predicted transition location was approximated by a laminar separation point given by the laminar boundary layer method and is located too far upstream.

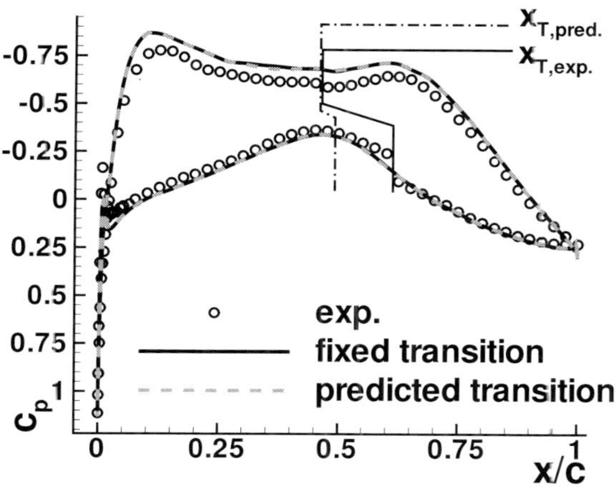

Fig. 3.12. Pressure distributions (experimental, fixed and predicted transition) and transition locations (experimental and predicted) of the CAST 10 airfoil, Wilcox k-ω model, $N_{x_{tr}} = 7.84$

In Fig. 3.13 the results of a test computation for the ONERA M6 wing, [26], for $M = 0.84$, $Re = 2.0 \cdot 10^6$ and $\alpha = -4.0°$ using the Baldwin-Lomax turbulence model are shown. The limiting N-factors were $N^{TS}_{x_{tr}} = 4.0$ and $N^{CF}_{x_{tr}} = 2.0$. The left part of the figure shows the initial and the predicted transition line on the upper and the lower sides of the wing. The transition points were predicted in three different wings sections applying the same procedure described above and the e^N-database methods in [15] and [19]. The character of the transition is marked with 'TS' for Tollmien-Schlichting wave, with 'CF' for crossflow instability and with 'ls' for transition due to a laminar separation.

3.4 Modeling of Transitional Flow

The easiest way to describe transitional flow regions in a RANS solver is the application of point transition, which means that turbulence quantities, which are suppressed in the laminar part of the flow, suddenly become active at the location of transition onset. This procedure results in a sudden change of the flow quantities in this area. Due to the effects of numerical dissipation a small transitional-like flow region is generated artificially in a computation without physical transition modeling. Nevertheless, the sudden change of the flow quantities is often strong enough to prevent the convergence of the iterative transition prediction process, [27]. In addition, the application of point transi-

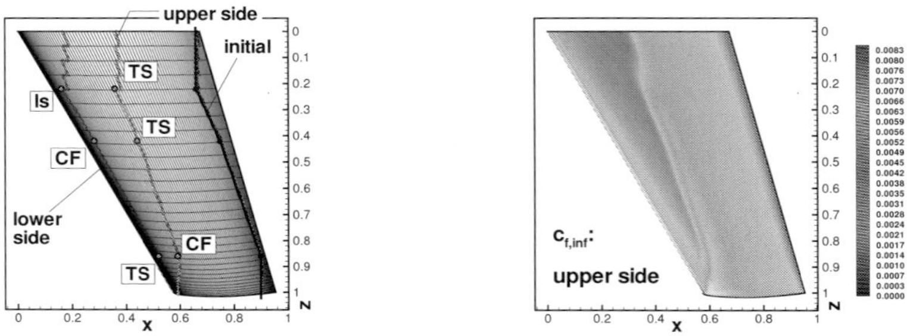

Fig. 3.13. ONERA M6 wing with initial and predicted transition lines (left), skin friction distribution on the upper side (right), $M = 0.84$, $Re = 2.0 \cdot 10^6$, $\alpha = -4.0°$, Baldwin-Lomax model, $N_{x_{tr}}^{\text{TS}} = 4.0$, $N_{x_{tr}}^{\text{CF}} = 2.0$

tion generates a strong upstream influence so that the transitional-like flow region starts considerably upstream of the transition location. In 2-dimensional airfoil flows an upstream influence up to 10% of the chord length of the airfoil can be observed. To overcome this limitation physical models for the modeling of transitional flow regions were incorporated into the FLOWer code. The physical models used for the modeling of transitional flow regions are based on empirical algebraic models for the transition length l_{tr}, [27] and [28],

$$l_{tr} = l_{tr}\left(\zeta_{tr}, \text{Re}_{\zeta_{tr}}\right) , \tag{3.4}$$

and on the intermittency function, [27] and [29],

$$\gamma = 1 - \exp\left(-0.412\,\zeta^2\right) , \quad \zeta = \zeta\left(l_{tr}\right) , \tag{3.5}$$

ζ being the longitudinal coordinate relevant for the laminar-turbulent transition. Details about the representation of the formulas for l_{tr} and γ and their implementation are given in [30]. The generation of transitional flow regions is done by the setting of the real value flag *ltflag* according to

$$\mu_t^{\text{code}}(P) = ltflag(P)\mu_t(P) = \gamma(P)\mu_t(P) \tag{3.6}$$

in the interval $\zeta_{tr} < \zeta < \zeta_{tr} + l_{tr}$ as described in [30].

Two transitional flow models were tested in [30] computing the 2-dimensional A310 3-element landing configuration, [31, 32, 33], at $M = 0.22$, $\alpha = 22.4°$, $Re = 4.1 \cdot 10^6$ using the Spalart-Allmaras turbulence model with Edwards modification (SAE), [34]. Fig. 3.14 shows the shape and extent of the transition regions according to one of the formulas (left) and the skin friction distribution with point transition and the application of physically modeled transitional flow (right).

For the test computations of this case the locations of laminar separation determined by the FLOWer code are supposed to represent the laminar-turbulent transition locations in a first step. In many cases this assumption

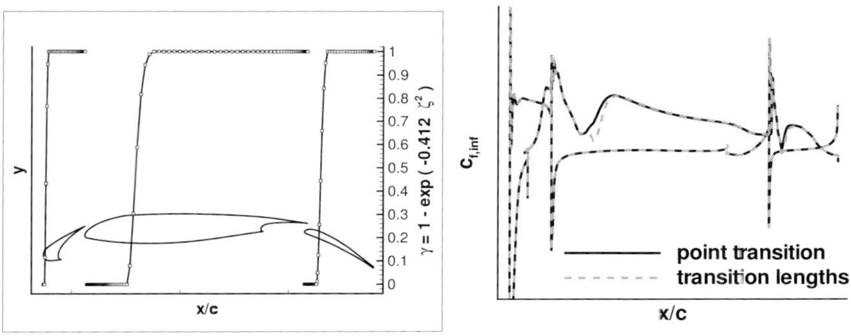

Fig. 3.14. A310 landing configuration — transition regions (left) and skin friction distribution (right), point transition and transitional flow, $M = 0.22$, $Re = 4.1 \cdot 10^6$, $\alpha = 22.4°$, SAE model

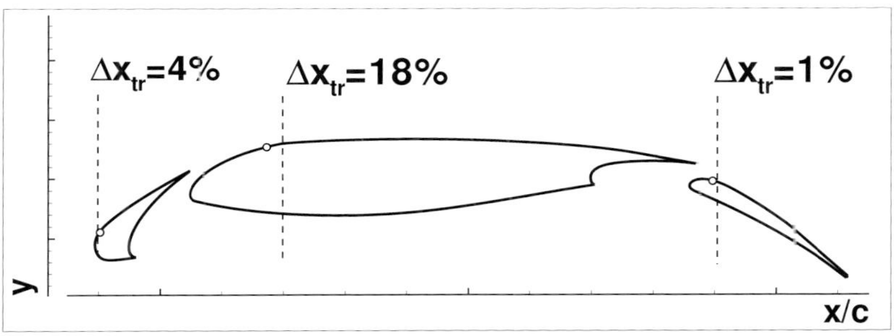

Fig. 3.15. A310 landing configuration — transition locations based on laminar separation, $M = 0.22$, $Re = 4.1 \cdot 10^6$, $\alpha = 22.4°$, SAE model

leads to a good approximation of the real transition point, particularly for low Reynolds number airfoil flows when transition does not occur before the laminar boundary layer separates. The transition locations determined in this way are shown in Fig. 3.15 as vertical lines, the transition locations which existed during the experimental measuring are plotted as circular symbols on the surfaces of the elements. The differences Δx_{tr} between the 'experimental' and the computed values of the transition locations are given in Fig. 3.15 as $\Delta x_{tr} = \left(x_{tr}^{\text{comp}} - x_{tr}^{\text{exp}}\right) / x_{tr}^{\text{exp}}$.

3.5 Conclusions

The DLR RANS solver FLOWer has been widely extended with regard to laminar-turbulent transition modeling techniques, that is to say transition prescription, transition prediction and the physical modeling of transitional flow.

A general transition prescription technique was developed and implemented in order to handle arbitrary transition lines on arbitrary surfaces. The technique is based on an automatic partitioning of the flow field into laminar and turbulent flow regions and generates an individual laminar zone for each element of a geometric configuration. The algorithm is independent of the grid structure and of the block topology in the case of a structured grid. The laminar-turbulent transition is emulated by a different numerical treatment of laminar and turbulent computational grid points and leads to an improved resolution of physical phenomena and better numerical results.

For the transition prediction on airfoils, wings and wing like components of an aircraft FLOWer was coupled with a transition prediction module consisting of a laminar boundary layer method and transition prediction methods, namely e^{N}-database methods for Tollmien-Schlichting waves and crossflow instabilities which are based on the local linear stability theory and the parallel flow assumption. The use of database methods guarantees an automatic transition prediction procedure. The underlying transition prediction strategy leads to an iteration of the transition locations within the solution process of the RANS equations. For a number of test cases highly accurate values of the transition locations were obtained. The prediction procedure is independent of the grid structure and of the block topology in the case of a structured grid. Presently, the transition prediction module of FLOWer can be applied to 2-dimensional multi-element configurations and to 3-dimensional single-element wings.

Different transitional flow models can be used in combination with the transition prescription and the transition prediction techniques.

The next steps are the extension to 3-dimensional multi-element wing configurations, the validation with an improved e^{N} crossflow database for 3-dimensional wing configurations and the application of the complete system to 3-dimensional high lift systems of aircraft.

References

1. FLOWer . Installation and User Handbook, Release 116, Doc.Nr. MEGAFLOW-1001, Institut fur Entwurfsaerodynamik, Deutsches Zentrum fur Luft- und Raumfahrt e.V., 2000
2. Becker, K.; Kroll, N.; Rossow, C. C.; Thiele, F., "The MEGAFLOW project", *Aerosp. Sci. Technol.* 4, 2000, pp. 223-237
3. Krumbein, A., "AVTAC Advanced Viscous Flow Simulation Tools for Complete Civil Aircraft Design - Transition Prescription and Prediction", Deliverable Task 3.2, AVTAC/DEL/DLR/D3.2C5, 1999
4. Krumbein, A., Stock, H. W., "Laminar-turbulent Transition Modeling in Navier-Stokes Solvers using Engineering Methods", ECCOMAS 2000, Barcelona (e), 11.-14. September 15 2000, *ECCOMAS 2000 - CD-Rom Proceedings*, editor:

International Center for Numerical Methods in Engineering (CIMNE), 2000, ISBN: 84-89925-70-4, Depósito Legal: B-37139- 2000

5. HELIFUSE-Helicopter Fuselage Drag, Assessment report, Task 2 . Navier-Stokes Calculations, Assessment of Blind Test Calculations, Deliverable of subtask 2.1, HELIFUSE/C/1/DLR/03/A, 1997
6. Wilcox, D. C., "Reassessment of the Scale-Determining Equation for Advanced Turbulence Models", *AIAA Journal*, Vol.26, No. 11, 1988, pp. 1299-1310
7. Becle, J. P., "Essai de la demi-maquette AS28 dans la soufflerie S1Ma . Partie Effets Reynolds et Partie TPS", Rapport d'Études ONERA no. 0962GY100G et 3423 AY043G, 1985
8. Moir, I. R. M., "Measurements on a Two-Dimensional Aerofoil with High-Lift Devices", *AGARD Report* No.303, pp. A2-1 - A2-12, 1994
9. Wild, J. W., "Direct Optimization of Multi-Element-Airfoils for High-Lift using Navier- Stokes Equations", Computational Fluid Dynamics 98, Proceedings of the Fourth European Computational Fluid Dynamics Conference, pp. 383, 1998
10. Radespiel, R., Graage, K., Brodersen, O., "Transition Predictions Using Reynolds-Averaged Navier-Stokes and Linear Stability Analysis Methods", AIAA Paper 91-1641, 1991
11. Smith, A. M. O., Gamberoni, N., "Transition, Pressure Gradient and Stability Theory", Douglas Aircraft Company, Long Beach , Calif. Rep. ES 26388, 1956
12. van Ingen, J. L., "A suggested Semi-Empirical Method for the Calculation of the Boundary Layer Transition Region", University of Delft, Dept. of Aerospace Engineering, Delft, The Netherlands, Rep. VTH-74, 1956
13. Stock, H. W., Haase, W., "A Feasibility Study of e^{N} Transition Prediction in Navier-Stokes Methods for Airfoils", *AIAA Journal*, Vol 37, no. 10, 1999, pp. 1187-1196
14. Horton, H. P., Stock, H. W., "Computation of Compressible, Laminar Boundary Layers on Swept, Tapered Wings", *Journal of Aircraft*, Vol.32, No. 6, 1995, pp.1402-1405
15. Stock, H. W., Degenhardt, E., "A simplified e^{N} method for transition prediction in twodimensional, incompressible boundary layers", *Zeitung fur Flugwissenschaft und Weltraumforschung*, Vol.13, 1989, pp.16-30
16. Warren, E. S., Hassan, H. A., "Transition Closure Model for Predicting Transition Onset", *Journal of Aircraft*, Vol.35, 1998, pp. 769-775
17. Czerwiec, R. M., Edwards, J. R., Rumsey, C. L., Bertelrud, A., Hassan, H. A., "Study of High-Lift Configurations Using k-ζ Transition/Turbulence Model", AIAA Paper 99-3186, 1999
18. Edwards, J. R., Roy, C. J., Blottner, F. G., Hassan, H. A., "Development of a One-Equation Transition/Turbulence Model", *AIAA Journal*, Vol.39, no. 9, 2001, pp. 1691-1698
19. Casalis, G., Arnal, D., "ELFIN II Subtask 2.3: Database method . Development and validation of the simplified method for pure crossflow instability at low speed", ELFIN II - European Laminar Flow Investigation, Technical Report no. 145, ONERA-CERT, Département d'Études et de Recherches en Aérothermodynamique (DERAT), R.T. DERAT no. 119/5618.16, 1996
20. Krumbein, A., "Coupling of the DLR Navier-Stokes Solver FLOWer with an e^{N}-Database Method for laminar-turbulent Transition Prediction on Airfoils", *New Results in Numerical and Experimental Fluid Mechanics III*, Notes on Numerical Fluid Mechanics - Vol.77, Berlin, Heidelberg, New York, Springer Verlag, 2002, pp. 92-99

21. Somers, D. A., "Design and Experimental Results for a Natural-Laminar Flow Airfoil for General Aviation Applications", NASA Technical Paper 1861, Scientific and Technical Information Branch, 1981
22. Baldwin, B. S., Lomax, H., "Thin-Layer Approximation and Algebraic Model for Separated Turbulent Flows", AIAA Paper 78-257, 1978
23. Stock, H. W., "Airfoil Validation Using Coupled Navier-Stokes and e^{N} Transition Prediction Methods", *Journal of Aircraft*, Vol.39, No. 1, 2002, pp.51-58
24. Mignosi, A., "Fundamental Reflections on Cryogenic Testing", *AGARD Report* No.722, 16 1985, pp. 7-1 - 7-25
25. Arthur, M. T., Dol, H., Krumbein, A., Houdeville, R., Ponsin, J., "Application of Transition Criteria in Navier-Stokes Computations", GARTEUR AD(AG35), TP-137, 2003
26. Schmitt, V., Charpin, F., "Pressure Distributions on the ONERA-M6-Wing at Transonic Mach Numbers", *AGARD Advisory Report* No.138, 1979, pp. B1-1 . B1-44
27. Stock, H. W., Haase, W., "Navier-Stokes Airfoil Computations with e^{N} Transition Prediction Including Transitional Flow Regions", *AIAA Journal*, Vol.38, no. 11, 2000, pp. 2059-2066
28. Walker, G. J., "Transitional Flow on Axial Turbomachine Blading", *AIAA Journal*, Vol.27, No. 5, 1989, pp. 595-602
29. Dhawan, S., Narasimha, R., "Some properties of boundary layer flow during the transition from laminar to turbulent motion", *Journal of Fluid Mechanics*, Vol.3, 1958, pp. 418-436
30. Krumbein, A., "On Modeling of Transitional Flow and its Application to a High Lift Multielement Airfoil Configuration", AIAA Paper 2003-724, 2003 (accepted at *Journal of Aircraft*)
31. Dargel, G., Schnieder, H., "GARTEUR AD (AG08) Final Report", GARTEUR High Lift Action Group AD (AG08), TP043, MBB Transport- und Verkehrsflugzeuge, Bremen, 1989
32. Thibert, J. J., "The GARTEUR High Lift Research Programme", AGARD Conference Proceedings 515 - High-Lift System Aerodynamics, 1993, pp. 16-1 - 16-2[illegible]
33. Brodersen, O., Ronzheimer, A., Ziegler, R., Kunert, T., Wild, J., Hepperle, M., "Aerodynamic Applications using MegaCads", Proc. of 6th International Conference on Numerical Grid Generation in Computational Field Simulation, editor: M. Cross, publisher: ISGG, NSF Eng. Research Center, Mississippi State University, 1998, pp. 793-802
34. Edwards, J.R., Chandra, S., "Comparison of Eddy Viscosity - Transport Turbulence Models for Three-Dimensional, Shock-Separated Flowfields", *AIAA Journal*, Vol.34, No. 4, 1996, pp. 756-763

4

Turbulence Models in FLOWer

B. Eisfeld

DLR, Institute of Aerodynamics and Flow Technology, Lilienthalplatz 7, D-38108 Braunschweig

Summary. The turbulence models implemented into the FLOWer code are briefly characterized, considering their basic equations and emphasizing their differences with respect to their aimed at field of application. The influence of the models on the flow solution is demonstrated for two simple test cases, the flow around the RAE 2822 airfoil, representing transonic conditions, and the flow around the Aérospatiale A airfoil, representing high-lift conditions. Results for industrially more relevant test cases of the flow around two different wing-body configurations and a three-element airfoil are presented, confirming the findings for the simple geometries at least under transonic conditions. From this, recommendations for the choice of suitable models are derived.

4.1 Introduction

Turbulence modelling is still an important issue for the development of modern CFD codes and will probably remain important for a long time. Practically all flows of technical interest take place at Reynolds numbers where turbulence naturally occurs.

In principal, this phenomenon is already covered by the three-dimensional, unsteady Navier-Stokes equations which, for sufficiently simple cases, can be solved directly (Direct Numerical Simulation, DNS). However, in technically important flows the ratio between the geometrical scale of the problem and the smallest occurring scale of turbulence is so large that an accurate resolution of all of these scales is unaffordable. Therefore, modelling techniques are applied that economically allow computations of turbulent technical flows. In general, these approaches rely on mathematically manipulated forms of the exact Navier-Stokes equations, shifting unresolved details of turbulence into a representation of their net effect.

In Large Eddy Simulations (LES) the Navier-Stokes equations are spatially filtered, leading to equations that describe the flow in terms of the resolved scale quantities. The unresolved scales enter via additional terms which are modelled in terms of the resolved scale quantities [22]. This approach has made some progress towards technical aerodynamic flows [4], but since it still requires a three-dimensional, unsteady computation at a rather fine spatial resolution, it is not yet practicable for routine industrial use [10].

Alternatively, the Navier-Stokes equations can be averaged, preferably imagined with respect to time, thus, removing all information on the turbulent fluctuations. Their effect enters the so-called Reynolds averaged Navier-Stokes (RANS) equations for the mean flow via unknown correlations for the fluctuating quantities [22]. Exact equations can be derived for these Reynolds stresses and the turbulent heat flux, involving again higher order correlations. For the latter, in principle, further equations could be derived and so forth.

Today, Reynolds stress transport models (second moment closures) are the highest level of approximation of that type [22]. Unfortunately, their in principal higher physical accuracy is accompanied by a reduced numerical stability [11]. Moreover, for three-dimensional problems, seven additional transport equations have to be solved, compared to five equations for the mean flow.

Therefore, for industrial aerodynamic CFD simpler, numerically more forgiving models are favoured. The preferred modelling approach is based on Boussinesq's hypothesis [22], assuming the tensor of the specific Reynolds stresses $\widetilde{R}_{ij}$ to be proportional to the traceless strain tensor $\widetilde{S}^*_{ij}$, i. e.

$$\widetilde{R}_{ij} = \mu_t \widetilde{S}^*_{ij} - \frac{2}{3}\overline{\rho}\widetilde{k}\delta_{ij} \tag{4.1}$$

where μ_t is a model coefficient called eddy viscosity, and

$$\widetilde{S}^*_{ij} = \frac{1}{2}\left(\frac{\partial \widetilde{U}_i}{\partial x_j} + \frac{\partial \widetilde{U}_j}{\partial x_i}\right) - \frac{1}{3}\frac{\partial \widetilde{U}_k}{\partial x_k}\delta_{ij} \tag{4.2}$$

with $\widetilde{U}_i$ the mean flow velocity components and x_i the cartesian coordinates. $\widetilde{k}$ is the specific kinetic turbulence energy which is only introduced to yield the correct trace of the Reynolds stress tensor.

Thus, turbulence is imagined to effectively increase the fluid's viscosity which considerably simplifies the numerical treatment of such models. The increase in numerical effort is basically determined by the number of the equations needed for computing the eddy viscosity and the grid resolution necessary for their solution.

Nevertheless, due to Boussinesq's hypothesis the predictive capabilities of eddy viscosity models is restricted [22]. Therefore, in order to increase their range of applicability, non-linear extensions of such models have been developed, taking the form

$$\widetilde{R}_{ij} = \left(\sum_n \beta_n T^{(n)}_{ij}\right) - \frac{2}{3}\overline{\rho}\widetilde{k}\delta_{ij} \tag{4.3}$$

The tensors $T^{(n)}_{ij}$ are functions of the traceless strain tensor $\widetilde{S}^*_{ij}$ and the rotation tensor

$$\widetilde{W}_{ij} = \frac{1}{2}\left(\frac{\partial \widetilde{U}_i}{\partial x_j} - \frac{\partial \widetilde{U}_j}{\partial x_i}\right) \tag{4.4}$$

and can be rigorously derived [7]; the first term of the series expansion, then, corresponds to the eddy viscosity formula. The coefficients β_n can be either determined by calibration, leading to "non-linear eddy viscosity models", or derived from a full Reynolds stress transport model, leading to "Explicit Algebraic Reynolds Stress Models" (EARSM) [7]. The latter can be considered a compromise between the predictive capabilities of full Reynolds stress transport models and the more favourable numerical behaviour of eddy viscosity models.

4.2 Turbulence models implemented in FLOWer

Within the MEGAFLOW project various turbulence models have been implemented into the FLOWer code and made operational with support of TU Berlin. All implemented models are based on or related to the eddy viscosity approach and can be grouped into four different classes.

4.2.1 Algebraic models

Algebraic or zero-equation models describe the eddy viscosity directly as a function of the mean flow quantities. Such models typically are based on Prandtl's mixing length hypothesis [13], stating that

$$\mu_t = \overline{\rho}\ell_{mix}^2 \left|\frac{d\widetilde{U}}{dy}\right|$$

in a two-dimensional flow in the $x - y$ plane where ℓ_{mix} is the mixing length that has to be computed from the mean flow.

In the FLOWer code the Baldwin-Lomax model [1] is implemented. This model has been developed for wall-bounded flows, dividing the boundary layer into an inner and an outer part for which the following relations hold

Inner part

$$\mu_t^{(inner)} = \ell_{mix}^2 \Omega \tag{4.5}$$

$$\ell_{mix} = \kappa y \left[\exp\left\{-\frac{y^+}{26}\right\}\right] \tag{4.6}$$

where y^+ is defined by

$$y^+ = \frac{y}{\sqrt{\overline{\mu}/\left(\overline{\rho}\Omega_{max}\right)}}.$$

Outer part

$$\mu_t^{(outer)} = 0.02688\overline{\rho} F_{wake} F_{Kleb} \tag{4.7}$$

$$F_{wake} = y_{max} \min\left(F_{max}, \frac{\widetilde{U}^2}{F_{max}}\right) \tag{4.8}$$

$$F_{Kleb} = \frac{1}{1 + 5.5\left(0.3\frac{y}{y_{max}}\right)^6} \tag{4.9}$$

$$F_{max} = \frac{1}{\kappa} \max_y \left(\ell_{mix} \Omega\right) \tag{4.10}$$

In these equations the distance to the surface, y, is taken along grid lines, emanating from walls, and the value of y_{max} is given by the location corresponding to F_{max}. Using the modification of Degani and Schiff [5], the local maximum closest to the wall replaces the global maximum along y, in order to allow computations of wall bounded flows, involving free vortices. Finally, $\Omega = \sqrt{2\widetilde{W}_{ij}\widetilde{W}_{ji}}$ is the absolute value of the vorticity, and the Kármán constant is assumed to take the value $\kappa = 0.40$.

The Baldwin-Lomax model [1] is suited for attached boundary layer flow [22], showing a very good convergence behaviour. Due to the required maximum search along grid lines the model is implemented in a block local fashion so that it is not recommended to be applied to complex multi-block geometries.

4.2.2 One-equation models

One-equation models provide a single transport equation, determining the eddy viscosity. The models implemented into the FLOWer code are based on a transport equation for the quantity $\widetilde{\nu}$ suggested by Spalart and Allmaras [19]

$$\frac{\partial\left(\overline{\rho}\widetilde{\nu}\right)}{\partial t} + \vec{\nabla}\cdot\left(\overline{\rho}\widetilde{\vec{V}}\widetilde{\nu}\right) = \vec{\nabla}\cdot\left[\frac{\overline{\mu}+\overline{\rho}\widetilde{\nu}}{\sigma_{SA}}\vec{\nabla}\widetilde{\nu}\right] + P_\nu - D_\nu \tag{4.11}$$

where the production and destruction terms are given by

$$P_\nu = c_{b1} C_\mu \widetilde{S}\overline{\rho}\widetilde{\nu} + c_{b2}\frac{\overline{\rho}\left(\vec{\nabla}\widetilde{\nu}\right)^2}{\sigma_{SA}} \tag{4.12}$$

$$D_\nu = c'_{w1} f_w \frac{\left(\overline{\rho}\widetilde{\nu}\right)^2}{\overline{\rho} d^2} \tag{4.13}$$

with

$$f_w = g\left(\frac{1+c_{w3}^6}{g^6+c_{w3}^6}\right)^{1/6} \tag{4.14}$$

$$g = r + c_{w2}\left(r^6 - r\right). \tag{4.15}$$

The eddy viscosity is related to the Spalart-Allmaras quantity $\widetilde{\nu}$ by

$$\mu_t = \overline{\rho}\widetilde{\nu}\frac{\chi^3}{\chi^3 + c_{v1}} \tag{4.16}$$

$$\chi = \frac{\overline{\rho}\widetilde{\nu}}{\overline{\mu}}, \tag{4.17}$$

and the closure coefficients are given by the following standard values

$$\sigma_{SA} = \frac{2}{3}; \qquad c_{b1} = 0.1355; \qquad c_{b2} = 0.622;$$

$$c_{v1} = 7.1; \qquad c_{w2} = 0.3; \qquad c_{w3} = 2.$$

Three different models of this type are implemented into the FLOWer code, the original Spalart-Allmaras model [19] (SA), its modified version of Edwards [6] (SAE) and a yet unpublished "strain adaptive linear" version (SALSA) of TU Berlin. These models can all be cast into the above form and differ only in the definitions of the parameters $\widetilde{S}$, r, C_μ, and c'_w.

The baseline Spalart-Allmaras model has been developed on empirical grounds along with physical reasoning [19]. It has been especially designed for exterior aerodynamics and is well suited for standard aeronautical configurations such as airfoils, wings, or wing-body combinations. The modification of Edwards [6] is mainly directed to increase the numerical stability of the baseline model, whereas the SALSA model is aiming at an improved prediction quality.

The major drawback of all models of this class is that they rely on the global minimum wall distance d, the computation of which is rather time consuming for complex multi-block configurations. Support has been given by TU Berlin to accelerate this part of the implementation and to stabilize the numerics of the Spalart-Allmaras type models in the FLOWer code.

4.2.3 Two-equation models

Two-equation models provide transport equations for a velocity scale, typically the specific kinetic turbulence energy $\widetilde{k}$, and a length scale [22]. The models implemented into the FLOWer code are based on equations for $\widetilde{k}$ and ω, the so-called specific dissipation rate [22], which take the following general form:

$$\frac{\partial\left(\overline{\rho}\widetilde{k}\right)}{\partial t} + \vec{\nabla}\cdot\left(\overline{\rho}\widetilde{\vec{V}}\widetilde{k}\right) = \vec{\nabla}\cdot\left[\left(\overline{\mu} + \frac{\mu_t}{\sigma_k}\right)\vec{\nabla}\widetilde{k}\right] + P_k - D_k \tag{4.18}$$

$$\frac{\partial\left(\overline{\rho}\omega\right)}{\partial t} + \vec{\nabla}\cdot\left(\overline{\rho}\widetilde{\vec{V}}\omega\right) = \vec{\nabla}\cdot\left[\left(\overline{\mu} + \frac{\mu_t}{\sigma_\omega}\right)\vec{\nabla}\omega\right] + P_\omega - D_\omega + C_D. \tag{4.19}$$

The production and destruction terms are defined by

$$P_k = \mu_t S^2 - \frac{2}{3}\overline{\rho}\widetilde{k}\left(\vec{\nabla}\cdot\vec{\widetilde{V}}\right) \tag{4.20}$$

$$P_\omega = \beta_{\omega 1} P_k \frac{\omega}{\widetilde{k}} \tag{4.21}$$

$$D_k = \beta_k \overline{\rho}\widetilde{k}\omega \tag{4.22}$$

$$D_\omega = \beta_{\omega 2}\overline{\rho}\omega^2 \tag{4.23}$$

where $S = \sqrt{2\widetilde{S}^*_{ij}\widetilde{S}^*_{ij}}$ is the strain-rate. The cross-diffusion term

$$C_D = \sigma_d \frac{\overline{\rho}}{\omega}\left(\vec{\nabla}\widetilde{k}\right)\cdot\left(\vec{\nabla}\omega\right) \tag{4.24}$$

results from the formal transformation of the well-known equation for the dissipation rate ϵ to the specific dissipation rate ω and is present only in part of the implemented models.

From the above equations the eddy viscosity is computed according to

$$\mu_t = C_\mu \frac{\overline{\rho}\widetilde{k}}{\omega}. \tag{4.25}$$

Currently, five different $\widetilde{k}-\omega$ type turbulence models are implemented into the FLOWer code which differ only in the definition of the closure coefficients σ_k, σ_ω, σ_d, β_k, $\beta_{\omega 1}$, $\beta_{\omega 2}$, and C_μ. They are briefly described in the following.

- **Wilcox $\widetilde{k}-\omega$ model (1988) [21]**
  The $\widetilde{k}-\omega$ model of Wilcox is implemented in its 1988 version [21]. It has only constant closure coefficients, especially $C_\mu \equiv 1$. The model is numerically rather robust, but its results are known to be dependent on the freestream value of ω [8], due to the lacking cross diffusion term ($\sigma_d = 0$). Nevertheless, from a practical point of view this does not pose a severe problem, if the far field is placed far enough away, so that ω decays to its lower limit.
- **Kok TNT $\widetilde{k}-\omega$ model [8]**
  Kok's TNT model [8] has been developed to resolve the freestream dependence of the Wilcox model [21] by introducing a cross diffusion term ($\sigma_d = 0.5$) and re-defining the values of σ_k and σ_ω. Computations with the FLOWer code have shown only little influence on the results with respect to the baseline Wilcox model.
- **Menter SST $\widetilde{k}-\omega$ model [12]**
  Menter's SST model [12] resolves the problem of freestream dependence by combining the $\widetilde{k}-\omega$ model of Wilcox in the near-wall area with the standard $\widetilde{k}-\epsilon$ model in the off-wall region. Transformation of ϵ into ω introduces the cross-diffusion term. Furthermore, the eddy viscosity is limited to cope with Bradshaw's assumption [12].
  Menter's model has shown to improve the predictions, especially for separated flows. However, like the Spalart-Allmaras type models, it suffers from its dependence on the global minimum wall distance.

- **LLR $\widetilde{k} - \omega$ model [18]**
  The LLR $\widetilde{k} - \omega$ model [18] is a development of TU Berlin, aiming at a linear eddy viscosity model in strictly local formulation that satisfies a number of realizability constraints. It does not contain the cross diffusion term ($\sigma_d = 0$), but all other closure coefficients, except $\sigma_{\widetilde{k}}$ and σ_ω, are functions of the mean flow.
  The original version of the model appeared to be numerically rather unstable due to the ω-production term. Therefore, the near-wall treatment of the corresponding coefficient $\beta_{\omega 1}$ has been re-formulated [16], making the model operational at least for simple test cases.
- **LEA $\widetilde{k} - \omega$ model [17]**
  The LEA $\widetilde{k} - \omega$ model [17] is the linear truncation of an Explicit Algebraic Reynolds Stress Model. It has been developed especially for transonic flows and differs from the $\widetilde{k} - \omega$ model of Wilcox [21] only by a variable coefficient C_μ.

4.2.4 Explicit Algebraic Reynolds Stress Models (EARSM)

As already stated, EARSMs can be interpreted as non-linear tensor expansions of linear eddy viscosity models. As representative of that class the model of Wallin and Johannson [20] is implemented into the FLOWer code, based on the $\widetilde{k} - \omega$ model of Wilcox. If the Reynolds stresses of the linear eddy viscosity model are given by

$$\widetilde{R}_{ij}^{(EVM)} = \mu_t \widetilde{S}_{ij}^* - \frac{2}{3}\overline{\rho}\widetilde{k}\delta_{ij}, \tag{4.26}$$

then the Wallin and Johansson formulation [20] can be written as

$$\widetilde{R}_{ij}^{(EARSM)} = \widetilde{R}_{ij}^{(EVM)} - \overline{\rho}\widetilde{k}a_{ij}^{(ex)} \tag{4.27}$$

In this equation $a_{ij}^{(ex)}$ denotes the extra anisotropy tensor which is defined by

$$\begin{aligned} a_{ij}^{(ex)} = \beta_3 &\left(\widetilde{W}_{ik}\widetilde{W}_{kj} - \frac{1}{3}\widetilde{W}_{kl}\widetilde{W}_{lk}\delta_{ij}\right) + \beta_4 \left(\widetilde{S}_{ik}^*\widetilde{W}_{kj} - \widetilde{W}_{ik}\widetilde{S}_{kj}^*\right) \\ +\beta_6 &\left(\widetilde{S}_{ik}^*\widetilde{W}_{kl}\widetilde{W}_{lj} + \widetilde{W}_{ik}\widetilde{W}_{kl}\widetilde{S}_{lj}^* - \widetilde{W}_{kl}\widetilde{W}_{lk}\widetilde{S}_{ij}^* - \frac{2}{3}\widetilde{S}_{kl}^*\widetilde{W}_{lm}\widetilde{W}_{mk}\delta_{ij}\right) \\ +\beta_9 &\left(\widetilde{W}_{ik}\widetilde{S}_{kl}^*\widetilde{W}_{lm}\widetilde{W}_{mj} - \widetilde{W}_{ik}\widetilde{W}_{kl}\widetilde{S}_{lm}^*\widetilde{W}_{mj}\right) \end{aligned} \tag{4.28}$$

Finally, it should be noted that the eddy viscosity of the Wallin-Johansson EARSM [20] is based on a variable coefficient C_μ.

There exist two different formulations of that model, one for two- and one for three-dimensional flow, that are both implemented into the FLOWer code. Moreover, the implementation allows the linear truncation of the model, leading to a linear eddy viscosity model that is related to the above LEA $\widetilde{k} - \omega$ model [17].

Its relation to the LEA $\widetilde{k} - \omega$ model [17] suggests the EARSM of Wallin and Johansson [20] being applied to transonic flow. It should be noted that its non-linear terms have not yet shown to significantly influence the results for such type of external aerodynamic problems. However, they may reduce the numerical robustness of the code, although this is not necessarily the case.

4.3 Comparison of the models

In order to assess the predictive capabilities of the different turbulence models implemented into the FLOWer code within a verification phase, two simple test cases have been computed. They are regarded typical for representing transonic and high-lift flow conditions.

4.3.1 Transonic flow: RAE 2822 airfoil

The flow around the RAE 2822 airfoil has been computed for conditions referred to as "Case 9“ ($Ma = 0.73$, $Re = 6.5 \cdot 10^6$, $\alpha = 2.80°$) and "Case 10“ ($Ma = 0.75$, $Re = 6.2 \cdot 10^6$, $\alpha = 2.80°$) [3]. Case 9 is characterized by a moderate shock that does not induce separation, whereas Case 10 shows a small separation bubble at the foot of a rather strong shock. According to the measurements, transition was fixed at 3% chord length.

The FLOWer computations were carried out on a rather fine grid, consisting of 736x176 cells. It is shown in Fig. 4.1. As requested, the first wall spacing yielded a value of $y_1^+ < 0.8$ for the Wilcox $\widetilde{k} - \omega$ model. Grid independence of the solution was checked with the same model by successively removing every second grid line in i- and j-directions. A central discretization scheme with Jameson type artificial dissipation and dissipation coefficients of $k^{(2)} = 1/2$ and $k^{(4)} = 1/64$ was used.

Fig. 4.2 depicts the pressure distributions for the baseline turbulence models of each class implemented into FLOWer. As one can see, all predictions agree fairly well with the experiments, except of the shock position and, for Case 10, the region behind. The Baldwin-Lomax model [1] predicts the shock always downstream the measured position, especially for Case 10. The Spalart-Allmaras model [19] shows good agreement with the experimental shock position of Case 10, but yields a too early shock for Case 9. In contrast, the Wilcox $\widetilde{k} - \omega$ model shows just the opposite behaviour with respect to the predicted shock positions.

Fig. 4.3 shows the influence of the modifications to the baseline one-equation models on the pressure distribution close to the respective shock positions for Cases 9 and 10. As one can see, the Edwards modification [6] and the SALSA model both generally move the shock position downstream with respect to the baseline model, while narrowing the overall deviation from the experiments. Concerning the results in Fig. 4.3, the SALSA model of TU Berlin seems to be the best compromise for the test cases considered.

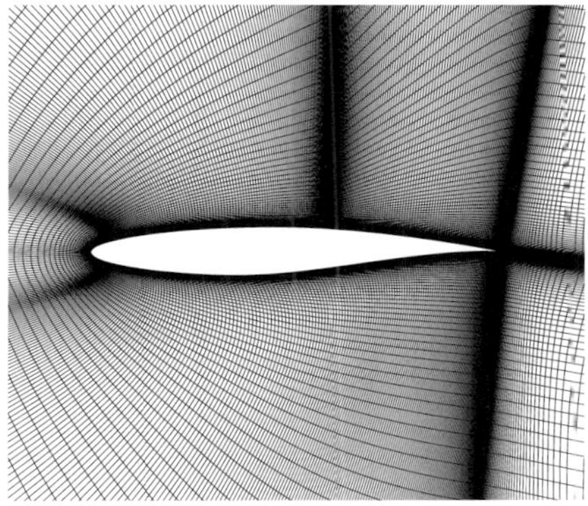

Fig. 4.1. RAE 2822, detail of the grid.

Fig. 4.4 shows the same close-up of the pressure distributions as Fig. 4.3, but for different two-equation models and the Wallin and Johansson EARSM [20]. As one can see, all more advanced two-equation models generally move the shock position upstream with respect to the baseline Wilcox $\widetilde{k}-\omega$ model [21]. Once again the overall deviation from the experiments is reduced by these models where the LEA $\widetilde{k}-\omega$ model [17] and the Wallin and Johansson EARSM [20] seem to be the best compromise for the tested cases. It should be noted that both models are based on similar ideas, supporting the rationale behind Explicit Algebraic Reynolds Stress Modelling at least for transonic flow.

4.3.2 High-lift flow: Aérospatiale A airfoil

The Aérospatiale A airfoil serves as a test case for high-lift flow conditions, showing a small laminar separation bubble at approximately 12% chord length and trailing edge separation [2]. The flow around that configuration has been computed for four different incidences, ranging from $\alpha = 7.2°$ to $\alpha = 17.1°$ at $Ma = 0.15$ and $Re = 2.0 \cdot 10^6$. For all computations transition has been fixed at 30% chord length on the pressure side and at 12% on the suction side. Only two-equation models, including the Wallin and Johansson EARSM [20], have been considered, because they represent the standard application with the FLOWer code.

The FLOWer computations were carried out on a standard grid, consisting of 352x64 cells which is shown ind Fig. 4.5. The first grid spacing resulted in greater values than $y_1^+ = 1$ in some regions, the highest value of $y_{1,max}^+ \approx 2.3$ being obtained with the LEA $\widetilde{k}-\omega$ model [17] at $\alpha = 17.1°$. The numerical parameters were kept identical with those for the RAE 2822 test case.

The left part of Fig. 4.6 shows the obtained lift coefficients C_L versus the angle of attack compared to the measurements in the ONERA F1 wind tun-

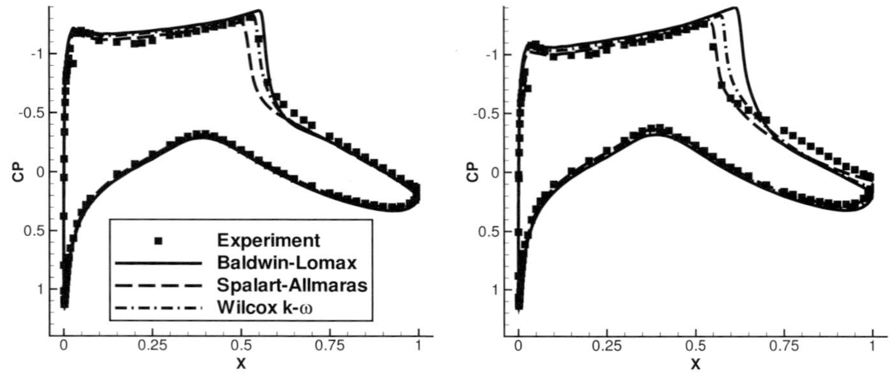

Fig. 4.2. RAE 2822, Cases 9 and 10. Pressure distributions of baseline models.

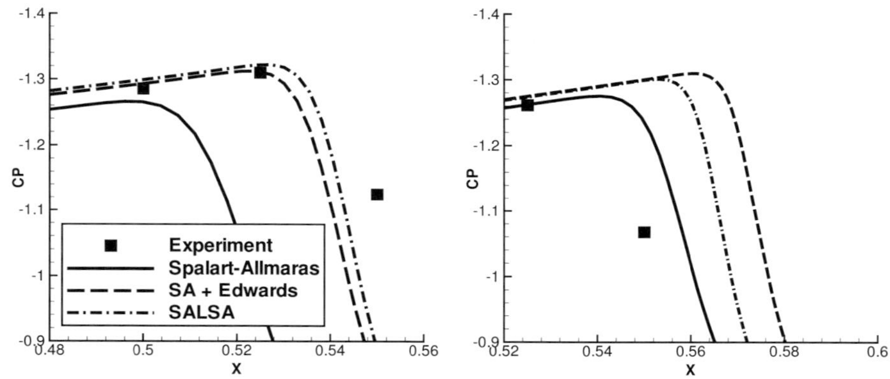

Fig. 4.3. RAE 2822, Cases 9 and 10. Pressure distributions of one-equation turbulence models near the shock.

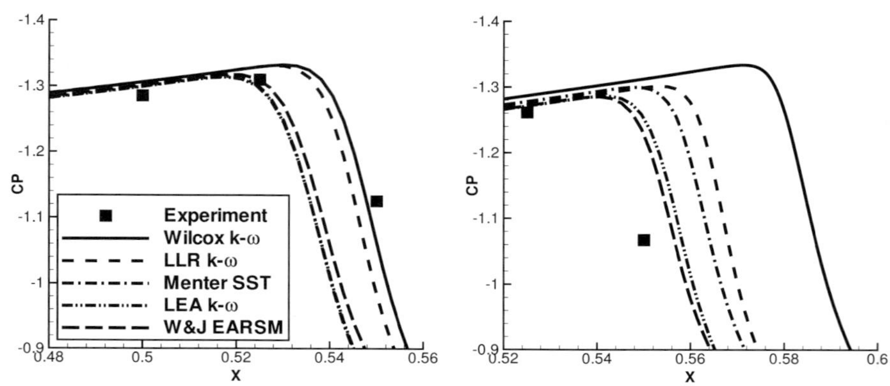

Fig. 4.4. RAE 2822, Cases 9 and 10. Pressure distributions of two-equation turbulence models and EARSM near the shock.

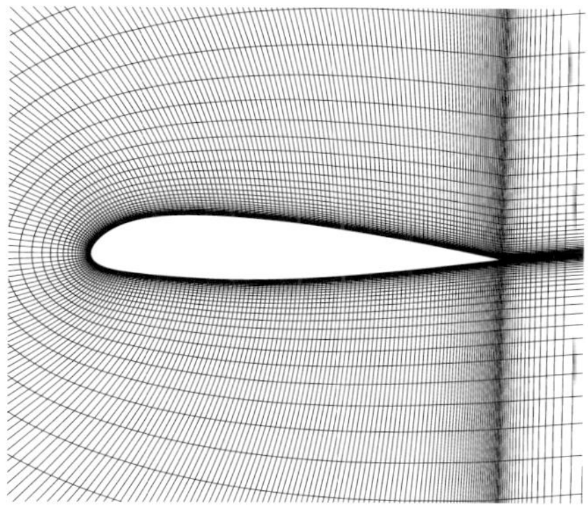

Fig. 4.5. Aérospatiale A airfoil, detail of the grid.

nel. As one can see, for $\alpha = 7.2°$ all models are in very good agreement with the experimental value. However, with increasing incidence, the differences between the models grow, and deviations from the measured curve become visible. In general, all models fail to predict the non-linearity close to maximum lift correctly. Once again, the LEA $\widetilde{k} - \omega$ model [17] and the EARSM of Wallin and Johansson [20] achieve the best agreement with the experiments which may partly be due to a cancelation of errors at higher angles of attack. Nevertheless, the comparison with the measured pressure distribution at $\alpha = 13.3°$ near the suction peak on the left hand side of Fig. 4.6 seems to confirm the improved predictive capabilities of the LEA and EARSM. Moreover, it should be noted that the non-linear Wallin and Johansson model [20] showed an excellent convergence behaviour for this test case.

4.4 Realistic applications

In the last section turbulence models were tested with respect to simple two-dimensional flows, inhibiting phenomena that were regarded characteristic of industrial applications. Therefore, in this section examples are given for realistic flow computations around more complex configurations.

Fig. 4.7 shows two pressure distributions at an inboard and an outboard wing section of a wing-body combination at $Ma = 0.85$, $Re = 32.5 \cdot 10^6$ and $C_L = 0.67$ investigated in the EU-project HiReTT [9]. As one can see, at $\eta = 0.216$ there is not much difference between the Wilcox [21], Menter SST [12] and LEA [17] $\widetilde{k} - \omega$ models, all being in good agreement with the experiments. However, towards the wing tip the predictions deviate considerably. At $\eta = 0.837$ the Wilcox model [21] gives a shock position too far downstream, whereas the Menter SST model [12] results in a shock position slightly upstream of the measurements. In contrast, the LEA $\widetilde{k} - \omega$ model predicts the

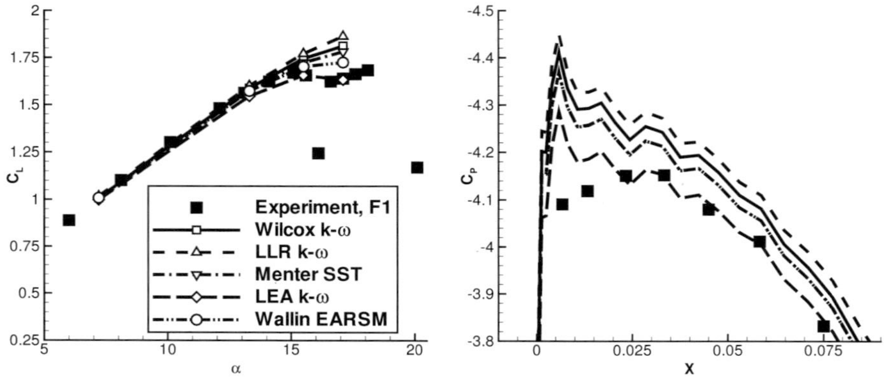

Fig. 4.6. Aérospatiale A airfoil. *Left*: Lift coefficient vs. incidence for two-equation turbulence models and EARSM. *Right*: Pressure distribution at $\alpha = 13.3°$ near the suction peak.

pressure distribution in good agreement with the experiments all along the span, thus, confirming the findings with the RAE 2822 airfoil.

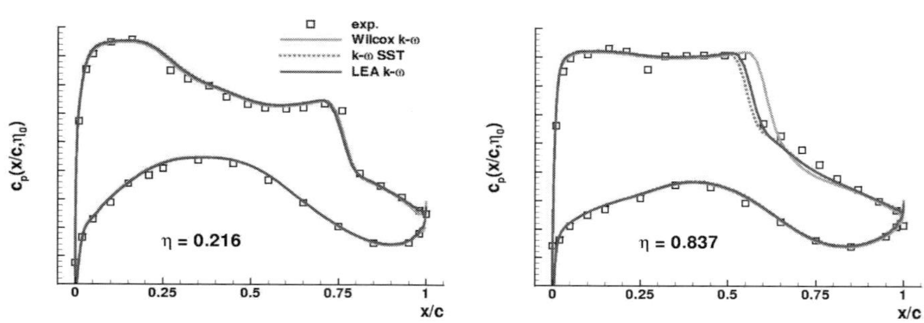

Fig. 4.7. HiReTT wing-body combination. Pressure distributions at an inboard ($\eta = 0.216$) and an outboard wing section ($\eta = 0.837$). Comparison of Wilcox [21], Menter SST [12] and LEA [17] $\widetilde{k} - \omega$ models.

Another wing-body combination has been investigated in the 2001 AIAA Drag Prediction Workshop [14]. The DLR-F4 configuration was computed at six different angles of attack, ranging from $\alpha = -3°$ to $\alpha = 2°$ at $Ma = 0.75$ and $Re = 3 \cdot 10^6$. Fig. 4.8 shows the results for the lift versus incidence and the moment versus lift curves for the Spalart-Allmaras model with Edwards modification [6], the Wilcox [21], and the LEA [17] $\widetilde{k} - \omega$ models. As one can see, the Wilcox model [21] consistently over-estimates the lift at a given incidence, while the SAE model [6] yields a too low lift at $\alpha > 0°$. Only the results obtained with the LEA model [17] follow the experimental curve throughout the whole incidence range, including the slight non-linearity at $\alpha > 1.5°$. The superiority of the LEA model [17] is even more pronounced, considering the moment versus lift coefficient. Noticing that there is quite some scatter in the experimental results, anyway, the LEA results are the only ones, following closely the S-shape of the DRA measurements.

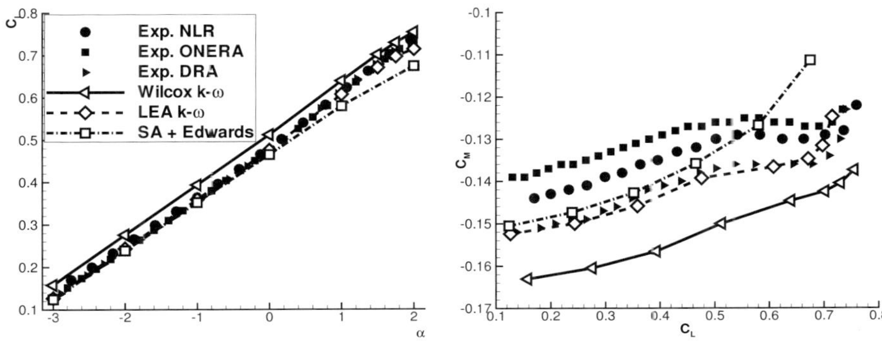

Fig. 4.8. DLR-F4 wing-body combination. *Left*: Lift coefficient vs. incidence. *Right*: Lift vs. moment coefficient.

Finally, Fig. 4.9 shows results for the three-element A-310 airfoil in landing configuration, as investigated in the EU-project EUROLIFT [15]. Computations have been carried out for seven angles of attack between $\alpha = 12°$ and $\alpha = 25°$ at $Ma = 0.22$ and $Re = 4 \cdot 10^6$. The results show that the Spalart-Allmaras [19], Wilcox [21] and LEA models [17] similarly follow the experimental curve up to an incidence of $\alpha \approx 22°$. However, with the Spalart-Allmaras model [19] the breakdown of lift is predicted prematurely, whereas the results obtained with the two $\widetilde{k}-\omega$ models remain close to the experiments up to the highest angles of attack. Moreover, unsteady computations indicate that with the Wilcox model [21] even the lift breakdown can be predicted correctly.

4.5 Conclusions

In the FLOWer code a variety of turbulence models is available, ranging from algebraic to two-equation models and EARSM. In general, it can be stated that the more advanced models yield improved predictions, but it should be noted that they might require a higher grid resolution because they rely on powers of spatial derivatives of the mean flow velocity components. Based on the experience gained so far, the Wilcox $\widetilde{k}-\omega$ model [21] is recommended as a reliable standard model. However, for transonic flow the LEA $\widetilde{k}-\omega$ model [17] and the EARSM of Wallin and Johansson [20] give the best results. The latter

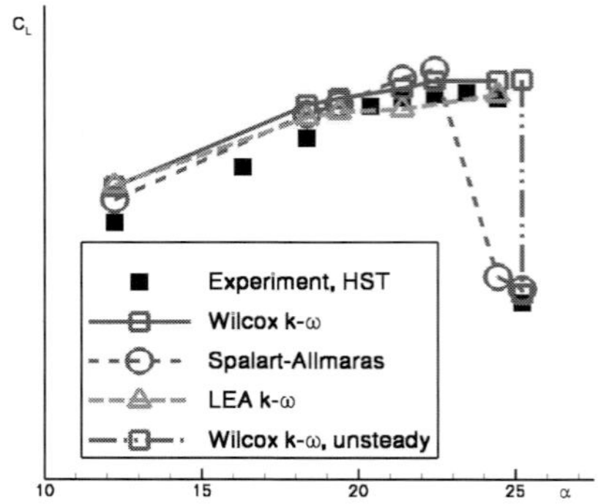

Fig. 4.9. A-310 airfoil in landing configuration. Lift coefficient vs. incidence.

models are also promising, concerning the results for the Aérospatiale A airfoil. Nevertheless, for high-lift flow the question on the best suited turbulence model is still open. Maybe full Reynolds stress transport models can provide some improvement due to their inherent ability of capturing rotational and curvature effects.

References

1. Baldwin, B. S., Lomax, H., "Thin-Layer Approximation and Algebraic Model for Separated Turbulent Flows", AIAA Paper 78-0257, 1978
2. Chaput, E., "Aérospatiale-A Airfoil", In: Haase, W., Chaput, E., Elsholz, E., Leschziner, M. A., Müller, U. R., "ECARP – European Computational Aerodynamics Research Project: Validation of CFD Codes and Assessment of Turbulence Models", Notes on Numerical Fluid Mechanics, Vol. 58, Vieweg, 1997
3. Cook, P. H., McDonald, M. A., Firmin, M. C. P., "Aerofoil RAE 2822 – Pressure Distributions, and Boundary Layer and Wake Measurements", In: J. Barche (Ed.), "Experimental Data Base for Computer Program Assessment", AGARD-AR-138, 1979
4. Davidson, L., Cokljat, D., Fröhlich, J., Leschziner, M. A., Mellen, C., Rodi, W., (Eds.), "LESFOIL: Large Eddy Simulation of Flow Around a High Lift Airfoil", Springer, to appear 2003
5. Degani, D., Schiff, L. B., "Computation of Turbulent Supersonic Flows around Pointed Bodies Having Crossflow Separation", Journal of Computational Physics, **66** (1986) 173-196
6. Edwards, J. R., Chandra, S., "Comparison of Eddy Viscosity-Transport Turbulence Models for Three-Dimensional, Shock-Separated Flowfields", AIAA Journal **34** (1996) 756-763
7. Gatski, T. B., Rumsey, C. L., "Linear and Nonlinear Eddy Viscosity Models", In: Launder, B., Sandham, N., "Closure Strategies for Turbulent and Transitional Flows", Cambridge University Press, 2002

8. Kok, J. C., "Resolving the Dependence on Freestream Values for the k–ω Turbulence Model", AIAA Journal **38** (2000) 1292-1295
9. Krumbein, A., DLR, personal communication
10. Laurence, D., "Large Eddy Simulation for Industrial Flows?", In: Launder, B., Sandham, N. (Eds.), "Closure Strategies for Turbulent and Transitional Flows", Cambrigde University Press, 2002
11. Leschziner, M. A., Lien, F.-S., "Numerical Aspects of Applying Second-Moment Closure to Complex Flows", In: Launder, B., Sandham, N. (Eds.), "Closure Strategies for Turbulent and Transitional Flows", Cambridge University Press, 2002
12. Menter, F. R., "Two-Equation Eddy-Viscosity Turbulence Models for Engineering Applications", AIAA Journal **32** (1994) 1598-1605
13. Prandtl, L. "Bericht über Untersuchungen zur ausgebildeten Turbulenz", Zeitschrift für angewandte Mathematik und Mechanik **5** (1925) 136-139
14. Rakowitz, M., Sutcliffe, M., Eisfeld, B., Schwamborn, D., Bleecke, H., Fassbender, J., "Structured and Unstructured Computations of the DLR-F4 Wing-Body Configuration", AIAA Paper 2002-0837, Reno, 2002
15. Rudnik, R., Heinrich, R., Eisfeld, B. Schwarz, Th., "DLR Contributions to Code Validation Activities within the European High Lift Project EUROLIFT", STAB Symposium, München, 2002
16. Rung, T., Bombardier Transportation, personal communication, 2003
17. Rung, T., Lübcke, H., Franke, M., Xue, L., Thiele, F., Fu, S., "Assessment of Explicit Algebraic Stress Models in Transonic Flows", In: Proceedings of the 4th International Symposium on Engineering Turbulence Modelling and Measurements, Corsica, 1999, pp. 659-668
18. "Computational modelling of complex boundary-layer flows", In: Proceedings of the 9th International Symposium on Transport Phenomena in Thermal-Fluids Engineering, Singapore, 1996, pp. 321-326
19. Spalart, P. R., Allmaras, S. R., "A one-equation turbulence model for aerodynamic flows", La Recherche Aèrospatiale **1** (1994) 5-21
20. Wallin, S., Johansson, A. V., "An explicit algebraic Reynolds stress nodel for incompressible and compressible turbulent flows", Journal of Fluid Mechanics **403** (2000) 89-132
21. Wilcox, D. C., "Reassessment of the Scale-Determining Equation for Advances Turbulence Models", AIAA Journal **26** (1988) 1299-1310
22. Wilcox, D. C., "Turbulence Modeling for CFD", 2nd edition, DCW Industries, 1998

Part III

Hybrid Solver TAU

5

Overview of the Hybrid RANS Code TAU

Thomas Gerhold

DLR, Institute of Aerodynamics and Flow Technology, Bunsenstr. 10, D-37073 Göttingen

Summary. A brief introduction is given which first describes the history of frame in which the TAU code was developed before explaining the main advantages which were the drivers for the selection of the approach. In the following an algorithmic overview describes shortly the code functionality before a section about the code design gives some more insight about the implementation and its scripting capability.

5.1 The unstructured Navier-Stokes code TAU

5.1.1 History

The main developments for the unstructured Navier-Stokes code TAU were based on the national CFD project MEGAFLOW within the framework of the German aerospace program [1, 2], which combined activities from DLR, German universities and aircraft industry. Other DLR projects, like Aero-Sum/SikMa, Aerostabil, Amanda, 3FF, IMENS, contributed much with respect to specific algorithmic extensions as well as work on code verification and validation. Common CFD developments on the European level contained the TAU code as flow solver part, such as the DLR-NLR cooperation "CFD for complete Aircraft" and projects founded by the EC, like the FASTFLO I and FASTFLO II project as well as the project TAURUS. Furthermore, the TAU-code was/is exploited by DLR in a number of other European projects, like AVTAC, EUROLIFT, HiAer, HiRett, UNSI, and FLOWMANIA. A newly established activity, aiming for future extensions of the TAU-code is the DLR-ONERA cooperation for Common unstructured code development. Due to an intensive use of the code at DLR and in German Aerospace Industry in a wide range of applications [3, 4] the code has reached a high level of maturity and reliability. This holds especially for applications in the field of aircraft design and for analysis of external and internal flows.

5.1.2 Advantages of the method

At start, the driver of the development of a Navier-Stokes code for hybrid grids was to benefit from the flexibility of the unstructured methods with respect to grid generation, grid adaptation and grid partitioning.

The generation of unstructured grids allows for a much higher level of automation and geometrical complexity than it can be achieved for structured multi-block grids. This decreases the user interaction in the grid generation process and thus reduces the effort and time needed to obtain an initial grid for a given CAD geometry description significantly.

The possibility of local grid re- and de-refinement in unstructured grids allows for an automated adaptation of the cell sizes to local flow phenomena in a given initial grid during the computation, which is needed to reduce discretisation errors and thus increases the solution accuracy. This automation allows to further decrease the effort needed for the grid generation, because suitable grid resolution can be obtained, without generating several grids by hand depending on the results of previous calculations. Furthermore, the local refinement capability allows for fast transition from small to large cell sizes in regions where a coarse grid resolution is sufficient in order to reduce the number of grid points and thus the computational time required. The possibility of partioning unstructured grids in equally balanced parts allows in an optimal manner for a parallelisation of the algorithms based on domain decomposition and thus for highly optimised software for massive parallel computing. The potential to achieve a very high parallel efficiency, and thus a very low turnaround time for the computational problem, overweighs the disadvantage of unstructured methods that indirect addressing increases the computational time needed for the integration of the governing equations per grid node.

The above named aspects become important with increasing geometrical complexity, which makes the grid generation more difficult and time consuming. Furthermore together with geometrical complexity the range of different scales of flow phenomena to be resolved usually grows, too. Finally, this leads to an increased number of grid points, such that a parallel method becomes evident for a reasonable computational turnaround time. Fig. 5.1 illustrates the geometrical complexity of the today applications the TAU code is employed for the prediction of turbulent flows.

5.2 Algorithmic overview

The TAU-Code is an unstructured method based on a dual mesh approach, which is well suited for hybrid grids thus allowing the use of mixed-element meshes composed of tetrahedron, prisms, hexahedra, and/or pyramids. The combination of these elements is considered to allow for regular grids in the vicinity of walls in the combination with the more flexible tetrahedral elements in the remaining computational domain.

Fig. 5.1. Geometrical complexity of today applications of the TAU-code

Building the regular grid parts with hexahedra and/or prisms allows for high aspect ratio cells with their edges aligned to the wall-normal and the wall tangential directions. This near-wall grid topology is known from the established structured methods to be well suited for accurate and efficient resolution of boundary layers. With a given surface discretization composed of triangles and/or quadrilaterals the generation of structured-type subgrids over viscous aerodynamic surfaces can be done automatically by using a front method and is not too complex, because the extend of the subgrids can be restricted to thin overall heights estimated from a maximum boundary-layer thickness. Filling the remaining volume with tetrahedral elements can also be done automatically applying known algorithms as advancing front or Delaunay methods, when using pyramidal cells for the transition between quadrilateral and triangular cell faces. Thus the volume-grid generation process can be performed without intermediate user interaction, which allow for fast grid generation turnaround time.

5.2.1 Grid generation

Since the TAU code does not contain a grid generator a strategic cooperation has been established with the company CentaurSoft [5], which provides the hybrid grid generation package Centaur. The software consists of an interactive part to read in the CAD data of the geometry under consideration, to perform CAD cleaning if necessary and to set up the grid generation process. The second part is for the automatic surface and volume grid generation.

5.2.2 Preprocessing step

From the primary hybrid grid the TAU-Code computes the dual grid composed of general control volumes from the primary elements. They are stored in an edge based data structure (described in more detail in [6]), which makes the solver independent of the element types of the primary grid. All metrics are given by normal vectors, representing size and orientation of the faces, the geometric coordinates of the grid nodes and the volumes of the auxiliary cells. The connectivity of the grid is given by linking the two nodes on both sides of each face into the data structure for the face (=edge). In order to employ a multigrid technique the agglomeration approach is used to obtain coarse grids by fusing together the fine grid control volumes, which are again described by the same metrics. Therefore, the coarse grid solution can be computed with the same approach as on the finest grid. The transfer operators needed for the communication between the different grids are obtained directly during the agglomeration process.

5.2.3 Spatial discretization

Both inviscid and viscous fluxes are evaluated edgewise in a loop over all edges. By adding the flux contribution of each edge to its corresponding end points the flux balance for all control volumes is obtained as soon as the loop over all edges is finished. The inviscid fluxes along the edges are computed employing either one of several implemented upwind schemes or employing a central scheme. By considering whirl fluxes in flux balance the code is capable to account for arbitrary rigid body motion of the grids. Additional terms accounting for the geometric conservation allow also for free stream consistency when mesh deformation is applied. For the calculation of low-speed flows, preconditioning of the compressible flow equations is implemented.

5.2.4 Time stepping

The integration in time is done employing an explicit Runge-Kutta time stepping algorithm, which, as a part of a dual time stepping approach, can also be applied to transient flows. Alternatively to the Runge-Kutta scheme implicit techniques can be employed, which are under investigation.

For a high performance acceleration techniques are employed like the local time stepping concept and residual smoothing. A multigrid method is available to speed up convergence furthermore.

Optimisation techniques related to different computer platforms are employed, like colouring of the edges either for vector machines or for minimisation of cache loads.

5.2.5 Parallelization

With the explicit time stepping scheme, each domain can be treated as a complete grid when employing the domain decomposition method for parallel computing. During the flux integration data have to be exchanged between the different domains several times. In order to enable the correct communication, there is one layer of ghost nodes located at each interface between two neighbouring domains. Edges of the dual grid connect the ghost nodes with the regular ones. These edges are those, which have been cut by the grid partitioning algorithm. The cut edges are part of both domains. Since a ghost node of one domain is a regular node of the corresponding neighbour domain the ghost-node values are to be updated by the regular node values before an operation on cut edges is performed. In order to allow for the use of distributed memory computers MPI is employed for exchanging the data for the update of node values. The number of ghost nodes compared to that of the regular nodes depends on the relation between the number of domains and the size of the global (non-decomposed) grid. Since (massive) parallel computing is needed for reduced turnaround it is as more important as larger the size of the computational global grid is. For large (grid-) scale problems a reasonable number of points and edges remain in each grid partition and as a result the relation to the number of cut edges and ghost nodes remain small, such that the overhead introduced by parallelization does not dominate the performance. In any case, for moderate parallel computations (e.g. 8 domains or less) the overhead remains small or negligible.

Fig. 5.2 illustrates the parallel performance of the TAU-code on different computer architectures. The benchmark was performed with a one-equation turbulence model on a hybrid grid composed of about 2 million grid points employing a 4W multigrid cycle, a 3stage Runge-Kutta method and a central spatial discretization. The parallel efficiency decreases significantly using 64 processors, because the grid partitions become very small especially on the coarse multigrid levels. Applying the code for the same example on 64 processors in single grid mode the efficiency remains at 83%.

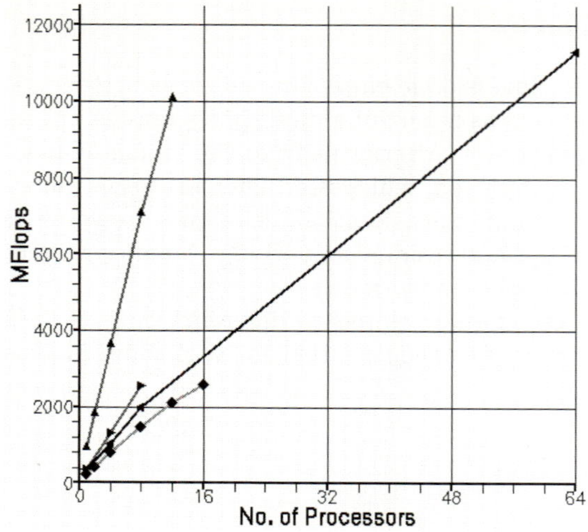

Architecture (left to right)	CPU/node	CPUs	parallel efficiency	sec. per step
NEC SX5-B	–	12	91 %	3.0
Xeon	1	8	92 %	11.0
Xeon	2	64	50 %	2.7
Athlon	2	16	65 %	12.0

Computing platforms:

NEC SX5-B	925 MFLOPS on 1 CPU
Xeon (2.0 GHz Myrinet)	350 MFLOPS on 1 CPU
Athlon (MP 16 GHz 1 GBit)	250 MFLOPS on 1 CPU

Fig. 5.2. Parallel performance on different architectures

5.2.6 Chimera

As the Chimera technique has been recognized as an important feature to efficiently simulate manoeuvring aircraft, it has been also integrated into the TAU-Code. In the context of hybrid meshes the overlapping grid technique allows an efficient handling of complex configurations with movable control surfaces or other bodies in relative motion. For the data exchange in grid overlap regions linear interpolation based on a finite element approach is used. The search algorithm for donor cells is based on the alternating digital tree data structure.

5.2.7 Grid adaptation

In order to efficiently resolve detailed flow features, a grid adaptation algorithm for hybrid meshes based on local grid refinement and wall-normal mesh movement in semi-structured near-wall layers is applied [7]. This algorithm also allow for de-refinement of earlier refined elements thus enabling the code to be used for unsteady time-accurate adaptation in unsteady flows. Fig. 5.3 gives a simple example of the process for viscous airfoil calculation. First a flow solution is calculated on a basic grid (a). After some refinement an adapted grid/solution is obtained (b). Changing the flow parameters and specifying e.g. that the number of mesh points should not increase any further, the de-refinement interacts with the refinement (c) and finally the new shock position is resolved (d).

5.2.8 Turbulence modeling

For turbulent flows different one- and two-equation turbulence models are implemented, which are described in detail in a later section [8].

5.3 Code design

5.3.1 Modules

The implementation of the algorithms follows a modular software concept in order to reach a high level of flexibility and to ease maintenance. The main modules are the grid partitioner, the preprocessor, the solver, the adaptation and the deformation. Furthermore, data postprocessing tools and converters exists as separate software parts.

The functionality of the main modules is:

- The grid partitioner is used to compute load balanced domains of the primary grid on which the preprocessing-algorithms (see below) can be employed in parallel. If the grid connectivity remains unchanged, i. e. no grid re- or de-refinement is performed, and the number of domains, i. e. the number of parallel processes, is kept constant during a computation, the partitioner needs to be employed only once for a complete simulation.
- The preprocessing contains all algorithms needed to compute the dual grid data from the primary grid cells. It needs to be employed once for a given primary grid and at least parts of the algorithms need to be recomputed each time the primary grid is modified. Grid modifications can be due to adaptation, mesh deformation or changed chimera grid overlap when using grids in relative motion. The preprocessing contains e.g.:
  - the construction of dual grid control volumes,
  - the agglomeration of dual cells for the coarse multigrid levels,
  - partitioning of dual grids in case of running the pre-processor in sequential mode, computation of wall-distances for turbulence models,
  - computation of regions in which the production of turbulence is suppressed according to given arbitrary transition-line,

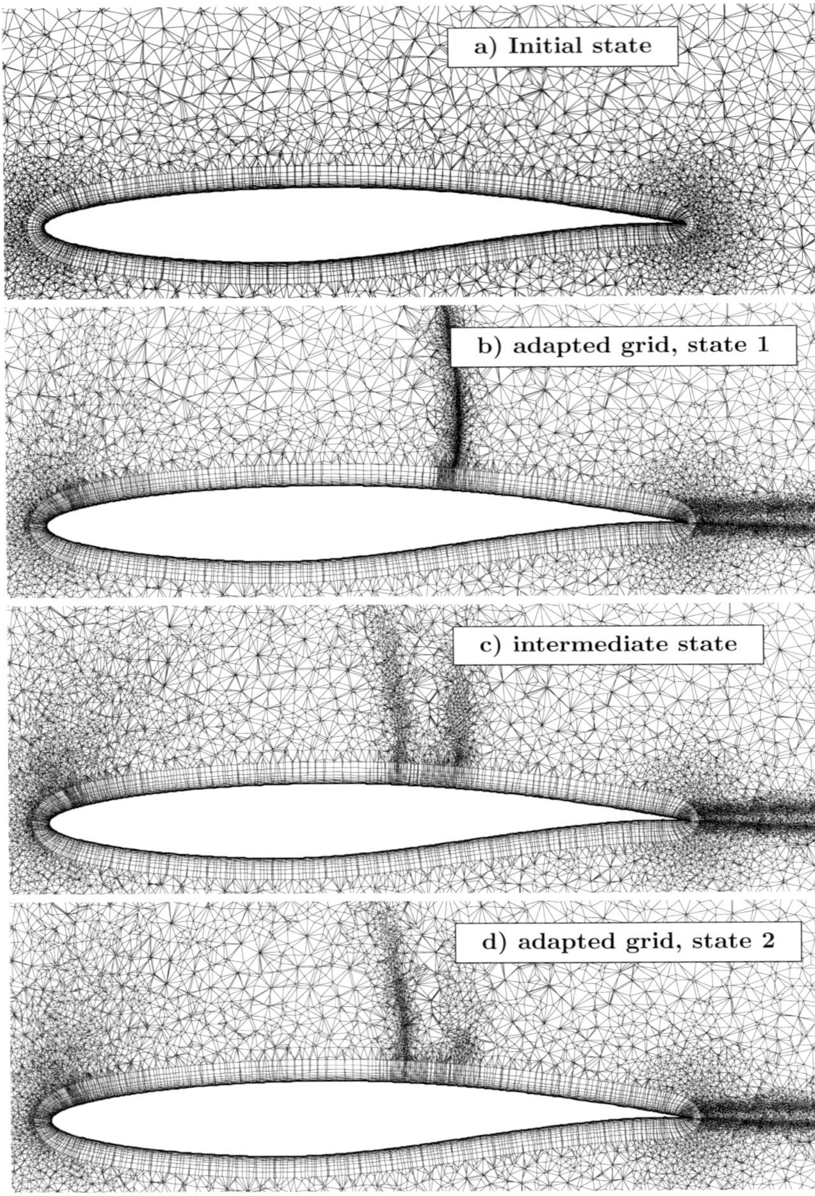

Fig. 5.3. Dynamic mesh adaptation

- computation of donor cell data for chimera overlap grids,
- coloring of edges for vector or cache computers,

- bandwidth optimisation of point numbers on the dual grid edges for memory access optimisation.
- The solver computes the flow solution on the dual grid provided by the preprocessor.
- The grid adaptation contains grid re- and de-refinement and point movement along wall-normal grid lines in turbulent flow regions for y+ adjustments as well as the interpolation of primitive variables to new or moved grid points.
- The grid deformation tool deforms the hybrid grid in the interior volume from given surface displacements, which are e.g. the result of a structural response or the result of a given surface motion in a fixed outer grid.

5.3.2 Data structures

The modules of the TAU-code are based on identical hierarchical data structures. The main data structures are a dual grid structure, a primary grid structure and a structure for the primitive variables, which contain the data and all data related attributes. Other data structures are e.g. sub-structures, like the primary grid connectivity, which is a part of the primary grid structure. The design and implementation of common data structures in all modules allows for a set of common basic functions in order to avoid code doubling.

5.3.3 Libraries

The modules, the basic functions and the data structures are build in libraries. The storage of data inside the libraries allows using them as objects, which simplifies the data management as well as the interfaces to the existing methods. With this implementation the different functionalities can be combined like building blocks in an easy manor. This is explained in the following examples.

A first example is the coupling of different modules. Instead of using the modules in stand-alone mode with data exchange by file-IO the library functions of different modules can directly work with data of each other, because they are making use of the same objects. Coupling e.g. the preprocessing and the solver means to combine the building blocks of both sets in one main routine. Both modules work together without intermediate file-IO without further changes, since in both parts the IO-routines for the dual grids need not to be called. The dual grid data structures are filled during the preprocessing and stored inside the dual grid library. Thus the data is available for the solver as it is in stand-alone mode, in which the data structures are filled by the dual grid IO-routines.

If the grid spacing needs to be adapted during a computation only parts of the preprocessing need to be repeated before the solver restart, because the grid connectivity remains constant, but some cell volumes and grid point coordinates change. Because the different parts of the preprocessor are separated in different building blocks, the parts to be re-computed can be selected by choosing the correct set of building blocks. Thus, as in the above example, the required sequence of functions can be set up to get a dynamic adaptation for y+ during solver run time. Running the same simulation employing the stand-alone modules would mean to run the pre-processor and to write the dual grid to disc, then to start the solver and to write restart data, then to apply the adaptation, to read the grid and restart data and to write a new primary grid and the adapted solution, then to restart the complete pre-processor, read data and to write a new dual grid and finally to restart the solver again.

5.3.4 Scripting capability

Many combinations of different building blocks in varying sequences are possible and strongly dependent on the simulation requirements for a specific scenario. In order to allow for the tailoring of the simulation code for specific types of such scenarios without compiling specific versions a scripting capability is introduced into the TAU-code.

A script-interpreter allows for calling the building block library functions directly from inside a script. Because the interpretation of the script is performed during runtime, no recompiling is needed. The script is thus, like a user programmable main-routine, which can be set up depending on the specific problem type. This provides increased flexibility, since the building block functionality is more granular than the functionality provided by the stand-alone modules, which are composed of fixed (compiled) building block sequences. Especially, for standard simulations to be performed several times with different input, it is obvious, that problem specific scripts are of advantage.

The script interpreter is implemented in python language [9]. Explaining all aspects of the usage of TAU and the scripting capability would be too much, here. However, without explaining each line, the following python script shows how to compute a polar for different angles of attack at different Mach numbers. For each defined Mach number or angle of attack some solver parameters are set. Once the computation for one polar point is finished, the final data is written out before continuing the computation for the next polar point based on the current data as initial solution.

```
Prep       = PyPrep.Preprocessing(parameter_file)
Solver     = PySolv.Solver(paramater_file)
mach_list  = [0.5, 0.7, 0.82]  # Mach numbers to be computed
alpha_list = [2.0, 4.0, 8.0]   # Alpha for each Mach number

Prep.run()
Solver.init(online_monitoring=xmgrace)
```

```
for m in range(len(mach_list)):
  for a in range(len(alpha_list)):
    file_prefix = "Restartfile_Mach=" + str(mach_list[m])
                  + "_Alpha=" + str(alpha_list[a])
    para_list = {}   # init dictionary for parameter update
    para_list['Reference Mach number']  = mach_list[m]
    para_list['Output files prefix']    = file_prefix
    para_list['Angle alpha (degree)']   = alpha_list[a]
    Para.update(para_list)

    Solver.init(cmd='para_update')
    Solver.inner_loop()
    Solver.output()
    alpha_list.reverse()  # restart for new mach
                          # with last angle of attack
Solver.finalize()
tau("exit")
```

The above script performs a sequence of computations with dynamic change of parameters during run time, which is (at least this is taken for granted here) not too complicate to understand and thus adjustable for other simulation sequences also by the users. This allows for easy code tailoring according to the requirements of the specific simulation problem. For more complex work flows in special vertical applications, it is possible at least for developers, to set up and provide scripts in a very short time frame.

Because it is possible to call other programs from inside a (python-) script, this provides in addition an open interface to couple them with the TAU code directly. A coupling can be done for example with visualisation software or post-processing tools from 3rd party in-house production chains or also with a finite element solver to compute the structural response on the aerodynamic loads computed by the flow solver.

5.4 Summary

The overview given above describes briefly the functionality and the structure of the DLR TAU-code in order to give a first idea of the capabilities and the performance of the code, which enable for accurate and efficient simulations of complete aircrafts including all geometrical complexity which needs to be considered in the industrial aerodynamic aircraft design process. For details about specific features of the code it is referred here to the following sections, which describes single aspects of the code more precisely.

References

1. N. Kroll, C.C. Rossow, K. Becker and F. Thiele: *MEGAFLOW — A Numerical Flow Simulation System*, 21st ICAS Congress, Melbourne, paper 98-2-7.3, 1998.
2. N. Kroll, C.C. Rossow, K. Becker and F. Thiele: *The MEGAFLOW Project*, Aerosp. Sci. Technol., Vol. 4. pp. 223-237, 2000.
3. P. Aumann, H. Barnewitz, H. Schwarten, K. Becker, R. Heinrich, B. Roll, M. Galle, N. Kroll, Th. Gerhold, D. Schwamborn and M. Franke: *MEGAFLOW: Parallel Complete Aircraft CFD*, Parallel Computing, Vol. 27, pp. 415-440, 2001.
4. N. Kroll, C.C. Rossow, D. Schwamborn, K. Becker and G. Heller: *MEGA-FLOW: A numerical flow simulation tool for transport aircraft design*, 23rd ICAS Congress, Toronto, paper 2002-1.10.5, 2002.
5. *CentaurSoft, http://www.centaursoft.com/.*
6. T. Gerhold, O. Friedrich, J. Evans and M. Galle: *Calculation of Complex Three-Dimensional Configurations Employing the DLR TAU-Code*, AIAA Paper 97-0167, 1997.
7. T. Alrutz: *Hybrid Grid Adaptation in TAU*, this volume.
8. K. Weinman: *Turbulence Model Implementation in TAU*, this volume.
9. *http://www.python.org/.*

6

Algorithmic Developments in TAU

Ralf Heinrich, Richard Dwight, Markus Widhalm, and Axel Raichle

DLR, Institute of Aerodynamics and Flow Technology, Lilienthalplatz 7, D-38108 Braunschweig

Summary. The paper describes a selection of algorithmic developments which have been implemented in the hybrid Navier-Stokes solver TAU during the MEGAFLOW II project. The paper concentrates on algorithms that help to improve the performance, the accuracy as well as the functionality. The algorithms presented are implicit MAPS-smoothing, low Mach number preconditioning, least square reconstruction in combination with a cell centered approach, the actuator disk boundary condition and a formulation for moving coordinate systems enabling steady solutions in a rotating frame. Results are presented in comparison to earlier versions of the TAU code, highlighting the improvements with respect to performance and/or accuracy. Comparisons with experimental data and results obtained with the FLOWer code are used to validate the new functionalities.

6.1 Introduction and Overview

The main motivation for implementing new algorithms into a simulation code is to achieve an improvement of the turn around time, the accuracy or/and the functionality. This article describes a selection of algorithms which have successfully been implemented in this context within the MEGAFLOW II project. As described in previous article, "TAU Overview", several techniques are implemented in TAU to reduce the turn around time. The convergence is accelerated by local time stepping, residual smoothing and multigrid. But compared to structured algorithms, like those implemented in the FLOWer code, there is usually a remarkable difference in the CPU time needed especially for viscous applications. To improve the convergence behavior the implicit MAPS-smoothing has been implemented in TAU. First results showing comparisons to the default explicit smoothing scheme are very promising. The algorithm and the numerical results are described in section 6.2.1.

An algorithmic development helping to improve performance, accuracy and functionality is the preconditioning technique. Through preconditioning flow computations down to very low Mach numbers like 10^{-5} are possible. For Mach numbers lower than 0.2 the convergence and accuracy behavior becomes nearly Mach number independent. Basic features of the implementation and results are described in section 6.3.1.

Within TAU several upwind schemes are implemented. To achieve second order accuracy a reconstruction of the flow variables at the cell interfaces is needed. Therefore the gradients of the variables are needed. By default an approximation based on the Green-Gauss theorem is used. But it was found that, on arbitrary hybrid meshes, even a linear function could not be reconstructed exactly. An outcome is promised by the so called least square reconstruction which is now available in TAU. First numerical experiments in two-dimensional flow show, that the accuracy of upwind schemes is improved. An additional improvement was found by switching from the dual mesh approach to a cell centered approach. The least square approach and results are presented in section 6.3.2.

The functionality is usually extended by new boundary conditions, for example the actuator disk boundary condition, which is now available in TAU. This boundary conditions helps to include the influence of a propeller on the aerodynamics of an airplane. Results achieved for a 4 propeller aircraft show good agreement with experimental data as shown in section 6.4.1. To enable an efficient and more accurate simulation of a real propeller or a helicopter rotor, in TAU the Navier-Stokes equations are now formulated in a moving coordinate system instead of using the inertial system. This enables the steady computation of flows in a rotating frame. Special effort was put into the formulation of the additional terms found in the flux balance including the interface velocity. As shown in section 6.4.2 results obtained for a hovering rotor show a good agreement to results of the FLOWer code, which is well validated for this type of flow.

6.2 Improvement of Performance

6.2.1 Implicit MAPS-smoothing

As default time integration scheme a Runge-Kutta scheme is applied in TAU. To accelerate the convergence to steady state an explicit residual smoothing in combination with local time stepping and multigrid is used. But unfortunately, especially for viscous, high Reynolds-number applications, the convergence rate is not comparable to state of the art structured codes like FLOWer. One possibility to improve the convergence is the usage of an implicit residual smoothing scheme. In the TAU code the so called "MAPS"-smoothing has been implemented, which was proposed by Rossow [1]. The MAPS scheme itself is an upwind scheme. The name stands for Mach number based advection pressure split scheme. In the following everything is explained in 1D for clarity. A time-implicit discretization of the one-dimensional Euler equations is given by

$$\frac{\Delta \underline{W}}{\Delta t} + \frac{\underline{F}^{n+1}_{i+1/2} - \underline{F}^{n+1}_{i-1/2}}{Vol} = \underline{0}. \tag{6.1}$$

$\underline{W} = \left(\rho \; \rho u \; \rho e_t\right)^T$ are the conserved quantities with density ρ, velocity u and total energy e_t. $\underline{F}^{n+1}_{i\pm 1/2}$ denotes the flux vector of the unknown state $t^{n+1} = t^n + \Delta t$. According the MAPS-scheme the fluxes are splitted into two parts, an advective and a pressure part.

$$\underline{F}^{n+1}_{i-1/2} = q^{(-)}_{ad,i-1/2}\underline{\Phi}_i + q^{(+)}_{ad,i-1/2}\underline{\Phi}_{i-1} + q^{(-)}_{P,i-1/2}\underline{P}_i + q^{(+)}_{P,i-1/2}\underline{P}_{i-1}. \tag{6.2}$$

The scalar values q are function of the Mach number normal to the cell face, see reference [1]. To simplify the implicit time integration the values of q are frozen. Additionally we assume that the rate of change of the conserved quantities $\Delta \underline{W}$ is approximately the same as the change of the advected quantities $\Delta \underline{\Phi}$ with $\underline{\Phi} = \left(\rho \; \rho u \; \rho h_t\right)^T$. h_t denotes the total enthalpy and $\underline{P} = \left(0 \; p \; 0\right)^T$ with the pressure p. Putting this into equation 6.2 and rearranging, the following Runge-Kutta time stepping scheme can be derived:

$$\epsilon_{i-1}\Delta \underline{W}_{i-1} + \epsilon_i \Delta \underline{W}_i + \epsilon_{i+1}\Delta \underline{W}_{i+1} = \underline{R}_i + \underline{S}_P\left(\Delta p^{\mu-1}_{i-1}, \Delta p^{\mu-1}_{i}, \Delta p^{\mu-1}_{i+1}\right) \tag{6.3}$$

On the left hand side are the changes of the conserved quantities at the point i and its neighbors. The pre multipliers ϵ are functions of q and so of the face normal Mach number. $\underline{R}_i$ is the explicit residual from the previous Runge Kutta stage $\mu - 1$. Within $\underline{S}_P$ we summarize all terms including changes of the pressure at the point i and its neighbors. We treat it explicitly in order to decouple the resulting set of equations. Equation 6.3 is a set of scalar implicit equations , which can be solved sufficiently with 2-5 point Jacobi iterations. This set of equations is very similar to other implicit smoothing schemes. The main difference is that the smoothing coefficients ϵ are derived based on the MAPS upwind scheme. For a detailed description of the different terms in equation 6.3 and the derivation of the scheme the reader is referred to [1]. The following examples show the effect of the MAPS-smoothing scheme on the performance compared to the standard explicit smoothing scheme. As a first example the inviscid flow around the ONERA-M6-wing has been computed. The mesh contains about 600000 tetrahedrons. The Mach number is 0.83 and the angel of attack is $\alpha = 3°$. As reference computation the standard TAU settings are used (3 stage Runge-Kutta combined with point explicit smoothing(PE)). Beside local time stepping, 4w-multigrid has been applied to accelerate the convergence. For the spatial discretization the AUSMDV scheme has been selected. Due to stability problems the CFL number has to be reduced to 1.2. The convergence history of the density residual and the lift is shown in figure 6.1 left. About 15000 CPU-seconds (424 multigrid cycles) are needed to achieve a convergence of 6 orders of magnitude for the density residual on a SGI workstation. The CPU time can be reduced to less than

4000 CPU seconds (70 multigrid cycles) using the MAPS-smoothing scheme which has been shown to perform well with 5 Rung-Kutta stages. Due to the implicit character of the smoothing and the higher number of stages the CFL number could be increased to a value of 6. A similar behavior can be achieved for the viscous flow around the RAE2822 airfoil as shown in figure 6.1 right. The Mach number is 0.73, the angle of attack is $\alpha = 2.83°$, Reynolds number is 6.5 million. The hybrid mesh contains about 22000 elements. What's even more interesting than the resulting density residuals is the behavior of the lift convergence. Less than 2000 CPU seconds are needed with the MAPS-smoothing to achieve a stable lift, compared to 10000 seconds using the default settings. So for the applications shown here the MAPS-smoothing helps to save between 60 and 80 percent.

6.3 Improvement of Accuracy

6.3.1 Preconditioning

For the numerical simulation of steady aerodynamic problems mainly time stepping algorithms are in use. Very efficient solution processes are enabled by convergence acceleration techniques like local time stepping, implicit residual smoothing and multigrid. But for applications under low Mach number conditions, the efficiency and the accuracy slows down, or even no convergent solution can be achieved. The reason for the bad convergence behavior is the growing stiffness of the system of equations with decreasing Mach number. The decreasing accuracy is due to an imbalance of the artificial dissipation [3] [2] terms for small Mach numbers, which are explicitly added for central schemes, or which are inherent in case of upwind schemes. These difficulties can be resolved by changing the time dependency in the equations without influencing the steady state solution. In literature this technique is known as time–derivative preconditioning. Following the work of Choi and Merkle [3], a preconditioner has been implemented and tested in the TAU code for a central scheme. The time derivative as well as the artificial dissipation $\underline{AD}$ is premultiplied by the preconditioning matrix $\underline{\underline{\mathbf{\Gamma}}}_P^{-1}$. For simplicity this is written down here for the two–dimensional Euler–equations in primitive variables $\underline{W}_P = \begin{pmatrix} p & u & v & T \end{pmatrix}^T$ with pressure p, velocity components u, v and temperature T:

$$\underline{\underline{\mathbf{\Gamma}}}_P^{-1} \frac{\partial \underline{W}_P}{\partial t} + \frac{\partial \underline{E}}{\partial x} + \frac{\partial \underline{F}}{\partial y} = \underline{\underline{\mathbf{\Gamma}}}_P^{-1} \underline{AD} \qquad \text{with} \qquad \underline{\underline{\mathbf{\Gamma}}}_P^{-1} = \begin{pmatrix} \frac{1}{a^2 M_r^2} & 0 & 0 & 0 \\ \frac{u}{a^2 M_r^2} & \rho & 0 & 0 \\ \frac{v}{a^2 M_r^2} & 0 & \rho & 0 \\ \frac{h_t}{a^2 M_r^2} & \rho u & \rho v & \frac{\gamma p}{\gamma - 1} \end{pmatrix} \tag{6.4}$$

$\underline{E}$ and $\underline{F}$ are the fluxes approximated by central differences, ρ denotes the density, h_t the total enthalpy, γ the ratio of specific heats and a the speed of sound, respectively. The preconditioning matrix includes a free parameter M_r^2, which is usually set to the square of the local Mach number M^2. To preserve the matrix from becoming singular near stagnation points or no–slip walls, M_r^2 is cut off by $K * M_\infty^2$, where M_∞ is the onflow Mach number and K a parameter, which can be specified by the user. Good results with respect to convergence and accuracy have been achieved by setting $K = 1$ for inviscid and $K = 1.5$ for viscous high lift test cases. To keep the good convergence properties of the unpreconditioned set of equations for supersonic flows, M_r^2 is set to 1 in supersonic regions.

$$M_r^2 = min(1, max(M^2, KM_\infty^2)) \tag{6.5}$$

In order to remove the stiffness of the equations, the elements of the preconditioning matrix are selected in such a way, that now all eigenvalues of the system of equations are of the same order of magnitude. The eigenvalue associated with the particle velocity remains unchanged. The premultiplication of the artificial dissipation ensures a good balance of the artificial dissipation [3] [2]. As pointed out in [2], different sets of variables can be used for preconditioning. Two different sets of variables can be chosen by the user of the TAU code. For preconditioner type I the above mentioned set of primitive variables $\underline{W}_P$ is used and for preconditioning type II the conservative variables are taken into account ($\underline{W}_C = \left(\rho\ \rho u\ \rho v\ \rho e_t\right)^T$ with the total energy e_t). A simple numerical experiment has been performed to test the implementation of the preconditioning algorithm. The inviscid flow around a NACA0012 airfoil has been calculated for a Mach number range form 0.1 down to 10^{-3}. For all computations the angle of attack is set to $\alpha = 2.0°$. The drag which is 0 for an inviscid, subsonic flow, can be used to "measure" the accuracy of the numerical solution. Table 6.1 summarizes results for the simulation with and without preconditioning. For the three Mach numbers taken into account the number of multigrid cycles and the final drag for a convergence of 7 orders of magnitude with respect to the L2-Norm of the density residual is printed. The convergence and accuracy becomes unacceptable without using preconditioning for Mach numbers approaching 0. The drag for the Mach number of 10^{-3} is almost 2 orders of magnitude higher, than for Mach number 0.1! For preconditioning method I and II, the convergence as well as the final drag is almost Mach number independent. Even for the best solution without preconditioning ($Ma = 0.1$), the drag-coefficient for all solutions using preconditioning is improved by a factor of 2.6. Using preconditioning method II, the convergence properties are improved compared to method I. A more detailed analysis of the results shows, that method II is slightly more dissipative compared to method I. Another well suited value for measuring the numerical quality of a scheme is the total pressure loss coefficient $cp_{tot,max} = (p_{tot} - p_{tot,\infty})_{max}/(0.5 * \rho_\infty \mathbf{v}_\infty^2))$. For Mach numbers ranging from 10^{-5} to 0.6 the maximum value of the coefficient on the profile is plotted in figure 6.2a for preconditioning method I and without preconditioning. Using preconditioning, it becomes visible, that the coefficient remains almost constant in the whole subsonic region. This shows, that the quality of the numerical solution is independent of the size of the Mach number for subsonic flows. Without preconditioning $cp_{tot,max}$ becomes unacceptable high. For a Mach number lower than 10^{-3} no convergent solution has been obtained. A similar behavior has been achieved for viscous flows. The flow around a three-element airfoil has been simulated for Mach numbers form 10^{-3} to 0.1, a Reynolds number of $3.52 \cdot 10^6$ and an angle of attack of $\alpha = 20.18°$. 6.2 shows the convergence of the total drag–coefficient without preconditioning (only for $Ma = 0.1$; the solutions for $Ma < 0.1$ are not visible in the range up to 3000 multigrid cycles!) and for preconditioning method I. Again the preconditioned solution is almost Mach number independent for the whole range of test cases. The drag of the preconditioned solution for $Ma = 0.1$ is slightly higher, than for the smaller Mach numbers. This is due to the compressibility effects, which have to be taken into account in the high lift regime locally. More example for applications using the preconditioned TAU code, especially in three-dimensional flow, may be bound in [4]. The result shown underline the advantages of preconditioning for nearly incompressible flows compared to standard schemes solving the compressible equations. The advantage of preconditioning compared to algorithms solving the incompressible equations is, that applications including incompressible and compressible regions are enabled. This is especially important for high lift applications.

Table 6.1. Number of multigrid cycles and drag for a convergence of 7 orders of magnitude

	no prec.		prec. type I		prec. type II	
	MG-cycles	drag	MG-cycles	drag	MG-cycles	drag
Ma = 0.1	381	0.000763	184	0.000289	133	0.000296
Ma = 0.01	1497	0.005384	184	0.000288	133	0.000293
Ma = 0.001	5432	0.042612	184	0.000288	133	0.000293

6.3.2 Least Square and Cell Centered

As described in the article "TAU Overview" in this book, the TAU code uses a node based finite volume discretization. The computational mesh called "dual mesh" as depicted in figure 6.3 b) and c) is constructed based on the so called primary mesh within a preprocessing step. The control volume associated to the point P_0 is constructed by connecting the centers of the surrounding elements with the centers of the edges connected to P_0. The vector $\vec{S}_{0,i}$ is the surface normal vector associated to the edge connecting the point 0 and

i. If upwind schemes are applied, usually a reconstruction of the quantities associated to the states on the left and right hand side of a cell face is used, to achieve second order accuracy. In the TAU code the gradient associated to the point P_0 is used to approximate the value of a quantity Φ on the left hand side hand side of face with normal vector $\vec{S}_{0,i}$ according

$$\Phi_L = \Phi_0 + (\nabla\Phi)_0 \, \vec{r}_{0,L}. \tag{6.6}$$

$\vec{r}_{0,L}$ is the direction vector pointing from point P_0 to the point P_L. In the TAU code the Green Gauss formula is used (abbreviation gg)

$$\int_V \nabla\Phi dV = \oint_{\partial V} \Phi d\vec{S}. \tag{6.7}$$

The discretized formula used within the TAU code is:

$$(\nabla\Phi)_0 = \frac{1}{V_0}\sum_{i=1}^{N_0}\frac{1}{2}(\Phi_i + \Phi_0)\,\vec{S}_{0,i}. \tag{6.8}$$

N_0 is the number of nodes connected to point P_0. A simple test to study the behavior of the approximation of the gradient is the following: Put the values of a linear function e.g. $\Phi(x,y)) = x + 2y$ on the nodes of a mesh. Using formula 6.8 and the mesh depicted in figure 6.3 b) we will get the analytical values for the gradient, so $\nabla\Phi_{0,gg} = \nabla\Phi_{0,exact} = \begin{pmatrix}1.0 \; 2.0\end{pmatrix}^T$. In general it can be shown, that a linear function can be reconstructed exactly on any triangular or tetrahedral mesh. But if we switch now to a hybrid mesh as depicted in figure 6.3 c) the approximation formula 6.8 fails. The result for the same test function is now $\nabla\Phi_{0,gg} = \begin{pmatrix}1.10 \; 2.11\end{pmatrix}^T \neq \nabla\Phi_{0,exact}$. It is obvious, that the error in the calculation of the gradient will influence the overall accuracy of the scheme. So we put effort on the approximation of the gradient posing that at least a linear function should be reconstructed exactly on any mesh. This is permitted by the so called least square approach [6], [7]. The approach is based upon a first order Taylor series approximation for each edge surrounding the point P_0. This results in an over-determined system of linear equations

$$\begin{bmatrix} \Delta x_{01} & \Delta y_{01} & \Delta z_{01} \\ \Delta x_{02} & \Delta y_{02} & \Delta z_{02} \\ \vdots & \vdots & \vdots \\ \Delta x_{0N_0} & \Delta y_{0N_0} & \Delta z_{0N_0} \end{bmatrix} \begin{bmatrix} \partial_x\Phi \\ \partial_y\Phi \\ \partial_z\Phi \end{bmatrix} = \begin{bmatrix} \Delta\Phi_{01} \\ \Delta\Phi_{02} \\ \vdots \\ \Delta\Phi_{0N_0} \end{bmatrix}. \tag{6.9}$$

For solving the linear system of the form $\vec{A}\vec{x} = \vec{b}$ matrix $\vec{A}$ is decomposed with a Gram-Schmidt process into an orthogonal matrix $\vec{Q}$ and an upper triangular matrix $\vec{R}$. More details of the solution procedure may be found in [5]. It can be shown, that this formulation leads to an exact prediction of the gradient of a linear function on arbitrary meshes.

Another problem beside the approximation of the gradient , which might influence the accuracy of the reconstruction, becomes visible in figure 6.3c. The node P_0 is in a general case not the center of the surrounding control volume. The point P_L is usually not the center of the cell face. Therefore the quality of the reconstruction will be reduced for dual control volumes if the surrounding primary mesh cells differ remarkably in their size. This can happen within hybrid mesh generation, especially in regions, where a prismatic layer is connected to a tetrahedral area. What can be done to bypass this problem is to switch from the dual mesh back to the primary mesh in connection with a cell centered approach. Additionally the location of the cell centers and faces are stored. Beside the improved quality of the reconstruction we learn from the structured FLOWer code, that a cell centered approach usually increases the robustness of a code.
The least square reconstruction as well as the cell centered approach has been implemented in the TAU code in order to investigate the influence on the accuracy [5]. As a first test case the inviscid subsonic flow around the RAE2822 airfoil is computed for a Mach number of 0.5 and an angle of attack of $\alpha = 2.9°$. As a measure of quality the pressure drag and the total pressure loss on the surface are good choices. Both values are 0 for an exact solution of the Euler equations in completely subsonic flows. The total pressure losses are shown in 6.4 (left). The solid line belongs to the total pressure loss using the node based approach in combination with Green-Gauss gradient for the reconstruction needed for the Roe upwind scheme. The corresponding pressure drag is of course close to 0, $c_D = 0.0046$. The situation can be improved slightly by using a central space discretization instead of the Roe scheme (dotted curve, $c_D = 0.0036$). A further improvement is achieved by using again Roe with Green-Gauss gradient computation, but now in combination with the cell centered approach (dashed curve, $c_D = 0.0028$). Using now the least square computation of the gradient instead of the Green Gauss approximation we come to the best solution for this test case (dashed-dotted curve, $c_D = 0.0022$).
As a first viscous test case a hybrid mesh has been generated around a profile with blunt trailing edge. For a Mach number of 0.2 and a Reynolds number of 6 million computations have been made for 3 different angles of attack $\alpha = 0°, 5°$ and $10°$. Fig. 6.4 (right) shows the polar of lift plotted over the pressure drag. What we expect is, especially for the two lower angles of attack, where no separation is expected, a curve close (and almost parallel) to the lift axis. Then only the shape drag contributes to the pressure drag of the profile, which does not vary much for low angles of attack. What we find (see figure 6.4) for the computation using the Roe scheme in combination with the node based approach and Green-Gauss computation of the gradient is a curve close to the lift axis for $\alpha = 0°$, but for the higher angles the pressure drag is increased remarkably (solid line with circles). The situation is improved switching now to the cell centered approach (dashed line with triangles pointing upward). A further improvement is made by using the central space discretization in combination with the node based approach. But again, the best solution is

obtained, by using the upwind scheme with least square computation of the gradient combined with the cell centered approach (dashed line with right pointing triangles). To prove, that this is really the best solution, a mesh convergence study has to be performed. This is ongoing work.

6.4 Improvement of Functionality

6.4.1 Actuator Disk

Operating propellers, especially wing mounted propellers, have significant and sometimes subtle effects on the aerodynamic of an airplane. What can be done to include the effects in the simulation is of course a detailed modeling of the propeller rotating relative to the wing. In that case the chimera technique has to be applied to enable the relative motion. The total number of mesh points will increase dramatically and additionally the flow field is unsteady. Such a simulation would be extremely expensive compared to a steady computation of the configuration without propellers. A way out of this situation is to make a simplification: The propeller can be included as a so called "actuator disk" [8]. The flow in the propeller area is defined using flow parameters on the inflow and outflow surfaces on the actuator disk. Such a model has been implemented and tested in the TAU code [9].

The inflow surface is the upstream facing side of the propeller, whereas the outflow surface is facing downstream, see figure 6.5. In order to simulate the effects of a propeller, the method allows a total pressure ratio (.i.e. the total pressure of the outflow surface in relation to the total pressure in the farfield) on the propeller outflow surface to be defined using a polynomial as a function of the propeller radius. Several options exist for controlling the direction of the flow both into and out of an actuator disc. If necessary, the direction of the flow leaving the outflow surface may be set via a vector, which is desirable for example, when trying to simulate the effect of an angle of attack. In addition to this a swirl angle as a function of the propeller radius can be defined as a polynomial function on the outflow surface. To find a good setting of the additional input parameters e.g. experimental data could be used as described in [9] or from separate tools for propeller design. Another good possibility is to derive these parameters from the simulation of an isolated propeller. A big advantage of the actuator disk model compared to the detailed modeling using the chimera technique is, that the simulations can be performed in steady mode. The actuator disk model has been tested on an actual four-engined transport aircraft in high-lift configuration ($Ma = 0.176$, $Re = 1.3 \cdot 10^6$. A hybrid mesh containing about 13 million nodes and 400k surface points has been generated including nacelles and deployed flaps, see figure 6.5. Computations have been performed for two angles of attack ($\alpha = 0°$ and $7.6°$). The resulting pressure coefficient distribution has been compared to experimental data in a section close to the inboard propeller (see figure 6.6

for $\alpha = 7.6°$) with power on and off. The good agreement to the experimental data shows, that the implemented actuator disc boundary condition is well suited for predicting the effects of the propellers on the aerodynamic behavior of an airplane. Additionally the significant influence of the propellers become visible in both, the wind tunnel and the numerical experiment.

6.4.2 Steady Computation in a rotating frame

For steady or unsteady computations of moving bodies, the balance equations have to be solved for a moving control volume. The momentum equation written for the inertial system in that case includes an additional velocity $\vec{v}_B$ in the flux balance. $\vec{v}_B$ is the velocity of the surface of the control volume.

$$\frac{d}{dt}\int_V \rho\vec{v}dV + \oint_{\partial V} (\rho\vec{v})(\vec{v} - \vec{v}_B)\,d\vec{S} - \oint_{\partial V} d\vec{T}d\vec{S} = \vec{0}. \tag{6.10}$$

$\vec{T}$ in equation 6.10 is the stress tensor. Instead of using the inertial system (x, y, z) the equations can be transformed into a moving coordinate system $(x\prime, y\prime, z\prime)$ as sketched in figure 6.7:

$$\frac{d'}{dt}\int_V \rho\vec{v}'dV + \oint_{\partial V} \left(\rho\vec{v}'\right)(\vec{v}' - \vec{v}'_B)\,d\vec{S}' - \oint_{\partial V} d\vec{T}'d\vec{S}' = -\int_V \rho\vec{\omega}' \times \vec{v}'dV. \tag{6.11}$$

In that case the velocity $\vec{v}'_B$ can be split into three contribution:

$$\vec{v}'_B = \vec{v}'_{trans} + \vec{v}'_{rot} + \vec{v}'_{flex} \tag{6.12}$$

$\vec{v}'_{trans}$ is due to a translation of a body, $\vec{v}'_{rot}$ has to be taken into account, if the body is rotating. In case of deforming meshes the deformation speed $\vec{v}'_{flex}$ has to be taken into account additionally. The extra term in 6.11 includes the time derivative of the unit normal direction vectors of the rotating frame, which are now a function of time. A big advantage compared to the formulation in the inertial system is, that the metric is kept constant over time in case of rigid body motions and it is possible to obtain steady solutions in a rotating frame. This is very useful to save computational time for applications in helicopter aerodynamic or in turbo machinery. So the formulation of equation 6.11 has been selected for the TAU code. One question arising is how to discretize the additional term $\oint_{\partial V} \vec{v}'_B d\vec{S}'$. The first idea was to compute this term in a straight forward and approximate manner. The contribution of the face associated to the edge connecting the point P_1 and P_2 (see figure 6.7) is

$$\oint_{\partial V} \vec{v}'_B d\vec{S}' \approx \frac{1}{2}\left(\vec{v}'_{B,1} + \vec{v}'_{B,2}\right)\vec{S}' = \frac{1}{2}\vec{\omega} \times (\vec{r}'_1 + \vec{r}'_2)\,\vec{S}' \tag{6.13}$$

The face normal vector $\vec{S}'$ (compare section 6.3.2) is calculated during the preprocessing as the sum of the normal vectors of the N facettes surrounding the edge, see figure 6.7. In two-dimensional meshes $N = 2$. To test the approximation a simple numerical experiment can be done, a so called "freestream consistency check". All nodes of a mesh are initialized with freestream quantities and zero velocity. Then you let rotate the mesh, containing only farfield boundaries. For a perfect solution every quantity should keep constant, for example the c_p value should be zero in the whole flow field. But what was found are disturbances of the order 10^{-3}! Additionally the density residual was far away from machine accuracy. So the approximate calculation of the additional term seems not to be the best choice. Fortunately it is possible to derive an exact formulation for a linear velocity field for $\vec{v}'_B$ (for a rigid body motion $\vec{v}'_B$ is always linear). Therefore the surface integral of $\vec{v}'_B$ on $\vec{S}'$ is split into its contributions of the different facettes. For a rotating frame we can write

$$\oint_{\partial V} \vec{v}'_B d\vec{S}' = \sum_{i=1}^{N} \oint_{\vec{n}'_i} (\vec{\omega} \times \vec{r}') \, d\vec{n}'_i = \vec{\omega} \sum_{i=1}^{N} \oint_{\vec{n}'_i} (\vec{r}' \times) \, d\vec{n}'_i = \vec{\omega} \sum_{i=1}^{N} (\vec{r}_i{}' \times \vec{n}'_i) \tag{6.14}$$

with the facette normal vector $\vec{n}'_i$ and the center of the facette located at $\vec{r}'_i$. The values of the components of the sum $\sum_{i=1}^{N} (\vec{r}_i{}' \times \vec{n}'_i)$ are only a function of the geometry of a dual control volume. These three values are computed in the preprocessing phase and stored additionally for each face. Using this exact computation of the additional term, the code passes the freestream consistent check with a residual close to machine accuracy. For a further verification of the implementation it is useful to compare numerical results of TAU to the FLOWer code, which is well validated for helicopter applications. Therefore a mesh has been generated around the HELI7A rotor blade. Because the geometry of the rotor is periodic, it's sufficient to simulate only a quarter part of the complete four-bladed rotor by applying periodic boundary conditions. The blade is rotating with a tip Mach number of 0.662. TAU c_p distribution have been compared to results of a FLOWer simulation in two slices, one on the mid of the wing and one quite close to the tip. The agreement between FLOWer and TAU is well, see figure 6.8.

6.5 Conclusions

Within MEGAFLOW II much effort has been successfully put into algorithmic improvements of the TAU code. For a large range of applications the CPU time can now be reduced up to a factor of 6 by using the implicit MAPS smoothing technique. The Choi-Merkle preconditioning now enable handling of nearly incompressible flows down to Mach numbers of 10^{-5}. For Mach numbers lower than 0.2 a nearly Mach number independent convergence behavior can be achieved. Due to a better scaling of the artificial dissipation of the central scheme the accuracy is improved. The advantage of the preconditioning technique compared to real incompressible codes is that flows with compressible and incompressible regions can be handled. This is especially important for high-lift applications. Using the least square reconstruction in combination with a cell centered approach, the accuracy of upwind schemes in the TAU code is improved. The range of application is enlarged by new boundary conditions. One of the new boundary conditions is the actuator disk, which enables the inclusion of the effects of propellers on the aerodynamics. Numerical results for a 4-propeller transport aircraft show good agreement with experimental data. The Navier-Stokes equations are now formulated for a moving coordinate system. This enables for example the efficient simulation of flows in a rotating frame. Special effort was put into the formulation of the additional terms, including the interface velocities in the flux balance, to ensure the so called freestream consistency. Numerical results for a helicopter in hover show good agreement with results obtained with the structured FLOWer code.

References

1. Rossow, C.-C.: "A Flux-Splitting Scheme for Compressible and Incompressible Flows". Journal of Computational Physics, volume **164**, pp104-122, 2000.
2. Radespiel, R.; Turkel, E.; Kroll, N.: "Assessment of Preconditioning Methods". DLR IB 95/29, 1995.
3. Choi, Y.-H.; Merkle, C. L.: The Application of Preconditioning to Viscous Flows. Journal of Computational Physics, volume **105**, 207-223, 1993.
4. Melber, S.; Heinrich, R.: "Low Mach-Number Preconditioning for the DLR-TAU Code and Application to High-Lift Flows". Contribution to the 13th AG STAB Symposium München 2002, to be published in Notes on Numerical Fluid Mechanics, Springer Verlag.
5. Widhalm, M.; Rossow, C.-C.: "Improvement of upwind schemes with the Least Square method in the DLR TAU Code". Contribution to the 13th AG STAB Symposium München 2002, to be published in Notes on Numerical Fluid Mechanics, Springer Verlag.
6. Anderson, W. K.; Bonhaus, D. L.: An Implicit Upwind Algorithm for Computing Turbulent Flows on Unstructured Grids. Computers & Fluids, volume **23**, No.1, pp. 1-21, 1994
7. Haselbacher, A.; Blazek, J.: On the Accurate and Efficient Discretisation of the Navier-Stokes Equations on Mixed Grids. AIAA Paper 99-33552, 1999
8. Yu, N. J.; Chen, H. C.: "Flow Simulation for Nacelle-Propeller Configurations using the Euler Equations". AIAA paper 84-2143, 1984
9. Hansing, J.; Sutcliffe, M., Kobloch, O.: "Numerical Simulation of the propeller flow around a four-engined aircraft in high-lift configuration". Contribution to the 13th AG STAB Symposium München 2002, to be published in Notes on Numerical Fluid Mechanics, Springer Verlag.

6.6 Figures

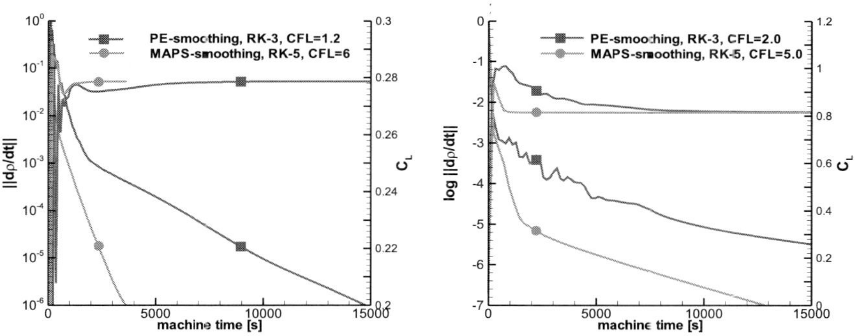

Fig. 6.1. Convergence histories for inviscid flow around ONERA-M6 wing (left) and viscous flow around the rae2822 airfoil (right) using point explicit smoothing and MAPS smoothing.

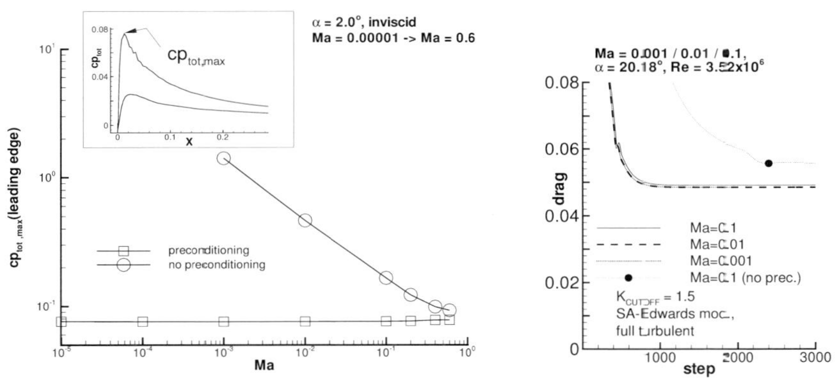

Fig. 6.2. Total pressure loss coefficient for inviscid NACA0012 (left); . Drag convergence for 3 element high lift application (right).

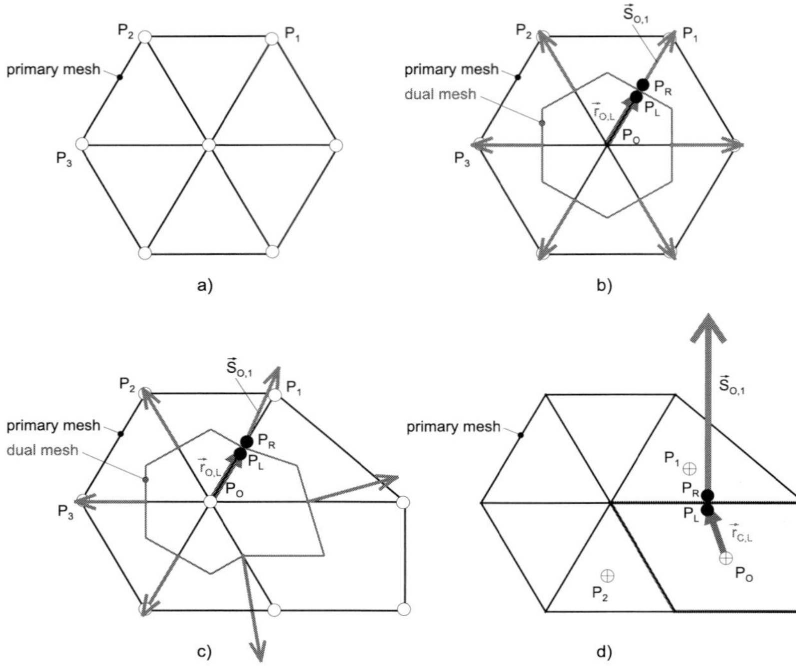

Fig. 6.3. Primary mesh and dual mesh.

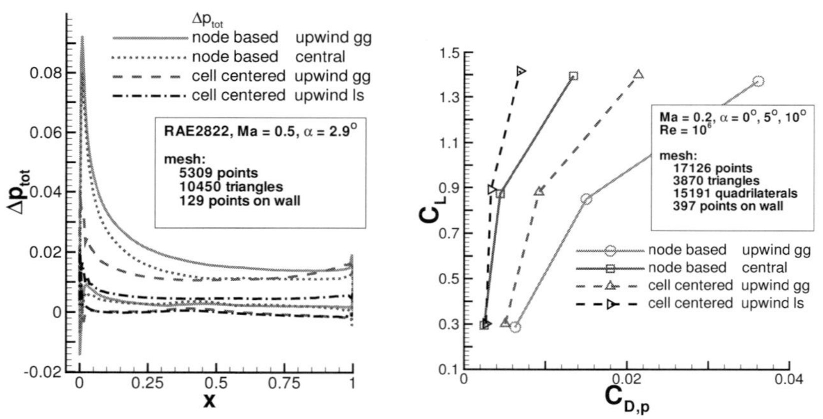

Fig. 6.4. Total pressure loss for inviscid, subsonic flow around rae2822 airfoil (left); Lift over pressure drag for vicous subsonic flow around an airfoil (right).

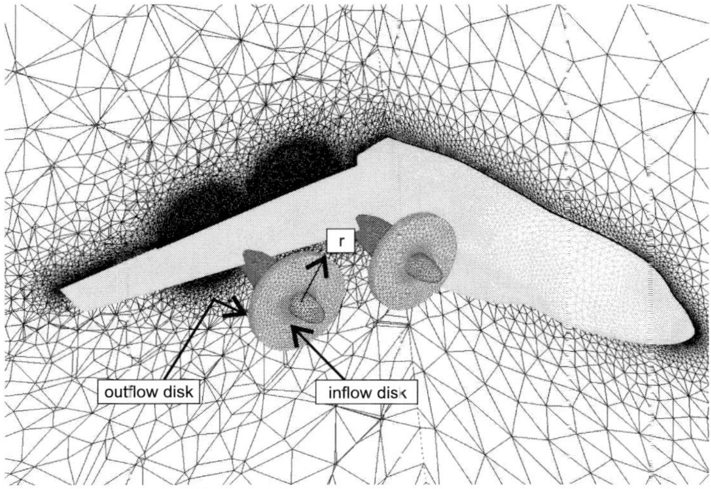

Fig. 6.5. Mesh of a 4 propeller transport aircraft.

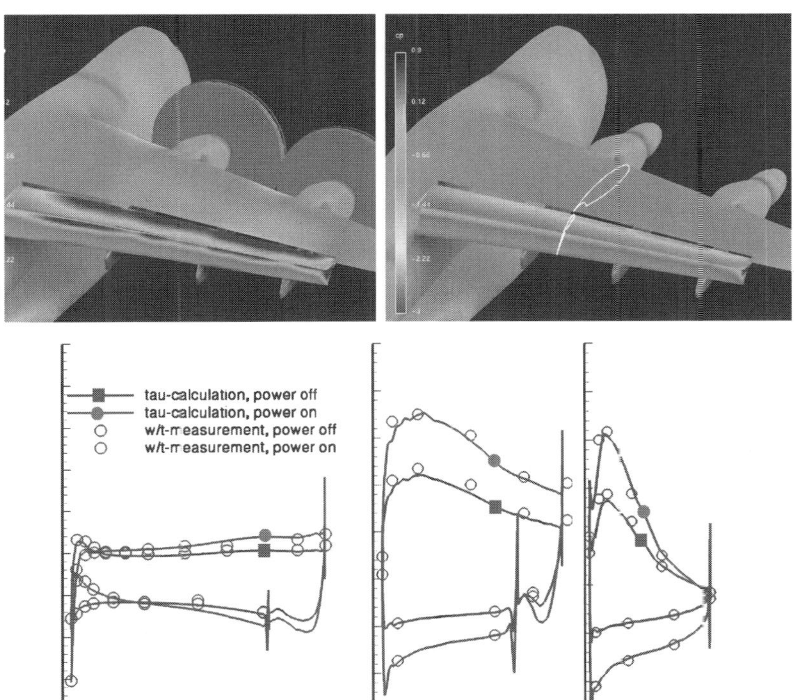

Fig. 6.6. Comparison of pressure coefficient distribution with experimental data (power on and off) for a 4 propeller transport aircraft.

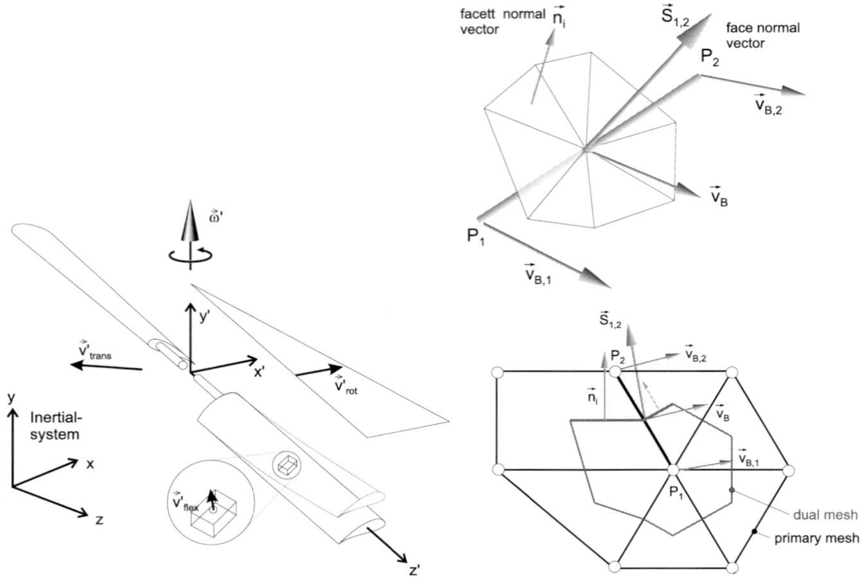

Fig. 6.7. Moving coordinate system (left); Cell face and facettes in 3D (right top) and 2D (right bottom).

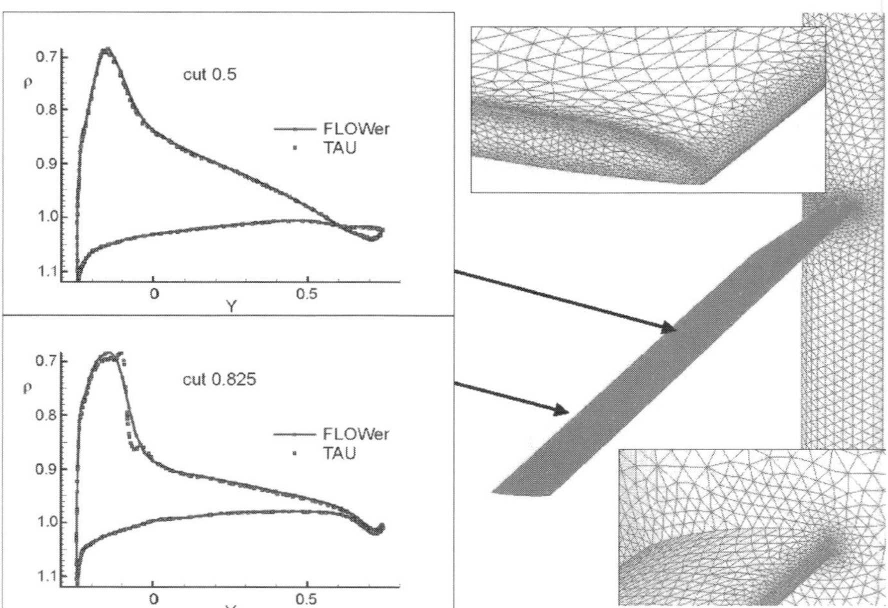

Fig. 6.8. Comparison of density distributions of FLOWer and TAU for a rotor in hover.

7

Hybrid Grid Adaptation in TAU

Thomas Alrutz

DLR, Institute of Aerodynamics and Flow Technology, Bunsenstr. 10, D-37073 Göttingen

Summary. Local grid refinement for unstructured meshes is a common approach to improve the accuracy of a solution for a given CFD problem or to reduce the amount of needed points for a numerical calculation. As part of the **DLR-TAU** code the **Adaptation** provides a hierarchical refinement and de-refinement tool for 2D and 3D hybrid grids. The object of this tool is to adapt the grid automatically to a given solution. This article described the major components of the tool, the grid refinement with surface approximation and the y_+ adjustment along wall-normal lines. Furthermore an overview of the used sensor functions is given, which are available for the detection of the local refinement and de-refinement.

7.1 Introduction

The need to solve complex and realistic CFD problems within a acceptable solution time has made it necessary to use local grid refinement contrary to grids global refined. The **TAU Adaptation** supports refinement and de-refinement for a different set of element types listed in table 7.1. In order to adapt the grid on a given solution, edge-indicator functions are used as sensors to bisect edges of elements, which may be refined or de-refined later on. In order to preserve surface fidelity, after introducing new vertices, a surface reconstruction tool is used to recompute the discretized geometry. Adaptation on hybrid grids usually used for viscous calculations, does require some further consideration. For example, bisecting an edge in wall normal direction would imply a whole new layer of elements and is therefore not allowed. In order to adjust the wall normal spacing with respect to a defined y_+ value, the points are moved along wall normal lines over surfaces with turbulent flow.

7.2 Functionality at the end of the MEGAFLOW project

The **Adaptation** consists mainly of two parts, one is the refinement, which includes the surface reconstruction, and the other is the y_+ point redistribu-

Table 7.1. Elements for refinement and de-refinement

	Volume Elements			Surface Elements	
2D	Hexahedra	Prims		Triangles	Quadrilaterals
3D	Tetrahedra	Prims	Pyramids	Triangles	Quadrilaterals

tion. Both tools are completely independent and are described in the following sections.

The refinement tool

As seen in the common literature (e.g. [1]) our algorithm works on a edge-based data structure. This means that a list of all edges for a given grid must to be calculated. The edgelist is then used to build up a structure that contains the element to edge connectivity.

7.2.1 The edge-based data structure

One of the reasons we use an edge-based data structure is that refining single edges of an element in all possible combinations allows more local refinement then does refining elements with a limited number of refinement cases (The same statement holds also for the de-refinement). The refinement algorithm in edge-based-formulation is described in the following:

1. Calculate the global edgelist from the grid connectivity
2. Check all edges by a sensor function and mark edges dependent on the selected strategy (see 7.2.2)
3. Check all elements for refinement by inspecting the corresponding edges and mark additional edges until grid consistency is achieved
4. Refine all elements which have one or more marked edges corresponding to the refinement case (7.2.3)
5. Reconstruct the surface geometry around a marked surface edge to project the new point to the "real surface geometry" (7.2.4)

7.2.2 The edge-indicator sensor functions

One approach for adaptation indicators is the use of gradients or differences of any suitable flow variable. The approximated gradient $G_{(V)}$ of a variable V in discrete form is $\Delta V/h$, with $\Delta V = V_{p_1} - V_{p_2}$ i.e. the difference between the point values of the two points p_1 and p_2 connected by one edge, where h is the length of the edge. We write the indicator function as:

$$I = \Delta V h^{\alpha} \tag{7.1}$$

A widely used formulation is $\alpha = 1$, i.e.: $G_{(V)}h^2$. The advantage of scaling the indicator with a positive value of α is that the adaptation stops automatically in the corresponding area after several cycles, and also when discontinuities are present in the flow field. Our three-dimensional numerical tests shows that values of $\alpha < 1$ may be of advantage in order to strengthen the grid refinement near surfaces where the cell sizes are typically some orders less than in the farfield region. We use $\alpha = 0.5$ as default value.

Our choice of ΔV for the indicator function is :

$$\Delta V_e = \max_{i=0,\dots N} \left(c_{\phi_i} \frac{(\Delta\phi_i)_e}{(\Delta\phi_i)_{max}} \right) \tag{7.2}$$

with N being the number of different flow variables considered and e the edges in the grid.

The weights c_{ϕ_i} are parameters which enable the choice of different combinations of the single parts of the indicator (to be set to zero in order to turn off ϕ_i).

The reference values $(\Delta\phi_i)_{max}$ are for an equilibrated scaling of each part of the indicator function with

$$(\Delta\phi_i)_{max} = \max_{e=0,1\dots} ((\Delta\phi_i)_e), \quad \text{for all edges } e \text{ in the grid.}$$

For the edge-indicator three differences we can use are

1. The differences (Δ_d) of the flow values

$$(\Delta_d\phi_i)_e = |\phi_i(x_{p_1}) - \phi_i(x_{p_2})| \tag{7.3}$$

2. The differences of the gradients (Δ_g) of the flow values

$$(\Delta_g\phi_i)_e = |grad(\phi_i(x_{p_1})) - grad(\phi_i(x_{p_2}))| \tag{7.4}$$

3. The differences of the reconstructed flow values (Δ_r) to the edge midfaces

$$(\Delta_r\phi_i)_e = |(\phi_i(x_{p_1}) + \frac{1}{2} x_e \cdot grad(\phi_i(x_{p_1}))) - (\phi_i(x_{p_2}) - \frac{1}{2} x_e \cdot grad(\phi_i(x_{p_2})))| \tag{7.5}$$

with ϕ_i the flow value, p_1 and p_2 the two edgepoints of edge e and $x_e = x_{p_1} - x_{p_2}$.

Any of the flow variables of the solver output can be used with the edge-indicators (see table 7.2 for a selected list). This means all primitive and all additional user defined flow variables can be applied.

7.2.3 Implemented refinement cases

In order to support a wide range of grids with mixed element types, we have implemented a set of refinement cases for multiple types of volume elements. The currently available refinement cases are listed in table 7.3. The implementation of the refinement cases for pyramids was necessary due to the fact

Table 7.2. Selected flow variables for sensor functions

sensor	pressure	density	x-, y-, z- velocity	total enthalpy	total pressure	vorticity	viscosity	cp
Δ_d	+	+	+	+	+	+	+	+
Δ_g	+	+	+	+	+	+	+	+
Δ_r	+	+	+	+	+	+	+	+

Table 7.3. Implemented refinement cases

Tetrahedra	Prims	Pyramids	Hexahedra	Triangles	Quadrilaterals
$1 \Longrightarrow 2$	$1 \Longrightarrow 2$	$1 \Longrightarrow 2$	$1 \Longrightarrow 2$	$1 \Longrightarrow 2$	$1 \Longrightarrow 2$
$1 \Longrightarrow 3$				$1 \Longrightarrow 3$	
$1 \Longrightarrow 4(3)$	$1 \Longrightarrow 4$			$1 \Longrightarrow 4$	
$1 \Longrightarrow 5$					
$1 \Longrightarrow 6(2)$					
$1 \Longrightarrow 7$					
$1 \Longrightarrow 8$		$1 \Longrightarrow 9$			

that one can have grids with chopped structured layers in 3D. To follow the strategy of local grid refinement, we also implemented all existing refinement cases for the tetrahedra. A closer discussion of the tetrahedra refinement cases and algorithm is given in [2].

To reduce the amount of needed grid points in order to save memory and computation time, the capability of removing points from the grid, which were introduced earlier due to refinement, is implemented as well. This feature is called de-refinement and it is also necessary for the calculation of time accurate flows.

Fig. 7.1. Parent with 2 refined edges recomputed from the children

The de-refinement uses a hierarchical child-parent data structure to store the refinement case, the actual level and the offset of the children for each refined element. With the hierarchical information, the algorithm can calculate the complete history of all refinement levels. This makes it possible to reconstruct the parent for every element that comes from an earlier refinement and to remove all vertices that were earlier introduced. With the same information, the global edgelist is extended by the edges of the actual parents in the grid. These are edges which have been refined by an earlier adaptation (see figure 7.2.3). The parent edges are used by the edge-indicator functions (7.2.2) to decide if a vertex have to be removed or not.

This part of the **Adaptation** was developed together within the Aero-Sum/SikMa project.

7.2.4 Surface reconstruction

For any calculation it is important to have a correct surface representation of the geometry, which requires a discretization depending on the surface curvature. Therefore using only linear interpolation for the new introduced surfaces points is not sufficient. A common approach is to use high order surface representation (CAD-data) to project the new surface points to the "real geometry", but this requires to have both grid and CAD-data. In cases where the CAD-data is missing no surface reconstruction can be done. In other cases where the geometry has to be deformed (e.g coupling CFD and CSM), the deformation of the high order surface representation is a difficult task.

Another method preferred here, is to use a local surface reconstruction with Bezier-splines for the surfaces and cubic-spline-curves for the boundaries of surfaces (e.g wing/body intersection). This method requires the extraction of the edges that are in the intersection of different surfaces. After this extraction the edges are sorted along lines which represents the boundary curves. For the reconstruction of those curve lines 3D cubic-spline-curves are used and for the remaining edges in the interior of a surface Bezier-splines are used.

To use a B-spline for interpolation the two old point coordinates and the normals on those points depending on the surrounding surface elements are needed.

The surface normal on a point p is then calculated through

$$n_p = \frac{1}{\| \sum_{i=1}^{N_p} \frac{\alpha_{e_i}}{|a_{e_i}+b_{e_i}|} n_{e_i} \|} \cdot \sum_{i=1}^{N_p} \frac{\alpha_{e_i}}{|a_{e_i}+b_{e_i}|} n_{e_i},$$

where N_p is the number of neighbouring surface-elements e_i of p, α_{e_i} is the angle at p of e_i, n_{e_i} is the unit normal on e_i and a_{e_i}, b_{e_i} are the edge-vectors of e_i from p.

To give the user more control over the process, the reconstruction of the surfaces can be turned on or off separately for every surface marker or mapping. In case of geometry problems (degenerated elements) due to point movement on the surface, the user can switch off the problematic area easily.

7.2.5 y_+ adaptation

Since a hybrid grid can have a variable number of structured layers, a major task is to support those grids. In order to preserve the grid consistency in the intersection of the structured/unstructured part of those grids, we do not allow point movement in the last layer. Therefore adjustment of the aspect ratio in the last layer of the structured part is only supported in grids with an equal number of layers.

To prevent collisions between two structural layers, an advancing front algorithm extracts the wall normal lines layer by layer. After the successful extraction of the wall normal lines the value of y_+ is determined.

First we calculate the vorticity with the aid of the gradient of the velocity vector V:

$$\omega(x) = \sqrt{[\partial_2 V_1(x) - \partial_1 V_2(x)]^2 + [\partial_3 V_2(x) - \partial_2 V_3(x)]^2 + [\partial_1 V_3(x) - \partial_3 V_1(x)]^2} \tag{7.6}$$

Now let x_W be a point on a solid wall, n the corresponding surface normal and x_N the near point of x_W in wall normal direction.

Then y_+ is defined by:

$$y_+ := \sqrt{\frac{\rho(x_W)\omega(x_W)}{\mu(x_W)}} \cdot \frac{n \cdot (x_W - x_N)}{|n|} \tag{7.7}$$

Now the points on the wall normal lines can be redistributed by the new value of the y_+ (see figure 7.2.5).

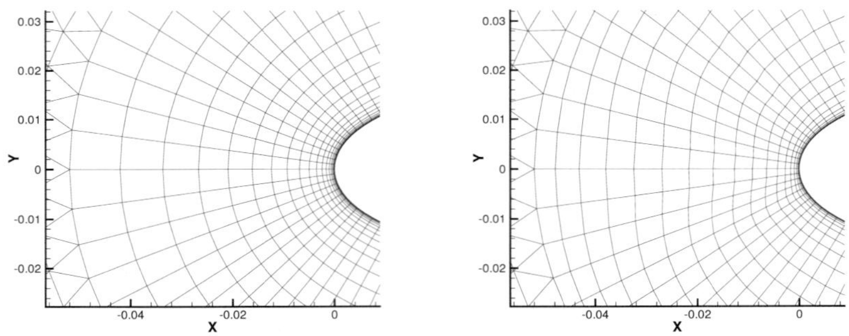

(a) $y_+ \approx 1.5$ (b) $y_+ \approx 0.5$

Fig. 7.2. adaptation of y_+

Another feature is, that the values of y_+ can be set for each turbulent surface separately. In cases where the user will have different values of y_+ (e.g.a slat, wing and flap configuration) this feature is of advantage.

In order to guarantee a consistent grid after the points are redistributed a full geometry check of the structured elements is performed to prevent hexahedra or prisms from degeneration.

7.3 Conclusion

The efficient solution of CFD problems requires local refinement and derefinement, which preserves grid consistency. At the current state, the **Adaptation** is able to handle most of the applications without user interaction. For very complex configurations, in particular with coarse surface geometry, it is still necessary that the user intervenes. Despite this, the results obtained from various applications [3] are encouraging.

References

1. J. Bey: *Finite-Volumen- und Mehrgitter-Verfahren für elliptische Randwertprobleme*. B.G. Teubner, 1998.
2. Thomas Alrutz: Erzeugung von unstrukturierten Netzen und deren Verfeinerung anhand des Adaptationsmoduls des DLR TAU-Codes. Master's thesis, Universität Göttingen, 2002.
3. N. Kroll and Th. Gerhold and S. Melber and R. Heinrich and Th. Schwarz and B. Schöning: Parallel Large Scale Computations for Aerodynamic Aircraft Design with the German CFD System MEGAFLOW. *Parallel CFD*, 2002.

8

Turbulence Model Implementation in TAU

Keith Weinman

DLR, Institute of Aerodynamics and Flow Technology, Bunsenstr. 10, D-37073 Göttingen

Summary. At the present time, aerodynamic analysis and numerical methods are inexorably linked. Thus, a quality control system for the aerodynamic simulation of a complete aircraft in varying configurations is very useful. For this purpose it is critical that suitable turbulence models for the various tasks in hand are available to a prospective user. These models should be robust, and yet sufficiently accurate to enable a proper evaluation of the flow in question. This paper provides an overview of the progress in turbulence model implementation in the TAU code during the duration of the MEGAFLOW project.

8.1 Introduction

The MEGAFLOW project is designed to provide a dependable, and efficient and quality controlled program system for the aerodynamic simulation of a complete aircraft in cruise and other configurations. It was essential that this system meet the standards of industrial implementation within the aircraft industry. Thus a principal goal of the MEGAFLOW project is to provide the basis of a development environment within which various modules of the TAU distribution could be easily modified and optimised. Exchange of modules between European partners was envisaged, and the general levels of performance of the codes comprising TAU were to be improved. This report discusses the significant influence of the MEGAFLOW project upon the development and maturation of the turbulence model implementation with the TAU code.

8.2 Status at start of MEGAFLOW II

Initially, only the basic one-equation Spalart-Allmaras and the two-equation Wilcox k-ω turbulence models had been implemented into the basic TAU solver. The modification due to Edwards had also been implemented into the Spalart-Allmaras model, while the Menter baseline and SST models were available as extensions to the standard Wilcox k-ω model. These extensions had been implemented as part of MEGAFLOW 1. The performance of the one-equation model implementation was robust and sufficiently accurate to allow this model to be used with confidence for the vast majority of design condition flows. Here it is meant that a design condition flow is a flow without phenomena which introduce strongly non-linear interactions (into the flow).

However, the performance of the base two equation Wilcox k-ω model was not totally satisfactory, and the implementation at this time could not be considered sufficiently reliable and robust for many test cases of interest. Of course, since the Wilcox k-ω model is, in some sense, a more complete model than the Spalart-Allmaras model, it was expected that this model would extend the range of computable flows from design flows to more interesting off-design condition problems where sources of non-linearity existed, such as shock/boundary layer interaction. However the more challenging problems were often computed only with the Spalart-Allmaras model. This practise, while usually generating stable solutions, resulted in flows with simply too high a level of eddy viscosity, and consequently unrealistic magnitudes of production and dissipation were computed. The classic symptom of this problem in an aerodynamic computation is the prediction of a shock position which is simply too far downstream in comparison with experimental observation.

Thus, the principal aims of the MEGAFLOW project with respect to the turbulence modelling in TAU, were to provide the basic structures by which improvements could be easily implemented and tested. Interest was in increasing the range of problems for which the satisfactory Spalart-Allmaras turbulence model family could be utilised, and to also improve the robustness and reliability of the Wilcox k-ω implementation. In a climate driven by user confidence, acceptance of the TAU code for the solution of more challenging flow problems can only be achieved when this confidence has been gained. A secondary, but by no means less important goal, was to foster the growth of strong links between the European Aeronautical community in the implementation and development of improved turbulence modelling strategies.

8.3 Progress during MEGAFLOW II: Turbulence Model Implementation

What is the current status of turbulence models implemented within the TAU code? And to what extent have the objectives, noted at the conclusion of the previous section, been achieved? It is useful to briefly overview the progress of turbulence model implementation over the duration of the MEGAFLOW project, shown in Table 8.1. A positive sign in the Robustness column of this table indicate that the code was quite robust, a negative sign has the converse meaning. Similarly a positive sign in the Verification column indicates that a significant number of tests have been performed. A negative sign denotes incomplete testing.

Table 8.1. Turbulence Model status at completion of MEGAFLOW 2. Projects abbreviations are as follows; MEGAFLOW 1 (M1), TAURUS (T), EUROLIFT (E)

Model	Robustness	Verification	Project	Comment
Spalart-Allmaras [1]	+	+	M1	recommended
Spalart-Allmaras + Edwards	+	+	M1	recommended
SALSA [2]	+	+	T	low skin friction
Wilcox k-ω [3]	+	+	M1	stable
Menter + SST [4]	+	+	M1	recommended
Menter + baseline [4]	+	+	M1	recommended
LEA+Wilcox k-ω[6]	+	+	E	recommended
RQEVM+Wilcox k-ω[7]	+	-	T	in testing
EARSM+Wilcox k-ω[8]	+	-	T	in testing

Examination of Table 8.1 illustrates that the robustness of the basic model implementations have been improved over the period of the MEGAFLOW project. This has been achieved by both continuous testing and reevaluation of the implementations within the code. The new data structures, which have been introduced as a direct consequence of MEGAFLOW project work, have contributed greatly to the ease with which existing implementations can be tested and with which new models can be implemented.

The non-linear Eddy Viscosity Models (EVM's) have greatly expanded the range of flows for which practical computation is now possible. This is, in part, due to general performance improvements within the solver. However, a significant contribution to this outcome has been made by technical partners within the European community. In particular, the non-linear EVM's have been introduced to TAU through a number of projects, such as TAURUS and EUROLIFT. In addition to the testing of model implementations within the MEGAFLOW project, some testing has also been achieved through other projects such as UNSI. Significantly, a culture has also arisen such that the day to day implementation and performance issues are usually treated and solved at an informal level between contributing partners. This highlights a significant aspect of the MEGAFLOW project; an environment under which code development and testing can be achieved with close co-operation between European partners. A example of the success in this approach is highlighted in the following section, where an old problem is revisited with the tools currently in hand.

8.3.1 The AGARD SSC test case

A typical example of the benefits obtained by the introduction of more advanced turbulence models, as well improved performance of the TAU code with respect to the basic solver and turbulence modelling, is seen in the AGARD SSC flow results presented at the closing MEGAFLOW meeting. The DLR first examined this flow within the UNSI project, where satisfactory computations were obtained with the one-equation turbulence model variants, but limited success was obtained with the two-equation model implementations for the steady problem. Note that computations were made for both a stationary and an oscillating NACA64A010 airfoil.

Fig. 8.1 illustrates that anisotropic eddy-viscosity models such as the RQEVM model [7] and the LEA model [8] offer much improved k-ω model based solutions. Here the RQEVM model has predicted the upper wing shock position well. It should be noted that the initial k-ω model implementation was incapable of computing a steady-state solution for this flow, and the use of global/dual time stepping strategies resulted in a quasi-steady shock position which was well downstream of the shock position as determined by experiment. Computed skin friction profiles are in good agreement, with of course the exception being in the predicted shock location. The error in the shock position is an indication of excessive turbulent viscosity in the base k-ω model, and the improvements noted in the LEA, and RQEVM models are not only due to a better resolution of the flow anisotropies, but also to improved implementation within the code.

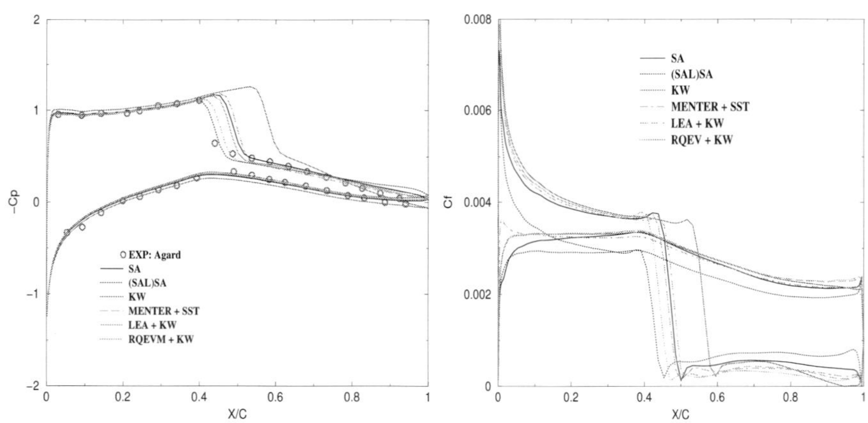

Surface pressure Skin friction

Fig. 8.1. AGARD SSC case: A comparison of C_p and C_f distributions for steady flow.

Examination of both the real and unsteady parts of the first harmonic in pressure for the oscillating airfoil, illustrated in Fig. 8.2, show that the non-linear eddy-viscosity models have clearly outperformed their linear counterparts. In particular, the response of the pressure field at both the shock position and at the airfoil trailing edge have been significantly improved. Consequently, the computed response frequencies, an important issue within the aeroelastic community, will be more accurate. This is due, in principle, to a better modelling of wall influences in the near wall regions of the solution.

The computational cost of the more complex models only slightly exceeds that of the linear EVM variants and, particularly for unsteady calculations, the preliminary evidence suggests that non-linear EVM computations may provide significantly superior results at only slightly more computational expense. Of course, a detailed examination of Figs. 8.1 and 8.2 shows that some subtle physical influences are still not properly modelled, even with more advanced RANS models. This suggests that the development and implementation of more advanced RANS models - or alternative approaches, such as LES and DES, will be required in the future.

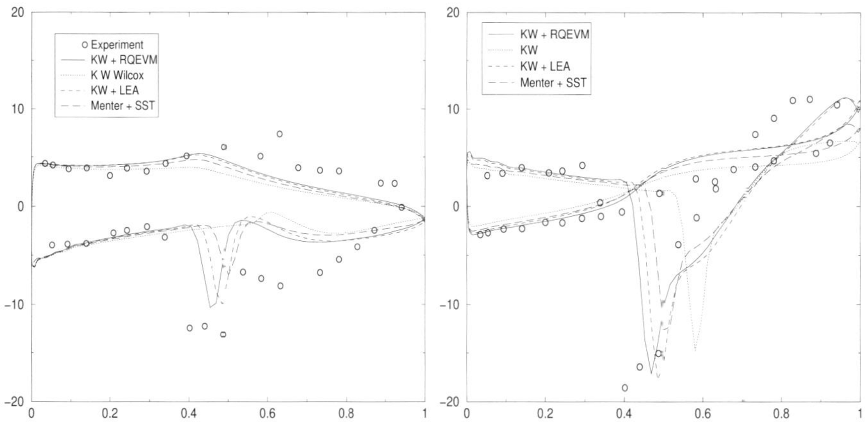

Real part of 1st Harmonic

Imaginary part of 1st Harmonic

Fig. 8.2. Comparison of the Linear and Non-Linear EVM response (in pressure) to forced unsteadiness in the motion of an airfoil

8.4 Future Ambitions and Conclusions

Further extensions to the range of turbulence models are envisaged through the new European project, FLOMANIA, which will be the vehicle for the introduction of DES/LES methods into TAU. Eventually DSM closure may be implemented. At the present time a variety of DES models have been implemented, and are in the process of verification. The intricacies of DSM implementation are being examined in detail within the FLOWer code, the structured and slightly older sister of TAU. Further work is underway in refining the data structures used for the implementation of the turbulence models within TAU and this work,which has arisen as a direct consequence of MEGAFLOW, should result in significant simplification and more efficient coding structures for turbulence model implementation in TAU.

In conclusion, it can be clearly stated that the principal improvement in the TAU code with respect to the modelling of complex and/or highly non-linear flows has been due to the introduction of more flexible and physically realistic turbulence models. The principal achievement of the MEGAFLOW project has been to provide a structure within which the various new models have been easily integrated into the TAU solver. This structure includes both more flexible code generation practices, as well as the equally important aspect of strengthening internal links within the European aerodynamic community in a highly competitive global environment.

8.5 Acknowledgements

The DLR is greatly indebted to both the Technical University of Berlin (TUB) and to the Aeronautical Research Institute of Sweden (FFA) for their work, advice and assistance in the initial implementation and testing of the non-linear EVM's. This work was completed as part of the EUROLIFT and TAURUS work packages. Similarly thanks are again due to the TUB for their implementation of the SALSA variant of the Spalart-Allmaras turbulence model into TAU as part of the TAURUS work package.

References

1. P.R. Spalart, S.R. Allmaras, "A One-Equation Turbulence Model for Aerodynamic Flows", AIAA paper **92-439**, Reno, Nevada
2. T. Rung: "Erweiterung von Eingleichungs-Turbulenzmodell für lokales Nichtgleichgewicht", Hermann-Föttinger-Institut, Technische Universität Berlin, Institutsbericht 4-98
3. D.C. Wilcox: "Reassessment of the Scale Determinign Equations for Advanced Turbulence Models", AIAA Journal **26**, No. 11, 1988, pp. 1299-1310.
4. F. Menter: "Improved Two-Equation k-ω Turbulence Models for Aerodynamic Flows", NASA-TM-103975, October, 1992
5. F. Menter: "Zonal Two-Equation k-ω Turbulence Models for Aerodynamic Flows", AIAA Journal **32**, No. 8, 1993
6. T. Rung, H. Lübcke, M. Franke, L. Xue, F. Thiele, S. Fu: "Assessment of Explicit Algebraic Stress Models in Transonic Flows", In: Proceedings of the 4th International Symposium on Engineering Turbulence Modelling and Measurements, Corsica, 1999, pp. 659-668

7. T. Rung, S. Fu, F. Thiele: "On the Realizability of Non-Linear Stress-Strain Relationships for Reynolds-Stress Closures". Flow, Turbulence and Combustion **60**, 1999, pp. 333-359.
8. S. Wallin, A. Johansson: "An Explicit Algebraic Reynolds Stress Model for Incompressible and Compressible Flows", Journal of Fluid Mechanics **403**, 2000, pp. 89-132.

9

G.I.G. — A Flexible User-Interface for CFD-Code Configuration Data

Uwe Tapper

DLR, Simulation and Software Technology, Lilienthalplatz 7, D-38108 Braunschweig

Summary. A possible solution for the growing demand for graphical user interfaces for simulation codes is shown. The situation at the German Aerospace Center is described and a set of resulting requirements for a flexible, configurable software system is collected. The main idea of separating the description of parameters and the editor is briefly summarized. Finally the current development status is presented, including some screenshots.

9.1 Situation and Requirements

At the German Aerospace Center (DLR), a number of simulation codes is existing which are covering different areas of application, such as computational fluid dynamics, structural mechanics, or flight dynamics. In addition to these numerical solvers, several supporting tools for pre- and post-processing, grid generation, and visualization are being used.

In the field of computational fluid dynamics (CFD), five major solvers are developed: FLOWer [5], TAU [6] and THETA by the Institute of Aerodynamics and Flow Technology, TRACE by the Institute of Propulsion Technology, and TRUST by the Institute of Combustion Technology. All these CFD codes have in common:

- They have a large number of input parameters.
- They are continuously improved and extended, which results in a constant change of existing and addition of new parameters.
- The more complex the use cases of the simulation programs become, the bigger and more complex the sets of input parameters get. In particular, not only the number of input parameters is increasing, but also their logical interdependencies.

Usually there are no or more or less rudimentary graphical user interfaces (GUI) supporting the user to edit parameters in a comfortable way and to ensure correct parameter sets. In any case the use differs from program to program, because of different file formats and also because of different parameters

Table 9.1. Requirements for a flexible parameter input editor

Situation	Requirement(s)
Use of different CFD-/simulation programs	Flexibility
Continuously changing input parameters	Maintainability
Demand for easy handling of input parameters Complex input Standard use cases	Defaults Online-help Guidance Templates
Parallel use of different programs	Single user-interface
Inhomogeneous hard- and software environment	Portability Platform independence

and the structure and hierarchy of parameter sets. Moreover the maintenance of existing user interfaces is expensive, due to the effort needed to integrate new parameters and to change existing ones.

Based on the task to develop a GUI for the FLOWer code being used and developed during the MEGAFLOW II project the idea was born to build a single, configurable GUI for both CFD codes (FLOWer and TAU) and other simulation codes. This common GUI was expected to become more useful because a growing number of users started to work with more than one simulation code to solve a problem. Examples for these tasks are the coupling of flight dynamics and CFD simulation codes or the coupling of structural simulations and CFD. The GUI should make the creating and editing of parameter sets as easy as possible and guide the user through this process. Support by an online help system and by checks of the input data should be self-evident. To meet the requirements of both users and developers a large number of different soft- and hardware environments (mainly UNIX-like systems) have to be supported, too.

In table 9.1 you can find a summary of these requirements.

9.2 Solution

One way to fulfill the given requirements is based on the separation of the description on the one hand and the editing of parameters on the other. All the parameters are described in a special markup language defined in XML. A flexible, configurable editor application displays the parameters, reads and writes files, and provides online-help and user-guidance and so on. Both the editor and the defined markup-language are independent from the actually used simulation code, though the development was of course influenced and driven mainly by the needs of the MEGAFLOW II project.

The use of XML helps to meet the demands for flexibility and maintainability. It is flexible since a new simulation code requires only a new XML-file, but not a new editor application. And it is easier to maintain as a "hard-wired" application, as the steady changes and additions of parameters don't require changes of the editor.

The editor itself is implemented in Java, so it is usable on all important Windows and UNIX-like systems. The runtime-system provides a system-independent GUI and the ability to use a number of needed tools and libraries.

XML and Java together build the common user interface for different simulation and CFD codes. Fig. 9.1 shows the general approach of the design and the embedding in a simulation environment.

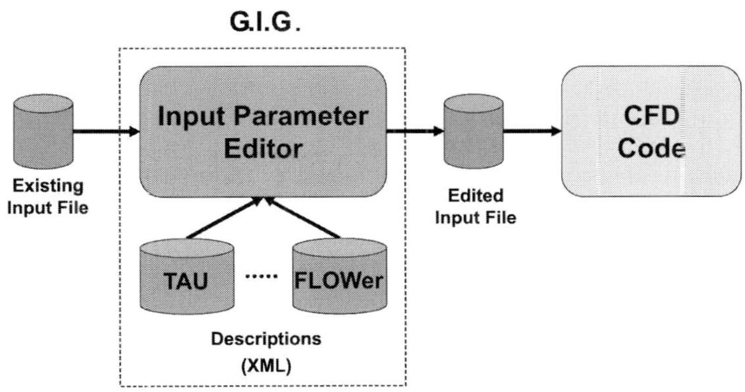

Fig. 9.1. Principle

More advantages are:

- The flexibility of the system makes it usable for applications outside the CFD domain. It may be used for a lot of non-interactive programs, getting their input data from files, e.g. pre- and postprocessors and grid generators.
- The maintenance of the system (editor and XML-files) can be distributed over the developers of the editor and the developers and users of the simulation codes more easily, because the latter can work on the XML-representation of the parameters without an in-depth knowledge of the editor's design and implementation.
- Improvements of the editor application aren't limited to a single application, but have a positive effect on all supported codes.
- A growing number of tools can be used for the creation and editing of XML files, e.g. XML-editors and syntax-checkers.
- The development of the editor takes advantage of the existing libraries (e.g. parsers) for a lot of programming languages.

- The editor might be exchanged without having an impact on the XML structures.
- Information stored in the XML structures may be used for different purposes, like generation of documentation or e.g. within the CFD codes to check data types and legal ranges of parameters.

9.3 Realization

The implementation of the described concept as a program called G.I.G. (generic input generator) was part of the MEGAFLOW II project. The goal was to get a single editor for both FLOWer and TAU. The editor was implemented in Java. It creates the graphical user interface dynamically, using the information stored in XML files. It checks the conformance of data according to the given constraints in the XML file, e.g. data types and legal ranges of values. Last but not least it reads and writes the input files of the CFD codes. G.I.G. supports the user in different ways during the process of editing:

- By marking incorrect input and data and by hiding unnecessary parameters it guides the user through the process of writing valid input files.
- It displays information on input parameters, so the need to lookup details in printed documentation is reduced.
- Useful, working default values for parameters are given.
- Predefined standard use cases reduce the number of required manual input.

9.3.1 Description of parameters in XML

The markup language describing the input parameters was specified in XML (eXtensible Markup Language [4], [1]). This language does not only describe single input parameters, but also their structure in groups and sub-groups, and also the format of in- and output. In the markup language the properties of parameters like data type, default value, legal ranges, the size of so called arrays, and a short description are specified. In a constraints section dependencies between parameters are defined, e.g. when a parameter has to be displayed or not or when a default-value has to be changed.

9.3.2 Editor

G.I.G. is implemented in pure Java, so it is platform-independent and can be used on several UNIX and Windows systems. It can be run as a stand-alone application, but it can also be integrated in software systems as a component (as a so called JavaBean [2]). As an example it is used as a part of TENT [3], a software integration and workflow management system.

9.3.3 Examples

In table 9.2 you can find examples for typical structures in parameter description files.

In the first two table rows the basic structures are shown: the definition of a single parameter and the grouping of parameters. The following two items are examples for the use of arrays. In arrays you can describe parameters consisting of a group of values, e.g. vectors, and you can describe sets of parameters that are read and written repeatedly, e.g. for every grid level in FLOWer. Another difference between the two examples is the definition of the array size. The size can be set to a static number (`length="3"`) or it can be determined by another parameter (`length="paraD"`).

In the last row you can see a very brief example of the implementation of dependencies between parameters. In this example parameter paraG is made visible ("enabled") or invisible ("disabled") in the user interface, depending on the value of parameter paraF. "enable" and "disable" are two members of a group of possible actions, being executed if a condition is fulfilled. E.g. the action "enable" is executed if paraF equals to "euler".

9.4 Conclusion

G.I.G. was developed as part of the MEGAFLOW II project as a single input editor for FLOWer and TAU. A first stable release is available. The second release is currently in a public test phase. G.I.G. has been tested and used at DLR, at the Technical University of Braunschweig and in industry. XML parameter descriptions for other codes, e.g. the flight mechanics simulation SIMULA (developed by the Institute of Flight Systems) are under development.

The work of two students at the Institute of Aerodynamics and Flow Technology has proven, that the XML descriptions of parameters can be changed and extended without detailed knowledge of XML and the editor application, and that the intended improvement in maintainability has been reached.

Further work is going to be done in the areas of user guidance, user interface design and file in-/output.

References

1. B. Bos. *XML in 10 points.* http://www.w3.org/XML/1999/XML-in-10-points
2. G. Hamilton (ed.). *JavaBeans API specification.* Sun Microsystems, 1997, http://java.sun.com/beans.
3. A. Schreiber. *The Integrated Simulation Environment TENT.* Concurrency and Computation: Practice and Experience 2002; 14; 1553-1568.
4. World Wide Web Consortium (W3C). *Extensible Markup Language (XML).* http://www.w3.org/XML/

Table 9.2. Examples of XML-structures (see section 9.3.3)

a single parameter	<PARAMETER name="paraA" datatype="integer" userlevel="expert" ... > <DEFAULT>1.0</DEFAULT> <CONSTRAINTS>... <DESCRIPTION>A brief description ... <SELECT type="Combobox">... <OUTPUT>paraA: %3i ... </PARAMETER>
a group of parameters	<PGROUP name="group1"> <PARAMETER name="paraA" ... > ... </PARAMETER> <PARAMETER name="paraB" ... > ... </PARAMETER> ... </PRGROUP>
array of a single parameter	<ARRAY length="3"> <PARAMETER name="paraC" ... > ... </PARAMETER> </ARRAY>
array of a group of parameters; array length is defined by another parameter (`paraD`)	<PARAMETER name="paraD" ... > ... <ARRAY length="paraD"> <PGROUP name=...> <PARAMETER name="paraE" ... > ... </PGROUP> </ARRAY>
visability of a parameter (`paraG`) depends on another parameter (`paraF`)	<PARAMETER name="paraF" ... > ... <PARAMETER name="paraG" ... > <CONSTRAINTS> <RULES> <RULE action="enable"> <CONDITION ref="paraF" relation="eq" value="euler"/> </RULE> <RULE action="disable"> ... </RULE> ...

5. Jochen Raddatz, Jens K. Fassbender. *Block Structured Navier-Stokes Solver FLOWer*, this volume.
6. T. Gerhold. *Overview of the Hybrid RANS Code TAU*, this volume.

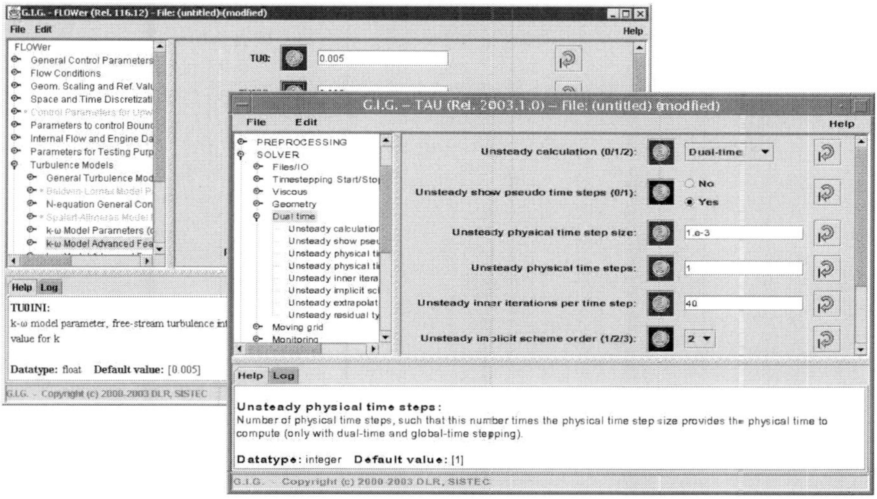

Fig. 9.2. Screenshots: G.I.G. in use for different codes (TAU and FLOWer), running on different operating systems

Part IV

Validation

10

Computation of Aerodynamic Coefficients for Transport Aircraft with MEGAFLOW

Mark Rakowitz[1], Sascha Heinrich[1], Andreas Krumbein[1], Bernhard Eisfeld[1], and Mark Sutcliffe[2]

[1] DLR, Institute of Aerodynamics and Flow Technology, Lilienthalplatz 7, D-38108 Braunschweig
[2] Airbus Deutschland GmbH, D-28183 Bremen, Germany

Summary. The accuracy of the DLR structured and unstructured computational fluid dynamics (CFD) codes in predicting aircraft forces and moments on several configurations at low and high Mach and Reynolds numbers is investigated. Using a combination of a high quality grid, i.e. a grid with sufficient resolution of important flow features, low levels of artificial dissipation and advanced turbulence models, the structured code (FLOWer) is able to both qualitatively and quantitatively predict the experimentally measured drag, lift, pitching moment and pressure distribution. Compared to the structured methods the total time for grid generation is significantly reduced with the unstructured approach (TAU). The quality of the flow solution is comparable to the structured method at significantly higher computational costs.

10.1 Introduction

Validation is an important aspect in the process of developing CFD software. There are many DLR efforts in this area, here a selection of some current projects is presented to demonstrate the capabilities of the MEGAFLOW [1] software. MEGAFLOW comprises the structured FLOWer [2] code, the unstructured TAU [3] code and the structured grid generator MegaCads [4] which have been primarily developed by the DLR in a cooperative effort of the DLR, Airbus Deutschland GmbH and several universities. The software is extensively used for aerodynamic design and research activities in the aircraft industry and the DLR.

For this purpose three typical wing-body geometries and a complex high-wing/high-lift geometry are used. The first set of results stems from the EU-project **HiReTT** (High Reynolds Number Tools and Techniques for Civil Transport Aircraft Design). Here structured CFD computations at Ma: 0.85 and Re: 32.5e6 including aeroelastic deformation are used. The second set includes structured and unstructured results at Ma: 0.75 and Re: 3e6 of the **DLR-F4** geometry which were computed in the framework of the first AIAA

Drag Prediction Workshop 2001 and later follow-up work. The third set of results comprises structured and unstructured results of a **Megaliner** geometry at Ma: 0.7-0.88 and Re: 37e6. The last set includes unstructured results of a complex **High-Wing/High-Lift** geometry at Ma: 0.2 and Re: 1.25e6.

10.2 Structured Flow Solution - FLOWer

The FLOWer code solves the 3D compressible Reynolds-averaged Navier-Stokes (RANS) equations in integral form. Turbulence is modelled by either algebraic or transport equation models. Here the k-ω [5] and the k-ω-LEA [6] turbulence models are used. The spatial discretization uses either a central cell-centered or a cell-vertex finite volume formulation. Dissipative terms are explicitly added in order to damp high frequency oscillations and to achieve sufficiently sharp resolution of shock waves. The dissipative operator is comprised of second and fourth differences scaled by the largest eigenvalue following Jameson et al. [7] and Martinelli and Jameson [8]. On smooth meshes, the scheme is second order accurate in space. Time integration is carried out by an explicit hybrid multistage Runge-Kutta scheme. For steady state calculations the integration is accelerated by local time stepping and implicit residual smoothing. These techniques are embedded in a multi-grid algorithm. The code allows two dummy layers around each block in order to maintain second order accuracy in space at block intersections.

10.3 Unstructured Flow Solution - TAU

The TAU code solves the 3D compressible RANS equations on hybrid grids. The initial/adapted grid is input into the pre-processing module, which computes dual grids using an edge-based data structure (independent of the element types in the hybrid grid). Coarse grids for the multi-grid algorithm are constructed recursively by agglomerating the control volumes at the finer grid level. For parallel computations the dual grids (fine grid and coarse grid levels) are partitioned into a number of domains, each corresponding to a processor. The flow solver is based on a finite volume scheme integrating the RANS equations. The flow variables are stored in the centers of the dual grid, i.e. the vertices of the primary grid. In this study the inviscid fluxes are calculated by employing a central method with scalar dissipation. The gradients of the flow variables are determined by a Green-Gauss formula. The viscous fluxes are discretized using central differences. The compressible flow solver employs an explicit multi-step Runge-Kutta scheme for the discretization of the temporal gradients. In order to accelerate the convergence to steady state, residual smoothing and a multi-grid technique are employed. The turbulence models used in this study are the one-equation Spalart-Allmaras model [9] and the Spalart-Allmaras model with Edwards modification [10]. The adaptation module allows for local refinement of the hybrid grid based on a sensor derived from the flow solution. The adaptation module also allows the redistribution of points in the structured prismatic/hexahedra sub-layers for an improved boundary layer resolution.

10.4 Structured Grid Generation - Megacads/INGRID

MegaCads is a parametric multi-block structured grid generation system providing interactive and batch capabilities for non-overlapping as well as overlapping grids. It enables the user to define block topologies, to distribute points, to create surface grids and to fill volume blocks. It supports a parametric design and restart concept using a script language. Grid spacings are defined as a fraction of a reference length so that grid generation processes can be reused for moderately different geometries. Trimming of curves and surfaces is performed relatively to their parameter values. Different elliptic and biharmonic grid generation techniques using multigrid acceleration are integrated to generate smooth grids.

INGRID is a single-block structured grid generation system developed and used by Airbus Deutschland GmbH. It is noninteractive and extensively used for wing-body geometries.

The structured multi-block grids of the HiReTT (left part of fig. 10.1) and the F4 (right part of fig. 10.1) wing-body geometries have a C-O-H topology for high resolution. The trailing edge of the wings is closed with Bézier splines from (x/c) = 90% onwards [11] and the fuselage plugs are slightly altered to allow the C-block around the wing. This geometry modification has a negligible effect on the results. The grid generation takes advantage of the replay capability of MegaCads, important parameters like cell numbers and spacings are controlled by variables which allow fast modifications to grids. The cell distribution in the 3.5e6 points meshes is chosen to have at least 20 cells normal to the wing surface in the boundary layer and a y^+ of maximal 1 in the first cell layer at a variety of flow conditions. The INGRID single-block structured grid of the Megaliner geometry (fig. 10.2) is provided by Airbus Deutschland GmbH. It contains a C-O block topology with 2.5e6 points.

10.5 Unstructured Grid Generation - *Centaur*^{TM}

Centaur^{TM} by CentaurSoft is a highly automated grid generation package for the generation of unstructured hybrid grids. The package consists of two parts. First, an interactive program reads in CAD data in IGES format and performs some CAD cleaning if necessary. It also allows the specification of boundary conditions and grid sources for the control of local element types and sizes. The interactive program also offers a variety of default settings for local grid clustering mainly based on curvature. The second program computes the

whole grid automatically. The boundary layer of hybrid grids is marched from the surface grid with prismatic elements, hexahedral elements are also possible. Chopping of prism layers can be applied automatically to maintain grid quality. The volume between the outer surface of the boundary layer grid and the farfield is filled by tetrahedral elements using an octree/advancing front approach. The transition between the boundary layer grid and the tetrahedral grid on quadrilateral faces is provided by pyramids.
All the unstructured grids here use prismatic elements close to the surface with at least 20 cell layers in the boundary layer and a y^+ of maximal 1 in the first cell layer. The unstructured Megaliner grid contains 3.4e6 points (fig. 10.2). The high-wing/high-lift geometry hybrid grid depicted in fig. 10.3 demonstrates the capabilities of the unstructured approach for complex geometries. The geometry contains a double-slotted high-lift system with flap-track-fairings and two nacelles. The grid was designed to be applicable for a variety of flow conditions for the computation of complete polars including maximum lift. To reduce the computational costs the number of grid points was kept as low as possible (7.4e6 points) without sacrificing grid quality.

10.6 FLOWer and TAU Solutions

10.6.1 HiReTT - FLOWer

The HiReTT results are computed fully turbulent with FLOWer using the k-ω - turbulence model at Ma: 0.85 and Re: 32.5e6 and are compared to measurements of the European Transonic Windtunnel (ETW). Two geometries are compared, the first geometry includes a wing shape with pre-estimated bend and twist, the second wing shape includes aeroelastic deformations (bend and twist) which are computed using a coupling of CFD and structural analysis (Timoshenko beam) at $\alpha \approx 1.5°$. The grid generation process takes full advantage of the replay capability of MegaCads, both multi-block structured grids use the same restart script. Fig. 10.4 includes three sets of curves which consist of the experiments, the FLOWer results with the pre-estimated wing shape and FLOWer results with the computed wing shape. The first set of curves on the left of fig. 10.4 shows $C_L(C_{D,net})$. The curve for the computed wing shape is very close to the experiment in a range from C_L : 0.32 - 0.68. The agreement of the computations with the pre-estimated wing shape with experiment is less favourable in this area. Above C_L : 0.68 both computational curves lie on top of each other, overpredicting the lift. The same applies for the middle set of curves for $C_L(C_D)$. The difference between the FLOWer computations with the computed and the pre-estimated wing shape is more pronounced in the third set of curves on the right side for $C_L(\alpha)$. In the range α : $1.2° - 3°$ the agreement of the computations with the computed wing shape with experiment is very good, while the computations with the estimated wing shape underpredict the lift noticeably. The same tendencies can be seen in fig. 10.5 which shows $C_D(\alpha)$ and $C_{D,net}(\alpha)$. The inserted enlargement of the region around α : 1.5° shows a close agreement of both computational total drag curves with experiment at α : 1.5°, but a noticeable difference of the pre-estimated wing shape computations to experiment for net drag. The difference between the computed wing shape drag curve and experiment is less than 3% for the whole α-range. Fig. 10.6 includes inboard and outboard wing pressure distributions at α : 1.57°. The agreement of the aeroelastic wing shape pressures with experiment is again excellent, the pre-estimated wing shape pressures have a forward shock position and an exaggerated re-expansion after the shock on the outboard wing. The anomaly of the experimental pressures on the inboard lower wing at $(x/c) \approx 0.24$ and on the outboard upper wing at $(x/c) \approx 0.25$ are due to blocked pressure tappings.

10.6.2 Megaliner Geometry - FLOWer vs. TAU

For the Megaliner geometry FLOWer computations using the k-$\omega - LEA$ - turbulence model on a single-block structured grid are compared to TAU computations using the Spalart-Allmaras turbulence model on an unstructured hybrid grid and to experiments (ETW) at Ma: 0.7 - 0.88 and Re: 37e6. Fig. 10.7 includes net drag rise curves, the FLOWer and TAU computations are quite close to each other and have an offset to lower drag as compared to the experimental drag rise below Ma: 0.86. The shape of all three curves is very similar and as for drag rise the gradient is important and not the absolute drag level, the prediction of drag rise by FLOWer and TAU is good. Fig. 10.8 depicts inboard and outboard wing pressure distributions at Ma: 0.8, Re: 37e6 and C_L : 0.54. The general agreement of the FLOWer and TAU computation which each other and experiment is quite good The rooftop of the TAU computation on the inboard wing is slightly high. The pressure niveau on the upper wing of both computations behind the shock is slightly above the experiment. On the upper side of the wing two blocked pressure tappings are visible in the experiment. On the outboard wing the computed FLOWer and TAU rooftop levels are similar, the shock of the TAU computation is slightly less steep and in a slightly more backward position. The tendency of the pressure level of both computations behind the shock to lie above the experiment has increased. The discrepancy of the computed pressures on the lower outboard wing trailing edge to experiment seems to be due to a blocked pressure tapping.

10.6.3 DLR F4 - FLOWer vs. TAU

In this section fully turbulent FLOWer computations on a multi-block structured grid are compared to fully turbulent TAU computations on the same grid converted to the necessary unstructured format. The FLOWer results stem from the DLR contribution to the first AIAA Drag Prediction Workshop 2001 and are achieved using a minimal amount of artificial dissipation and the k-ω-LEA turbulence model. The structured grid is converted into an unstructured grid to achieve a comparison solely based on the flow solver and turbulence model. The computations are compared to experimental results (including fixed transition on the wing and the fuselage nose) obtained in three different wind tunnels, the High Speed Wind Tunnel HST of the NLR, the ONERA-S2MA wind tunnel of the ONERA and the 8 foot Wind Tunnel of the DRA. The fully turbulent FLOWer and TAU computations overpredict the measured drag curve (upper left diagram of fig. 10.9) by approximately 20 drag-counts. The influence of transition accounts for approximately 14 drag-counts, reducing the drag overprediction to approximately 6 drag-counts. Finer grids [12] reduce the computational drag further, i.e. move the drag curve to the left. The upper right diagram of fig. 10.9 depicts $C_L(\alpha)$. The FLOWer lift curve matches the experiment very well, even in the non-linear region above $\alpha = 0.5$ degrees, with a slight deviation in the region $\alpha = 1.5$ to 2 degrees. In this area the TAU result is slightly below the experiment, the linear region compares very well to the experiment and the FLOWer result, too. The lower diagram of fig. 10.9 again shows good agreement between the FLOWer and experimental results for the pitching moment, when considering the experimental scatter between the different wind tunnel facilities. The changes in slope of the pitching moment curve are also captured very well. The TAU results agree well at lift coefficients below 0.5, too and have a deviation above C_L : 0.5 which corresponds to the deviation in the aforementioned diagrams. From turbulence model investigations with FLOWer in [12] it is known that this discrepancy between the two sets of computations can be mainly attributed to the different turbulence models. The good agreement between computational and experimental drag-, lift- and moment-curves is only possible when the computed pressure distributions agree very well with the experiment, as can be seen in figs. 10.10 and 10.11. On the inboard wing the agreement of the FLOWer and the TAU computation with each other and with experiment is good, on the outboard wing the TAU computation has a backward shock position which is mainly due to the turbulence model. One main insight of the whole F4 investigation is that the flow separations on the wing-root-fuselage intersection and on the upper wing trailing edge in the experiment (fig. 10.12) and in the computation (fig.10.13) have a significant impact on the surface pressure distribution. In figs. 10.12 and 10.13 it is obvious that the FLOWer computation captures these flow separations very well.

10.6.4 High-Wing/High-Lift Geometry - TAU

In this section fully turbulent TAU computations with a complex high-wing/high-lift geometry using the Spalart-Allmaras turbulence modell with Edwards modification at Re: 1.25e6 and Ma: 0.2 are compared to full- and half model experiments (ONERA F1 and Airbus Deutschland GmbH LSWT). On the left side of fig. 10.14 it can be seen that the computations overpredict the lift with a parallel shift in the linear range and that the maximum lift is underpredicted by 2%. The right side of fig. 10.14 depicts the corresponding $C_L(C_D)$-curve which is shifted parallel to lower drag (about 4% less than the experiment) except at maximum lift. Figs. 10.15 and 10.16 show wing pressure distributions on the inboard wing and in the outboard region of the double-slotted high-lift system. The agreement between computation and experiment is good on all geometric wing elements, indicating sufficient grid resolution. The overprediction of lift in fig. 10.14 might be explained through the influence of the fuselage.

10.7 Conclusions

From the preceding investigations the following prerequisites for high quality CFD results can be identified:

- The actual geometry has to be known, which is not a trivial matter in the case of highly loaded wind tunnel models like the HiReTT model at high Reynolds numbers. As measured aeroelastic deflections are usually not available, the deflections have to be estimated, most suitably with a coupled CFD/structual approach. Usually this is performed at one flow condition only and the geometry is used at other flow conditions, too. Strictly speaking, the aeroelastic deformation should be evaluated and incorporated into the geometry at each flow condition.
- For medium grid sizes ($\approx$ 4e6 grid points for wing-body and $\approx$ 8e6 for complex high-lift geometries) for industrial applications the grid quality is crucial. [12] shows, that the difference in drag can be up to 20% for computations with different grids of similar size using the same flow solver and parameter settings.
- To get the flow physics right, especially in case of flows with separations, advanced turbulence models (at least 1.5 and 2 equation models) are necessary.
- The best computational results are achieved with a minimum amount of artificial dissipation at the cost of robustness.

With these prerequisites given, the agreement of transonic CFD results for wing-body geometries of the drag polar with experiment can be less than 3% with a corresponding good agreement of lift, moment and pressure distribution. Unstructured computations at low speed for complex high-lift geometries show a good agreement of computational lift and drag gradients with experiment, except in the maximum lift area. Maximum lift is underpredicted by 2%, the drag is 4% too low for the geometry and flow condition at hand. The computational wing pressure distributions show a good agreement with experiment.

Acknowledgements

The support from CentaurSoft for the unstructured grid generation is appreciated. The aeroelastic wing deformations for the DLR contribution to HiReTT work-package 1.2 are provided by RWTH Aachen.

References

1. N. Kroll, C.-C. Rossow, K. Becker and F. Thiele: "MEGAFLOW — A Numerical Flow Simulation System", 21st ICAS Congress, Melbourne, paper 98-2.7.4, 1998.
2. N. Kroll, R. Radespiel and C.–C. Rossow: "Accurate and Efficient Flow Solvers for 3D Applications on Structured Meshes", AGARD R–807, 1995, pp.4.1–4.59.
3. T. Gerhold, O. Friedrich, J. Evans and M. Galle: "Calculation of Complex Three-Dimensional Configurations Employing the DLR TAU-Code", AIAA Paper 97-0167, 1997.
4. O. Brodersen, M. Hepperle, A. Ronzheimer, C.–C. Rossow and B. Schöning: "The Parametric Grid Generation System MegaCads", 5th Intern. Conference on Numerical Grid Generation in Computational Fluid Dynamics and Related Fields, edited by B. K. Soni, J. F. Thompson, J. Häuser and P. Eisenmann, 1996, pp. 353-362.
5. D. C. Wilcox: *"Turbulence Modeling for CFD"*,DCW Industries Inc., La Cañada, CA, 1993, pp. 84-87.
6. E. Monsen, M. Franke, T. Rung, P. Aumann and A. Ronzheimer:" Assessment of Advanced Transport-Equation Turbulence Models for Aircraft Aerodynamic Performance Prediction", AIAA Paper 99-3701, 1999.
7. A. Jameson, W. Schmidt and E. Turkel: "Numerical Solutions of the Euler Equations by Finite-Volume Methods using Runge-Kutta Time-Stepping Schemes", AIAA Paper 81-1259, 1981.
8. L. Martinelli and A. Jameson: "Validation of a Multigrid Method for the Reynolds-Averaged Navier-Stokes Equation", AIAA Paper 88–0414, 1988.
9. P.R. Spalart and S.R. Allmaras: "A One-Equation Turbulence Model for Aerodynamic Flows", AIAA Paper 92-0439, 1992.
10. J.R Edwards and S. Chandra: *"Comparison of Eddy Viscosity-Transport Turbulence Models for Three-Dimensional, Shock-Separated Flows"*, AIAA Journal, Vol. 34, No. 4, pp. 756-763.
11. E. Monsen and R. Rudnik: "Investigation of the Blunt Trailing Edge Problem for Supercritical Airfoils", AIAA Paper 95-0089, 1995.
12. M. Rakowitz, B. Eisfeld, D. Schwamborn and M. Sutcliffe: *"Structured and Unstructured Computations on the DLR-F4 Wing-Body Geometry"*, Journal of Aircraft , Vol. 40, No. 2, March-April 2003, pp. 256-264.

10.8 Figures

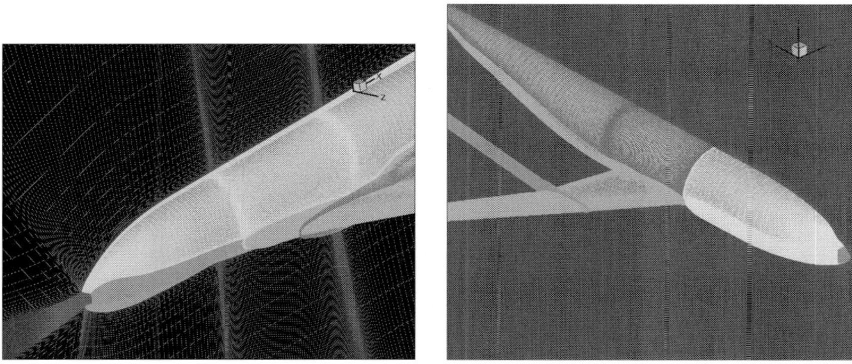

Fig. 10.1. Left: HiReTT MegaCads multi-block structured grid. Right: F4 MegaCads multi-block structured grid.

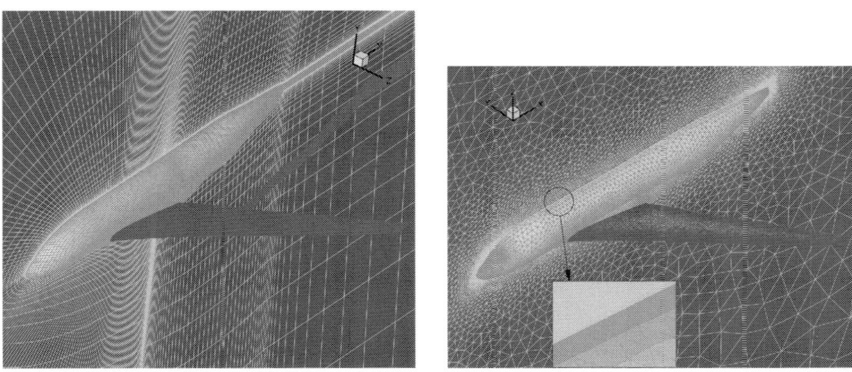

Fig. 10.2. Left: Megaliner INGRID single-block structured grid. Right: Megaliner Centaur hybrid unstructured grid.

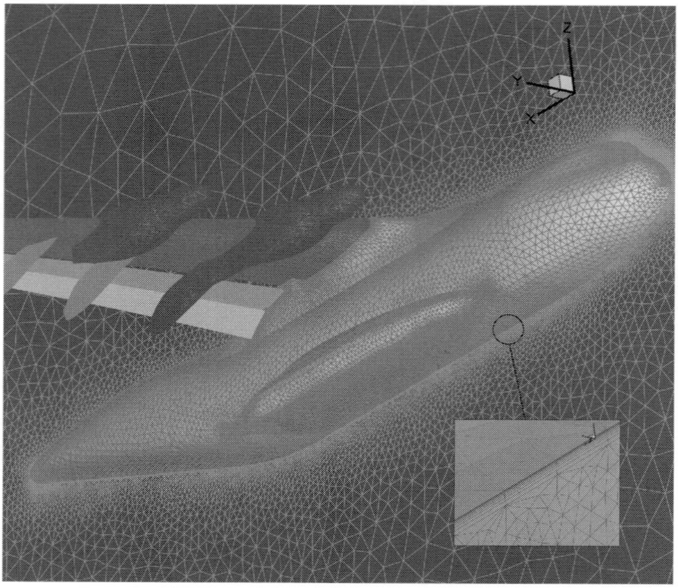

Fig. 10.3. High-wing/high-lift geometry Centaur hybrid unstructured grid

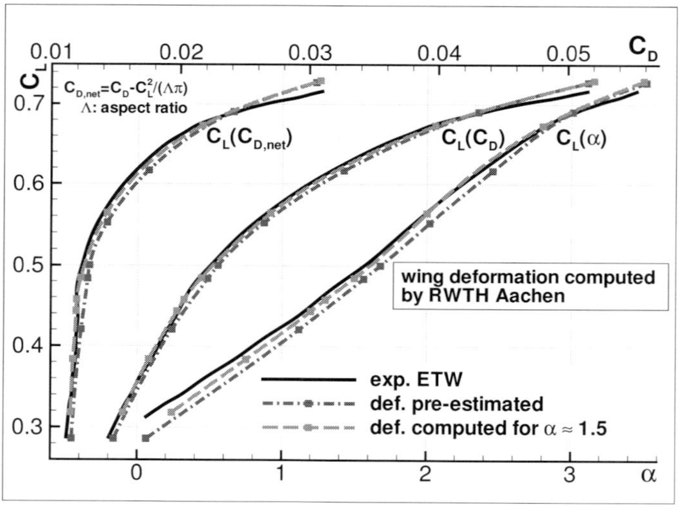

Fig. 10.4. HiReTT FLOWer $C_L(C_D)$, $C_L(\alpha)$ at Ma: 0.85 and Re: 32.5e6

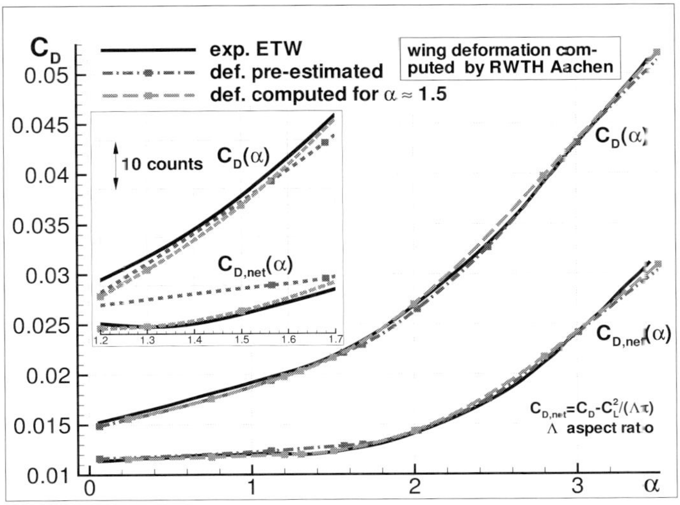

Fig. 10.5. HiReTT FLOWer $C_D(\alpha)$ at Ma: 0.85 and Re: 32.5e6

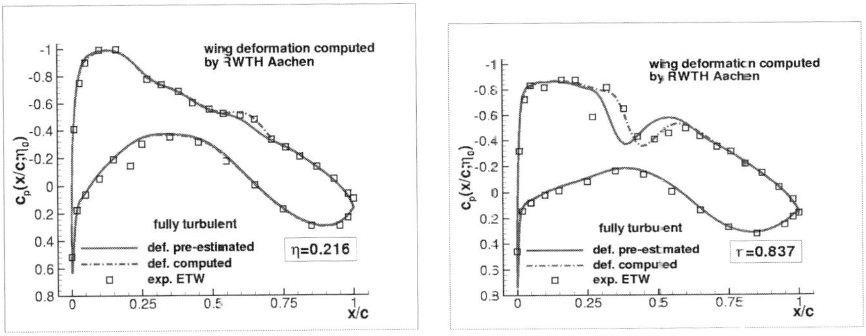

Fig. 10.6. HiReTT FLOWer $c_P(x/c)$ η : 0.216 and η : 0.837 at Ma: 0.85 and Re: 32.5e6.

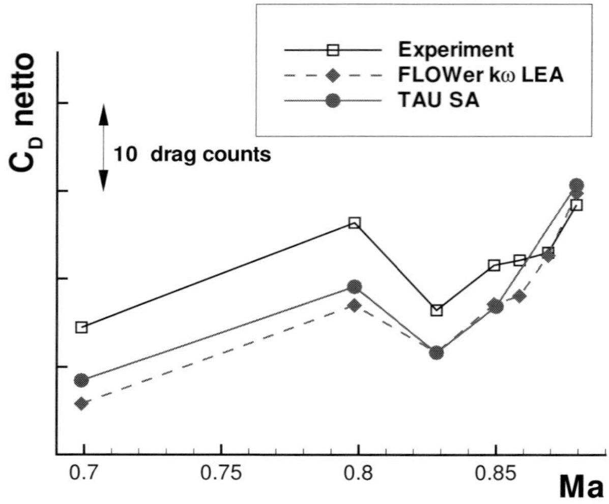

Fig. 10.7. Megaliner drag rise comparison FLOWer/TAU at Ma: 0.7 - 0.88 and Re: 37e6

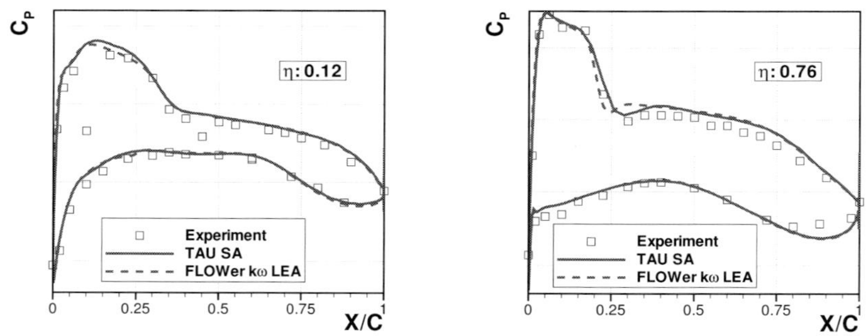

Fig. 10.8. Megaliner $c_P(x/c)$ η : 0.12 and η : 0.76 at Ma: 0.8, Re: 37e6 and C_L : 0.54.

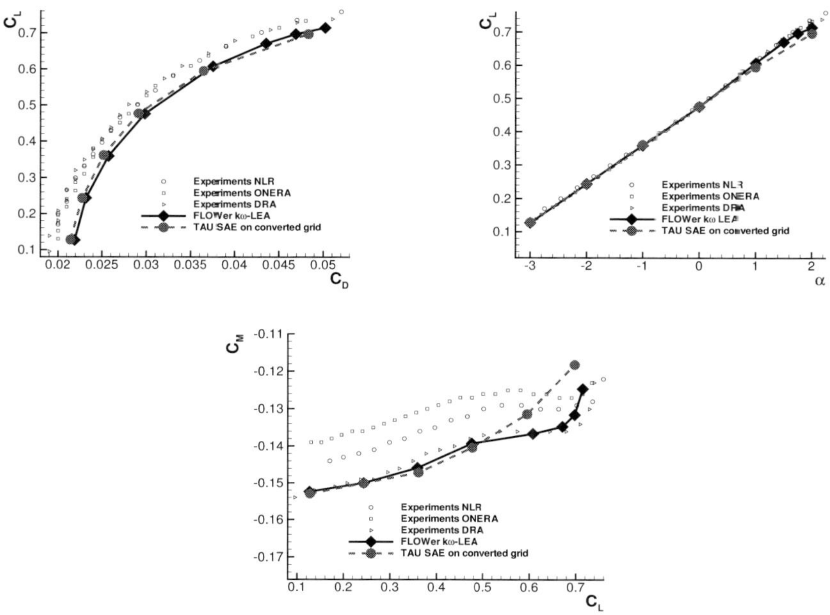

Fig. 10.9. F4 $C_L(C_D)$, $C_L(\alpha)$ and $C_M(C_L)$ FLOWer vs. TAU at Ma 0.75 and Re: 3e6.

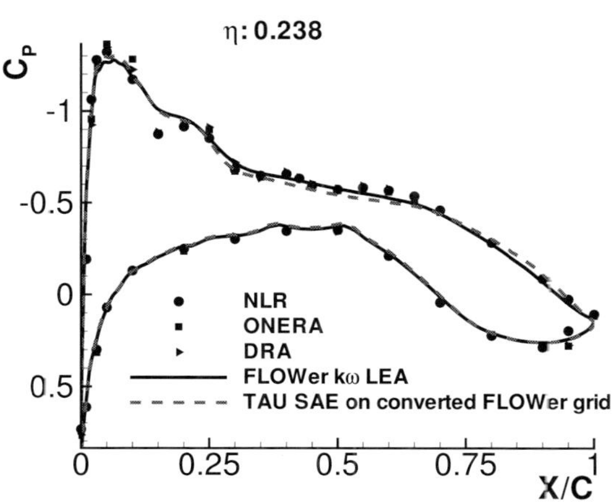

Fig. 10.10. F4 c_P at C_L : 0.5 η : 0.238 FLOWer vs. TAU at Ma: 0.75 and Re: 3e6

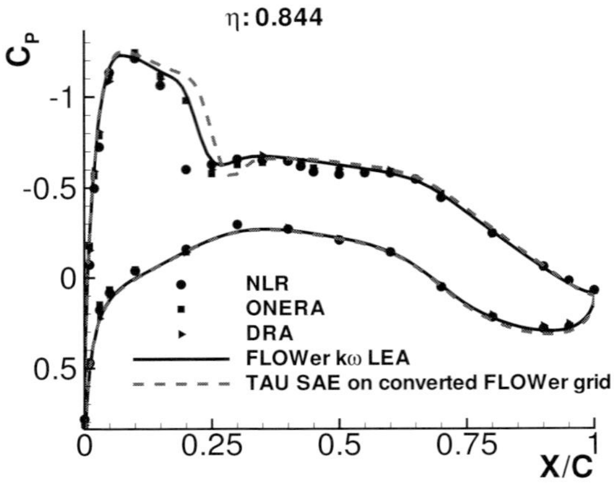

Fig. 10.11. F4 c_P at C_L : 0.5 η : 0.238 FLOWer vs. TAU at Ma: 0.75 and Re: 3e6

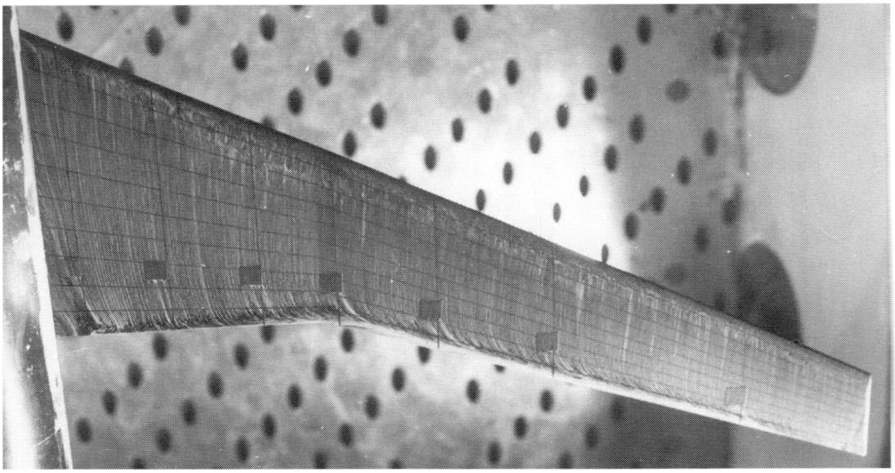

Fig. 10.12. F4 ONERA S2MA experimental surface streamlines (Ma: 0.75, Re: 3e6 and C_L : 0.5)

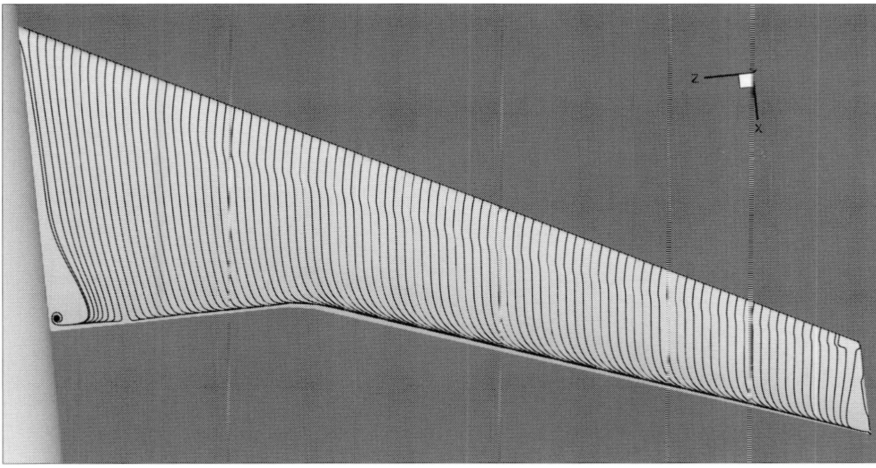

Fig. 10.13. F4 FLOWer k-$\omega - LEA$ computational surface streamlines (Ma: 0.75, Re: 3e6 and C_L : 0.5)

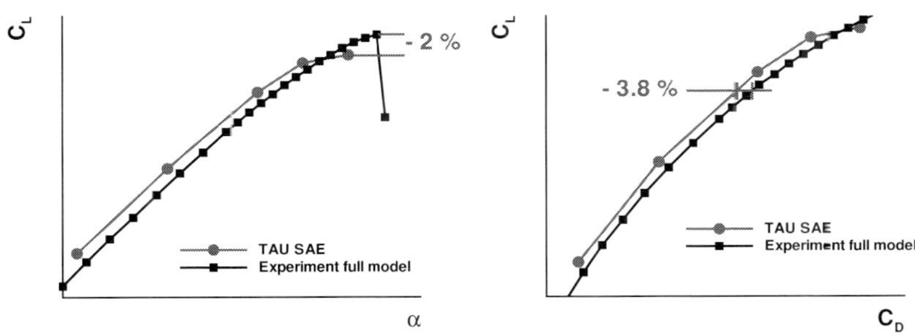

Fig. 10.14. High-wing/high-lift geometry TAU SAE $C_L(\alpha)$, $C_L(C_D)$ at Ma: 0.2, Re: 1.25e6

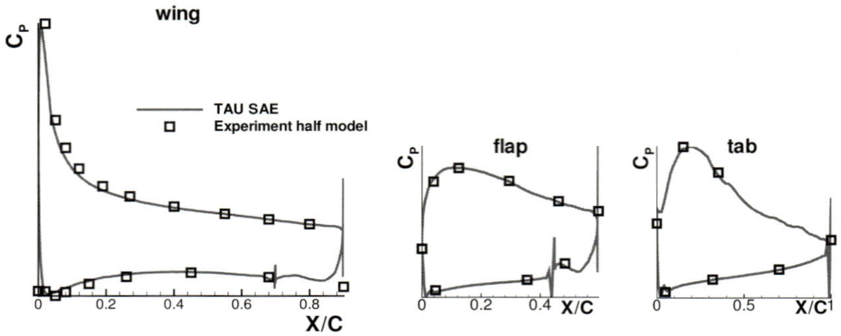

Fig. 10.15. High-wing/high-lift geometry TAU SAE $c_P(x/c)$ at η : 0.054 (Ma: 0.2, Re: 1.25e6, α : 10°)

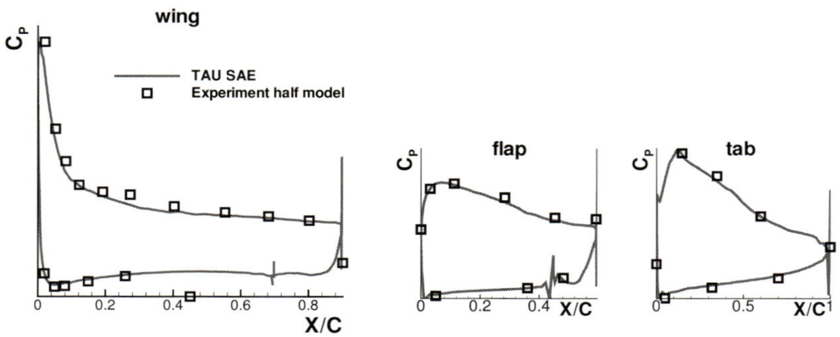

Fig. 10.16. High-wing/high-lift geometry TAU SAE $c_P(x/c)$ at η : 0.465 (Ma: 0.2, Re: 1.25e6, α : 10°)

11

Computation of Engine–Airframe Installation Drag

Olaf Brodersen

DLR, Institute of Aerodynamics and Flow Technology, Lilienthalplatz 7, D-38108 Braunschweig

Summary. It is demonstrated that the unstructured, hybrid method of the MEGAFLOW project is capable to compute the engine–airframe installation drag for several engine types although drag differences for varying engine positions can be very small. Results are presented for the DLR–F6 and ALVAST configurations with through–flow nacelles and powered engines. Verification by grid refinement as well as validation with wind tunnel data completes this investigation.

11.1 Introduction

The future of civil aircraft design will be characterized by more stringent environmental requirements. Great efforts are necessary to achieve a further significant reduction of the specific fuel consumption, noise emissions, and air pollution as demanded in [1]. An increase of the engine bypass ratios, which can be observed over the last decade, is one step to reach these goals. But an higher bypass ratio causes a geometrically larger engine which can influence severely the wing lift and pressure distribution and the overall drag.

For the success of a new aircraft the design process itself is of importance because decisions at the beginning of a project based on experimental and numerical performance predictions have a significant influence [2]. Accurate, fast, and reliable development methods can help to analyse more configurations and to find an optimised configuration.

The objective of the MEGAFLOW project was to develop and enhance the required theoretically based performance prediction capabilities by numerically solving the Reynolds–averaged Navier–Stokes (RANS) equations [3, 4]. This approach faces the difficulties that the viscous boundary layers have to be resolved adequately even for geometrically complex configurations, that accurate solution algorithms must be available, and that suitable turbulence models are necessary.

In the MEGAFLOW–I project mainly structured grid techniques have been applied for studying engine–airframe integration effects. Due to the lack of automatic structured grid generation software the manual driven process is very time–consuming for aircraft configurations. The unstructured method offers a very fast semi–automatic grid generation but questions still exist whether unstructured methods can be used for the accurate computation of flow features, of installation effects, and of the overall drag.

Due to the fact that since the beginning of the 1990s several investigations in the field of engine–airframe integration have been carried out, a good experimental data base and substantial experience exist so that these can be used to verify and validate the MEGAFLOW software. The data is based on results from an ONERA DLR cooperation using the DLR–F6 wind tunnel model with through–flow nacelles and on the results from the European projects DUPRIN, ENIFAIR, and AIRDATA wherein the DLR–ALVAST configuration with turbo–powered engines has been investigated [5, 6, 7, 8, 9, 10]. Wind tunnel testing as well as numerical investigations have been performed. The numerical solutions using structured grids are based on the Euler and RANS equations. Because of the structured grid technique only fundamental results could be achieved. Effects of engine position variations on the installation and total drag could not be computed.

In principle it has been shown in MEGAFLOW–I that the structured and unstructured approach are capable to compute solutions for analysing engine–airframe installation effects at transonic speed [11, 12]. As demonstrated in [13, 14] the results of the verification and validation of the unstructured method will be presented here. The calculations are focused on drag prediction for two configurations with a variety of different engine sizes and positions. Grid Refinement studies will be used to reduce discretisation errors and to verify the results.

11.2 Aircraft Configurations

The two wind tunnel models DLR–F6 and DLR–ALVAST exist for the investigation of engine–airframe interference effects. They represent a wide body twin engine aircraft. Through–flow nacelles can be mounted on the F6. ALVAST can be equipped with turbo–powered engine simulators. The configurations are very similar. The design Mach number is $M_\infty = 0.75$ at $C_L = 0.5$. The aspect ratio of the wing is $\Lambda = 9.42$ and the wing has a leading edge angle of $\varphi = 27.1°$. The dihedral is $\gamma = 4.8°$. The F6 is a full model (see Fig. 11.1) whereas the ALVAST model is larger and can be used as a full or half model. Fig. 11.2 shows the ALVAST with the conventional, the very high bypass ratio (VHBR), and the ultra high bypass ratio (UHBR) engine simulators in the ONERA S1 wind tunnel. For F6 three different axis–symmetric nacelles have been investigated.

1. A conventional engine, similar to a CFM-56 with a long nacelle. The nacelle and the wing lower side are overlapping. Three positions are available.

2. A conventional engine, similar to a CFM-56 with a short nacelle with an inner nozzle. The nacelle and the wing lower side are not overlapping. One position is available.
3. A very high bypass ratio engine (VHBR) with an inner nozzle. The nacelle and the wing lower side are not overlapping. Two positions are available.

Fig. 11.3 gives an impression of the F6 nacelles and table 11.1 lists their position. The ALVAST configuration is equivalent to F6 but approximately twice as large. The bypass ratios and sizes of the conventional, VHBR, and UHBR engine simulators can be found in table 11.2.

Table 11.1. F6 nacelle positions

Nacelle type	x_{TE}/l	z_{TE}/l
F6 CFM-56 long, position 1	0.49	-0.189
F6 CFM-56 long, position 2	0.30	-0.189
F6 CFM-56 long, position 3	0.30	-0.250
F6 CFM-56 short	-0.04	-0.160
F6 VHBR, position 1	-0.04	-0.160
F6 VHBR, position 1	-0.04	-0.160
ALVAST conventional	-0.027	-0.171
ALVAST VHBR	-0.017	-0.162
ALVAST UHBR	-0.446	-0.119

Table 11.2. ALVAST engine data, start–of–cruise (SOC)

	TF	VHBR	UHBR
Diameter	162.6 mm	198.0 mm	254.0 mm
Bypass ratio	6	9.2	15.7
ε_{Fan}	0.848	0.916	0.831
p_{Fan}/p_0	2.331	2.086	1.721
p_{Core}/p_0	2.067	1.708	1.297
T_{Fan}/T_0	1.141	1.103	1.042
T_{Core}/T_0	0.551	0.528	0.612

11.3 Wind Tunnel Experiments

The results of the extensive test program from several campaigns for both configurations can be used here. The DLR–F6 configuration has been tested in four wind tunnel campaigns. The model was equipped with pressure taps in eight spanwise sections. In addition the pressure has also been measured on the pylon and on the nacelle. Here, only the wing pressure distributions will be compared. Fig. 4(a) shows the location of the spanwise sections. The internal drag of the through–flow nacelle has been investigated in a calibration test so that the engine installation drag can be computed. The deviation of the drag is approximately one drag count ($1dc = 10^{-4}$) during a test campaign. Probably due to degradation of the model surface an increase of the drag of several drag counts has been observed over the years. The F6 model has been tested for cruise flight conditions only.

The test conditions of the ALVAST model vary only in relation to the engine condition because powered simulators are used. Here, only start–of–cruise (SOC) conditions are of interest. The fan and core pressure and temperature ratios are listed in table 11.2. The location of the wing pressure taps sections can be found in Fig. 4(b).

11.4 Solution Method

The solution of the RANS equations is performed with the TAU software based on unstructured grids. Details about the software and the solution algorithms can be found in the literature [12, 15]. For these investigations a central difference scheme with a Jameson–type of dissipation is used. The Spalart–Allmaras turbulence model with Edwards–Chandra modifications is applied [16, 17].

It is well known that a low discretisation error is of great importance when drag has to be determined. Grids with an adequate density and quality of the elements have to be used. Therefore, DLR applies the software Centaur from CentaurSoft for the generation of unstructured, hybrid grids [18]. As it is known, isosceles triangles should be used for unstructured grids to minimize discretisation errors. But this effects the number of points so that especially on wings a large number of points in spanwise direction will be located although this is not necessary for high aspect ratio wings. Here, a moderate stretching of the surface triangles of 2.5 is introduced to reduce the number of elements in wing spanwise direction. Within the 24 prismatic layers normal to all surfaces approximately 10–16 layers can be found in the boundary layer. Small tetrahedral elements are used to improve the wake resolution.

To ensure a comparable discretisation error for the different F6 and ALVAST configurations the grids are generated using the same parameter setting. The grids will be adapted based on pressure and velocity gradients of the solutions so that the grids can be refined and discretisation errors can be minimized. The tables 11.3 and 11.4 lists the initial and final adapted number of grid points for the F6 and ALVAST configurations. The Figs. 5(a) and 5(b) show a typical hybrid grid in the symmetry plane and partly the resolution of the aircraft surface.

Table 11.3. DLR-F6: number of grid Points (10^6 nodes)

Grid	Wing-Fuselage	CFM-long 1	CFM-long 2	CFM-long 3	CFM-short	VHBR-1	VHBR-2
Initial	2.58	3.79	3.97	3.73	4.03	4.70	4.55
Final	5.52	7.46	7.44	7.74	8.72	8.46	8.19

Table 11.4. DLR-ALVAST: number of grid Points (10^6 nodes)

Grid	Wing-Fuselage	Conventional	VHBR	UHBR
Initial	3.02	3.80	4.20	5.15
Final	6.16	6.97	7.60	9.31

11.5 Results

For both configurations calculations have been performed at $Ma_\infty = 0.75$. The target lift coefficient has been varied between $C_L = 0.2 \ldots C_L = 0.6$ for F6 and has been kept constant ($C_L = 0.5$) here for ALVAST. The Reynolds numbers are $3 \cdot 10^6$ and $4.3 \cdot 10^6$ for F6 and ALVAST, respectively. In [13, 14] it has been shown that the three–step grid adaptation influences only slightly the pressure distribution but has a stronger influence on the installation and overall drag. Although the discretisation error can be reduced using the adaptation a systematic deviation between the numerical and experimental results still exists. Figs. 11.6 and 11.7 present the result of the drag for all F6 engine configurations. For the short and VHBR nacelles only two adaptations instead of three as for the conventional long nacelles have been performed. Therefore a reduced deviation of the numerical and experimental installation drag for the short and VHBR nacelles can be expected for a further adapted grid. Fig. 11.8 demonstrates that the refinement of the grid causes a reduction of the differences between numerical and experimental results, so that it can be assumed that the deviations in Fig. 6(a) and 6(b) are mainly produced by systematic errors. Figs. 9(a) and 9(b) compare the pressure distributions at wing sections inboard ($\eta = 0.331$) and outboard ($\eta = 0.377$) of the engine close to the pylon. The matching of the results is good. But on the wing inboard lower side deviation are visible due to flow separation. The details have been presented in [13].

The computation of the flow for the ALVAST with powered simulators needs an higher effort because in addition jet effects have to be resolved. The analysis of the results presented here is not finished completely. Especially the thrust–drag book–keeping needs further investigations. Figs. 11.10 and 11.11 show the pressure distributions for ALVAST with the VHBR and the UHBR engines. For VHBR also results from the structured grid approach using FLOWer are included. Due to the grid adaptation the unstructured solution offers a stronger shock resolution. The UHBR wing lower inboard side shows an overshooting of the minimum pressure coefficient and deviations close to the trailing edge like it has been observed for F6. It is assumed that flow separations are responsible for this effect.

11.6 Conclusion

It has been demonstrated that the MEGAFLOW software is well suited for the investigation of engine–airframe interference effects and performance predictions. Now, a confidence exists at DLR that the unstructured technique using tetrahedral and prismatic elements can be applied to compute small drag differences due to engine position variations. Systematic deviations still exist when results are compared to experiments, but using adequate grids in combination with several steps of grid adaptation result in low discretisation errors. This guarantees that the MEGAFLOW software can be used for fast, accurate, and reliable performance predictions in the aircraft design process.

Acknowledgements

The author would like to thank the TAU development team for their support, especially T. Alrutz for the grid adaptation. Thanks earn also L. Lekemark for his CAD work, A. Ronzheimer for his CAD and grid generation support, and FrHr. H. v. Geyr for providing structured grid based solutions for the ALVAST configuration.

References

1. Busquin, Ph., et al.: "*Aeronautics: A Vision for 2020*"., European Commission, (2001)
2. Szodruch, J., Hilbig, R.: "*Building the Future Aircraft Design for the Next Century*". AIAA Paper 98–0135 (1998)
3. Kroll, N., Rossow, C.–C., Becker, K., Thiele, F.: "*MEGAFLOW – A Numerical Flow Simulation System*". Aerospace Science Technology, Vol. 4, (2000), pp. 223–237
4. Kroll, N., Rossow, C.–C., Schwamborn, D., Becker, K., Heller, G.: "*MEGAFLOW – A Numerical Flow Simulation Tool for Transport Aircraft Design*". 23rd ICAS Congress, Toronto, paper 2002-1.10.5 (2002)
5. Rossow, C.-C., Godard, J.-L., Hoheisel, H., Schmitt, V.: "*Investigation of Propulsion Integration Interference on a Transport Aircraft Configuration*". AIAA Paper 92–3097 (1992)
6. Hoheisel, H.: "*Aerodynamic Aspects of Engine–Aircraft Integration of Transport Aircraft*". Aerospace Science and Technology, November (1997), pp. 475–487

7. Burgsmüller, W., Rollin, C., Rossow, C.–C.: "*Engine Integration on Future Transport Aircraft – The European Research Programs DUPRIN/ENIFAIR*". ICAS Paper 98–5.6.2 (1998)
8. Laban, M.: "*Aircraft Drag and Thrust Analysis, Project Overview and Key Results*". Proc. of Workshop on EU–Research on Aerodynamic Engine/Aircraft Integration for Transport Aircraft. DLR Braunschweig, (2000), pp. 9.1–9.15
9. von Geyr, FrHr. H.: "*Key Results of Detailed Thrust and Drag Studies on the ALVAST Configuration*". Proc. of Workshop on EU–Research on Aerodynamic Engine/Aircraft Integration for Transport Aircraft. DLR Braunschweig, (2000), pp. 12.1–12.17
10. Rudnik, R., Rossow, C.-C., Frhr. von Geyr, H.: "*Numerical simulation of engine/airframe integration for high–bypass engines*". Aerospace Science and Technology, Vol. 6, January (2002), pp. 31–42
11. Brodersen, O., Monsen, E., Ronzheimer, A., Rudnik, R., Rossow, C.–C.: "*Computation of Aerodynamic Coefficients for the DLR–F6 Configuration using MEGAFLOW*". Notes on Numerical Fluid Mechanics, Ed. Nitsche W. et al., Vieweg Braunschweig, Vol. 72, (1998), pp. 85–92
12. Gerhold, T., Evans, J.: "*Efficient Computation of 3D–Flows for Complex Configurations with the DLR–Tau Code Using Automatic Adaptation*". Notes on Numerical Fluid Mechanics, Ed. Nitsche W. et al., Vieweg Braunschweig, Vol. 72, (1998), pp. 178–185
13. Brodersen, O.: "*Drag Prediction of Engine–Airframe Interference Effects Using Unstructured Navier–Stokes Calculations*". Journal of Aircraft, Vol. 39, No. 6, (2002), pp. 927–935
14. Brodersen, O., Hain, R., von Geyr, FrHr. H.: "*Hybrid Navier–Stokes Calculations for Transport Aircraft with Conventional and High–Bypass Engines.*" STAB–Tagung, Munich, (2002)
15. Galle, M.: "*Ein Verfahren zur numerischen Simulation kompressibler, reibungsbehafteter Strömungen auf hybriden Netzen*". PhD thesis, University of Stuttgart, DLR–FB 99–04, (1999)
16. Spalart, P.R., Allmaras, S.R.: "*A One–Equation Turbulence Model for Aerodynamic–Flows*". AIAA Paper, 92–0439, (1992)
17. Edwards, J., Chandra, S.: "*Comparison of Eddy Viscosity–Transport Turbulence Models for Three–Dimensional, Shock–Seperated Flowfields*". AIAA Journal of Aircraft, Vol. 34, No. 4, (1996), pp. 756–763
18. CentaurSoft: "*The Grid Generation Package Centaur*". http://www.centaursoft.com, (2003)

Fig. 11.1. DLR–F6 with VHBR type through–flow nacelles in the ONERA S2 wind tunnel

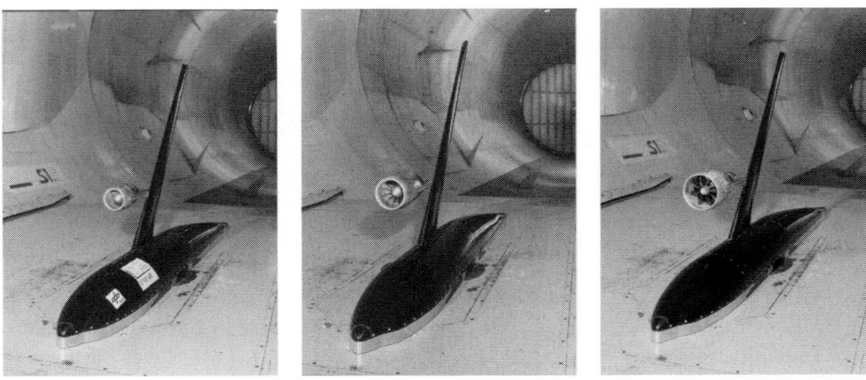

(a) Conventional (b) Very high bypass ratio (c) Ultra high bypass ratio

Fig. 11.2. DLR–ALVAST half model in the ONERA S1 wind tunnel

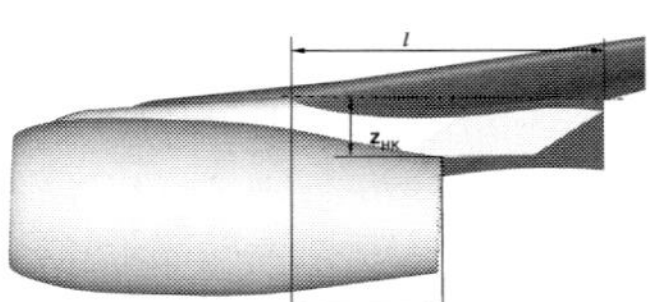

(a) CFM-56-long, position 1

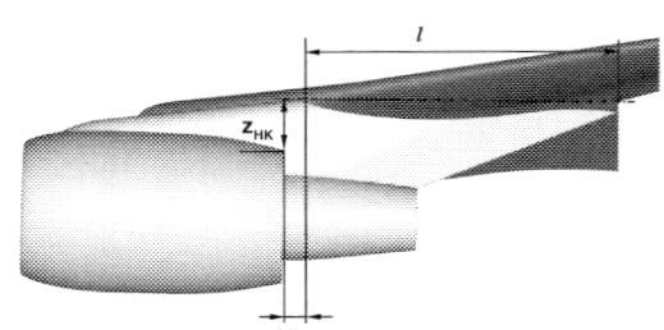

(b) CFM-56-short

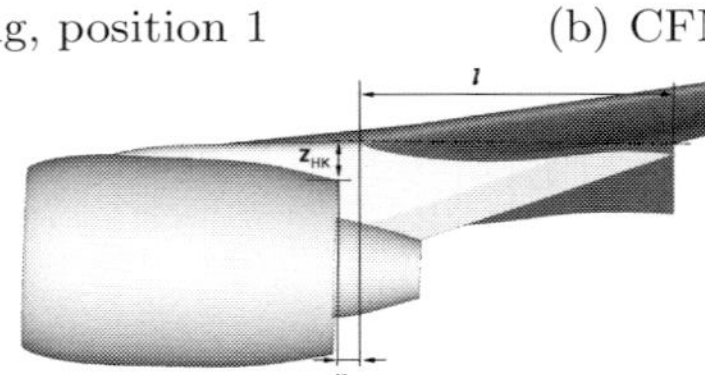

(c) VHBR, position 1

Fig. 11.3. DLR–F6 nacelles

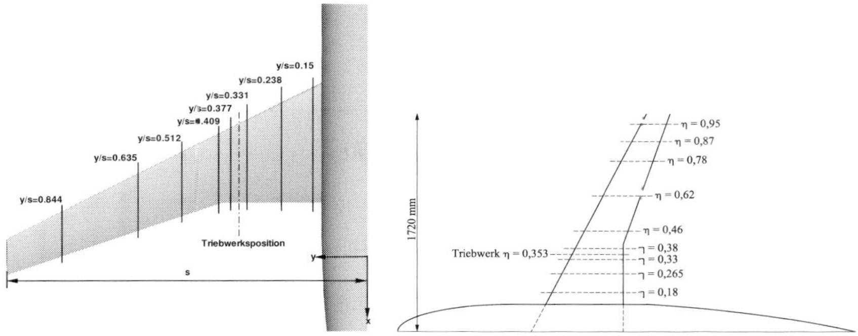

(a) DLR–F6 (b) ALVAST

Fig. 11.4. Wing pressure taps sections

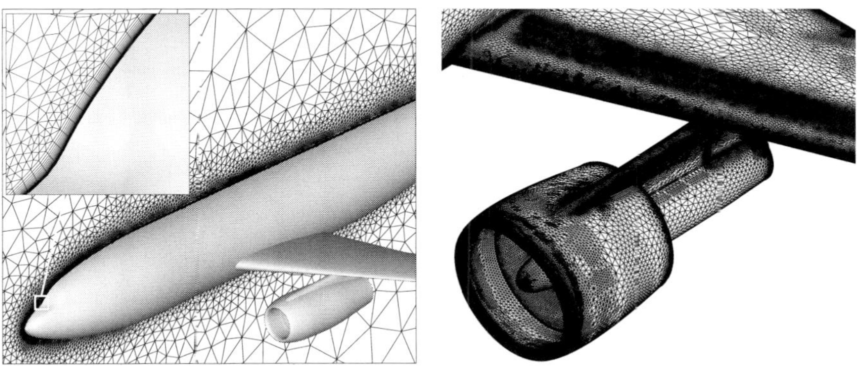

(a) F6 symmetry plane, initial grid (b) ALVAST-UHBR, adapted grid

Fig. 11.5. Hybrid grids

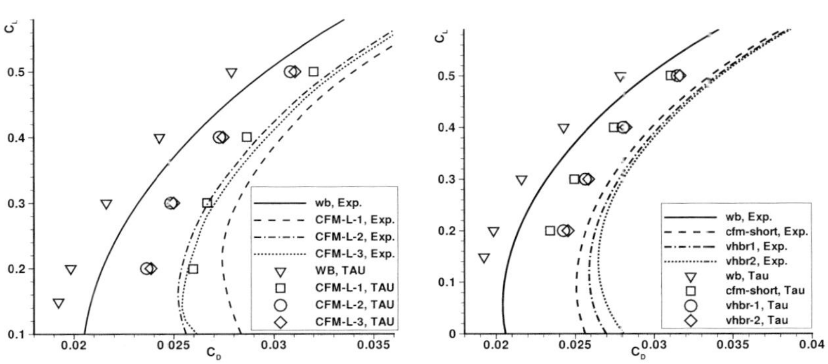

(a) Conv. long nacelles (b) Conv. short, VHBR nacelles

Fig. 11.6. DLR-F6: total drag

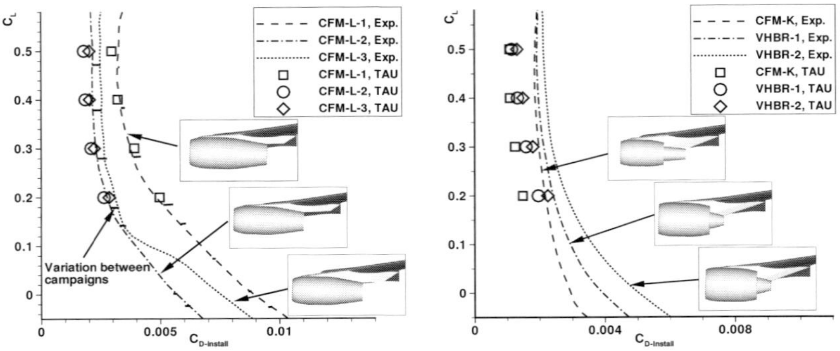

(a) Conv. long nacelles (b) Conv. short, VHBR nacelles

Fig. 11.7. DLR-F6: engine installation drag

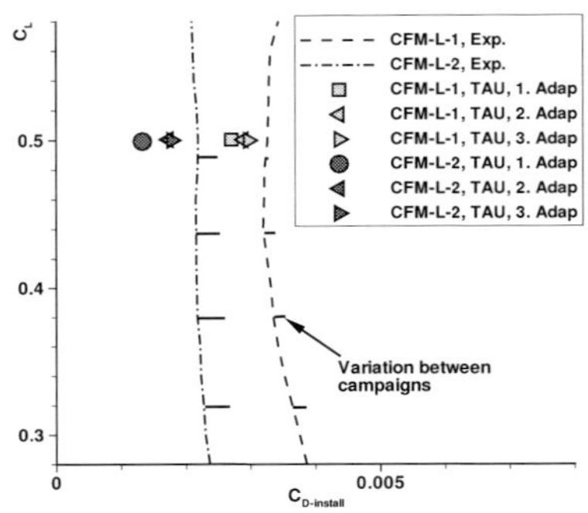

Fig. 11.8. DLR–F6: influence of grid refinement on the installation drag

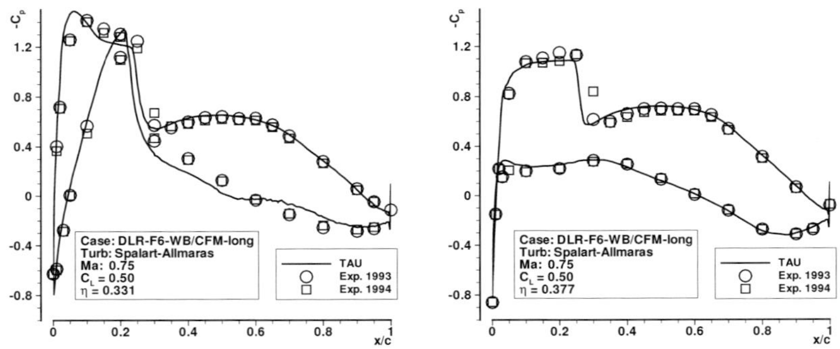

(a) Inboard of pylon (b) Outboard of pylon

Fig. 11.9. DLR-F6: pressure coefficient c_p distributions, CFM-56-long position 1, $C_L = 0.5$

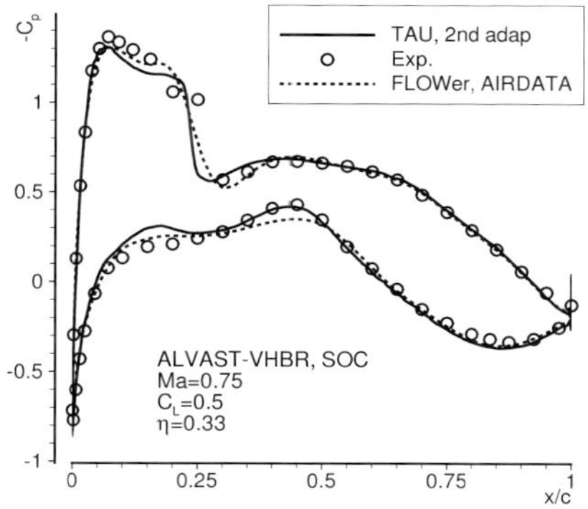

(a) Inboard of pylon

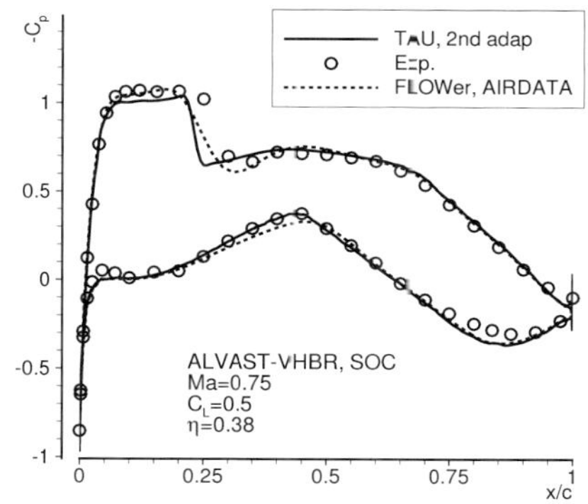

(b) Outboard of pylon

Fig. 11.10. DLR-ALVAST: pressure coefficient c_p distributions, VHBR, $C_L = 0.5$

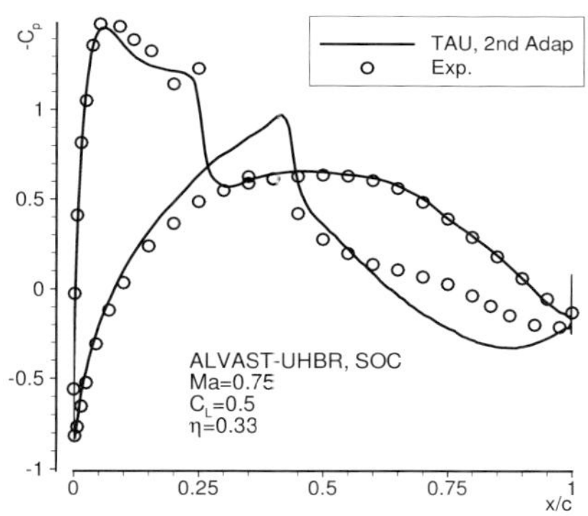

(a) Inboard of pylon

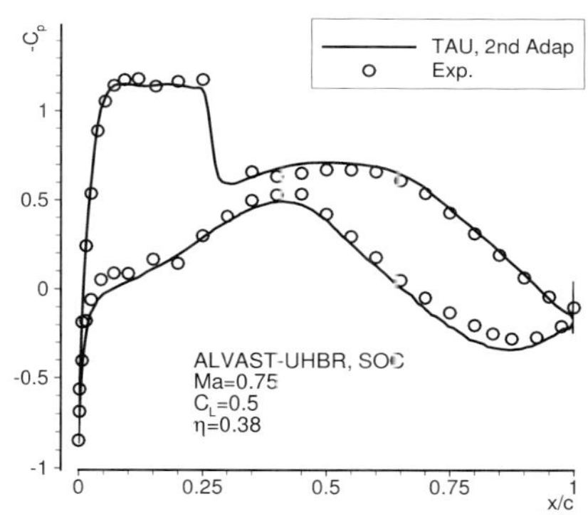

(b) Outboard of pylon

Fig. 11.11. DLR-ALVAST: pressure coefficient c_p distributions, UHBR, $C_L = 0.5$

12

Verification of MEGAFLOW-Software for High Lift Applications

S. Melber-Wilkending, R. Rudnik, A. Ronzheimer, and T. Schwarz

DLR, Institute of Aerodynamics and Flow Technology, Lilienthalplatz 7, D-38108 Braunschweig

Summary. This part of the MEGAFLOW project is concerned with the numerical simulation of the viscous compressible flow around transport aircraft high lift configurations and its validation against wind tunnel experiments. The investigations are based on the solution of the Reynolds-averaged Navier-Stokes equations (RANS) using the MEGAFLOW code system with a finite volume parallel solution algorithm. Both the block-structured (FLOWer) and the unstructured (TAU) code modules are used for this task. Based on the high lift configurations DLR-ALVAST and DLR-F11 lift polars are computed with both codes and compared against each other and against measurements. Further on the structured Chimera technique implemented in FLOWer is applied to a high lift configuration.

12.1 Introduction

The efficient design of a transport aircraft high lift configuration (Fig. 12.1a) with respect to low speed take-off/landing performance and handling qualities represents a complex aerodynamic problem. The flow field around the wing with deployed high lift devices at high incidences is characterized by the co-existence of flow phenomena such as large pressure induced separation, compressibility effects and strong wake/boundary layer interaction. In addition there are some critical areas with respect to the flow topology like the wing/fuselage junction or the engine/pylon/wing intersection, which have a strong influence on the overall aerodynamic performance.

Aside from continuous development and increasing experience in applying flow solvers for the solution of the RANS equations to 2D high lift problems during the last decade (e.g. [1] - [3]) only a comparatively limited number of national and international studies of three-dimensional applications for realistic aircraft configurations have been published in the near past (e.g. [4] - [9]). The reason is basically the high degree of geometrical complexity of deployed high lift systems in conjunction with a variety of relevant flow features, which the leads to a drastic increase in the required simulation performance even in the linear range of the $C_L(\alpha)$ curve.

Compared to established experimental investigations of such aerodynamic effects in wind tunnels or using flight testing the use of advanced CFD methods holds promise to substantially accelerate the aerodynamic design process and save costs, while providing a detailed insight in the complete flow field around the considered configuration.

The numerical simulation of the flow around a complete aircraft in high lift configuration up to maximum lift conditions is one of the major areas, where Navier Stokes methods can prove their superior potential compared to widespread and comparatively efficient coupled viscous/inviscid interaction methods. Due to the considerable point number needed for high lift configurations a considerable numerical effort is necessary to compute a lift and drag polar for a three-dimensional aircraft configuration. Consequently the prime usage for RANS simulations in the near future is likely to support the understanding of detailed flow phenomena, than an usage inside an inverse design or optimization procedure.

The present paper deals with the investigation of the numerical approach. In contrast to preceding studies, that concentrate on a single solution approach as the chimera-technique or the hybrid unstructured approach of [9], the MEGAFLOW software offers the possibility to compare the capabilities of the purely block-structured (FLOWer) and a hybrid unstructured approach (TAU). Consequently a comparison of the structured and unstructured grid generation techniques are carried out.

12.2 Flow Solution Method

12.2.1 Block-structured Flow Solver FLOWer

The FLOWer code [10] is characterized by a cell vertex finite volume formulation of the governing equations for compressible three-dimensional flows. The baseline features of the code, as used for the present computations, are a central differencing spatial discretization of the convective fluxes with Jameson-type scalar artificial dissipation operators based on second and fourth differences of the flow variables. Time integration to steady state is accomplished using an explicit five-stage Runge-Kutta time-stepping scheme. The convergence process is accelerated by local time stepping, implicit residual smoothing and full multigrid technique.

For the turbulence closure the k-ω model of Wilcox [11] in a slightly modified form [12] and the Spalart-Allmaras model [13] is used. The convective fluxes of the turbulence equations are discretized using a first order upwind scheme. Time integration is based on a point-implicit treatment of the turbulence source terms.

12.2.2 Hybrid Unstructured Flow Solver TAU

In contrast to the FLOWer code the DLR TAU code [14] is characterized by an unstructured data concept. As the data structure is based on the edges of the control volumes, the code is independent of the type of grids cells, allowing to handle either unstructured, structured or hybrid grids. The governing equations are solved on a dual background grid, which is determined directly from the initial grid. As in the FLOWer code, the flow variables are stored at the cell vertices. For the discretization of the convective fluxes several upwind or central discretization methods are available. The present computations are carried out using the second order accurate central differencing scheme with scalar dissipation. Time integration is accomplished using an explicit 3-stage Runge-Kutta scheme in conjunction with local time stepping and an agglomeration multigrid procedure.

Beside the k-ω turbulence model the Spalart-Allmaras model [13] is used for the present computations. As for the block-structured computations the convective fluxes of the turbulence equations are discretized using a first order upwind scheme. No implicit source term treatment is implemented.

12.3 Numerical Results

12.3.1 DLR - ALVAST Configuration

The following results are abstracts from [6] and [15], which both compares in more detail the results obtained with block-structured and hybrid unstructured approach against each other an against measurements on example of the ALVAST configuration.

Aircraft Configuration:

The DLR ALVAST wing/fuselage wind tunnel model [16] with deployed slat and flaps in take-off configuration is selected as an example of a complex high lift configuration, Fig. 12.1a. The geometry specifications are similar to an AIRBUS A320. For the present investigation the take-off configuration without engine is considered, characterized by a continuous slat with a deflection angle of $\delta_s = 20.0°$ and a single slotted flap with a deflection angle of $\delta_s = 19.5°$. The flap is departed in span-wise direction by a thrust gate.

Block-structured Grid Generation:

The generation of the block-structured grid is accomplished using the interactive grid generation module of the MEGAFLOW project called MEGA-CADS [17]. An important feature of this code is the parametric construction of multiblock grids using the script file technique. The present grid represents an extension of the grid around the simplified wing/fuselage junction used in [4], similar to that Fig. 12.3a. The geometry set-up and the generation of the baseline grid took about 3 man months for an experienced user. The parametric construction of the computational grid and the restart capability is appropriate to incorporate moderate changes in wing shape or the high lift setting, once the baseline grid has been generated.

The investigation of one polar-point at a free stream Mach-number of $M_{\infty} = 0.22$ and a Reynolds-number of $Re = 2.0 \cdot 10^6$ with the FLOWer code was one of two main milestones of the first project phase [4]. As depicted in Fig. 12.2a a good agreement with experiments was achieved for this complex 3D high lift configuration using the simplified geometry.

The comparatively straight forward inclusion of the slat and flap gaps at the fuselage intersection, Fig. 12.2b, demonstrates the benefits of parametric construction concept of MEGACADS. The extension of the script file to include the gaps has been accomplished in about one week compared to 3 month for the base grid [4]. With the gaps, the complexity of the configuration has been further increased towards the wind tunnel model. Considering the turn-around time a general outcome is that the degree of complexity for the purely block-structured approach is limited to wing/fuselage configurations because of the high effort needed for complex high lift configurations.

The computational grid of this complete configuration consists of about $9.2 \cdot 10^6$ grid points. The domain is decomposed into 50 blocks. The surface grid at the wing/fuselage junction is depicted in Fig. 12.3a. An H-topology is used in the spanwise direction with 128 cells on the wing. For the wing sections embedded C-type topology grids are wrapped around the single elements with 128 cells on the slat, 378 cells on the main wing, and 168 cells on the flap surface, respectively. The wing boundary layer is resolved with 25 cells in the normal direction. The first wall spacing amounts to 10^{-5} local chord length, corresponding to an average value $y^+ \approx 1$. 16 cells are placed on the blunt trailing edges of the wing elements.

Hybrid Unstructured Grid Generation:

The hybrid unstructured grid is generated with the code system CENTAUR [18]. The grid consists of two parts: a quasi-structured prismatic cell layer with 20 points is utilized in order to achieve an appropriate resolution of the viscous effects inside the boundary layer. In contrast to this, tetrahedral cells are used to fill the outer domain of the flow-field. A total of $4.1 \cdot 10^6$ grid points is used for the initial grid. The adapted grid consists of $5.8 \cdot 10^6$ grid points, Fig. 12.3b. The adaptation is based on a velocity gradient sensor, resulting in a higher grid resolution at the leading edges. The first grid spacing ensures, the maximum value is limited to $y+ = 1$ on all components. Thus an average $y+$ below 0.5 is obtained. Based on a well defined CAD description of the model, the grid generation process requires about 1 week, depending on the grid point number, adaptation steps, and hardware availability.

Lift- and Drag Polar:

The main emphasis of this investigation is a verification of FLOWer and TAU for 3D high lift flows. The reference experiments are conducted in the large low speed facility of the German-Dutch wind tunnel DNW at a free stream Mach-number of $M = 0.22$ and a chord Reynolds number of $Re = 2 \cdot 10^6$, [19].

The lift- and drag polars are presented in Fig. 12.5. The resulting lift coefficients C_L in the linear range of the $C_L(\alpha)$-curve are consistently over-predicted by about 5% computed with the unstructured TAU-Code. Beginning with an angle of attack of $\alpha \geq 22°$ the lift breaks down. However, the angle of attack for maximum lift agrees well with the measured data. A similar behavior can be observed for the drag coefficient, which is also under-predicted by the numerical method in the complete range of angle variations.

The aerodynamic coefficients computed with the block-structured FLOWer differ from the results computed with the unstructured TAU-Code. Whereas the agreement of structured computations with experimental data below an angle of attack $\alpha < 15°$ is quite good, the predicted lift coefficients between $15° < \alpha < 22°$ are 5% lower than the measured coefficients. Beginning with $\alpha = 22°$ the lift increases again, without predicting a clear maximum lift in the simulated range angle of attack. For a descriptive discussion of the lift breakdown, the subsequent section includes flow-field stream-line figures at angles of attack of 4°, 12° and 22°.

Streamlines off the Surface:

In Fig. 12.4 the ALVAST configuration is shown with stream lines in the flow-field for the structured and unstructured approach. As supposed by the remarkably linear character of the lift polar, the flow does not separate in the simulation with the unstructured TAU-code up to an angle of attack of $\alpha = 21°$. Beyond this point there is a separation in the area of the wing root, which extends over the whole wing chord. Adjacent to it there is a weak separation on the outer end of the outboard flap beginning at an angle of attack of $\alpha = 22°$. It extends to approximately 30 % of the local wing chord and which moves with increasing angle of attack to the inboard-side of the wing.

The simulation with the block-structured FLOWer-Code shows already weak separation on the side edge of the inboard flap in the area of the flap-cut out developing from an angle of attack of $\alpha = 4°$. This is not the case in the unstructured simulation. At $\alpha = 12°$ the separation disappears, coming back at $\alpha = 15°$ for the second time now spreading over the trailing edge in the whole flap cut-out. The separation sprawls out on the inboard flap with increasing angle of attack and shows up again at $\alpha = 21°$ in the area of the flap cut-out. The lift loss induced by this separation can clearly be found in a low lift coefficient in the lift curve starting at an angle of attack of $\alpha = 15°$.

Subsequently the differences between the numerical results from FLOWer and TAU caused by geometry influences, grid resolution, and the choice of turbulence model should be demonstrated and quantified as far as possible.

Turbulence Model:

The choice of the turbulence model plays an important role in the computation of viscous high lift flows. The prediction of the appearance and strength of flow separation is clearly linked to the turbulence model. While in the block-structured FLOWer-code the two-equation model k-ω of Wilcox [11] was used, the unstructured TAU-code uses the one-equation model of Spalart-Allmaras [13].

To investigate this influence the computation at $\alpha = 21°$ was done with the k-ω in TAU. This lead to a similar flow behavior as with the Spalart-Allmaras model. Obviously the differences between FLOWer and TAU are not directly related to the turbulence model.

CAD-Description, Wind Tunnel Model:

The CAD-description of the simulations was simplified compared to the wind tunnel model to simplify the block-structured grid generation: The slot between the slat and fuselage in the area of the wing-junction was not modeled with all details like slat-horn[1] and slat-stump at the leading edge, Fig. 12.6a.

To check the influence of geometrical differences between experiment and simulation an improved CAD-geometry was build. It captures many details of the wind tunnel model, like slat-end-plate and slat-stump at the leading edge, and the wing-root fairing on the upper-side of the wing/fuselage junction. The hybrid unstructured grid based on this CAD-geometry has an improved grid resolution with $12 \cdot 10^6$ points, compare Fig. 12.6b and Fig. 12.7a. The correspondingly numerical results are considerably improved, as depicted by the lift-polar in Fig. 12.7b for one incidence.

Grid Resolution:

The structured and unstructured grids were build from the same CAD-description and the surface grids of the configuration have a comparable resolution, Fig. 12.3. For the hybrid unstructured grid-generation this resolution is constant over the prismatic layer of the unstructured grid. In the adjacent tetrahedral volume grid the resolution decreases from the outer prism layer

[1] The slat-horn is a modification of the slat-end at the wing root, which produces a vortex on the upper side of the wing at high angles of attack (Fig. 12.1b). This eddy delays flow separation at the wing-junction and as a result it allows higher lift coefficients. This effect was shown in the experiment with the ALVAST configuration, [19].

to the farfield, Fig. 12.8a. In the structured grid the number of cells is constant over block-boundaries to the farfield, Fig. 12.8b. Correspondingly the structured grid has a higher resolution outside the boundary layer compared to the unstructured grid.

The adaptation of the unstructured grid increases the resolution where sensored. In contrast, the structured grid has a high number of points in the outer domains of the grid due to the necessary grid resolution to resolve flow phenomena. Consequently in the unstructured grid there are less points ($5.6 \cdot 10^6$ points after one adaptation) required then in the structured grid ($9.2 \cdot 10^6$ points) to have a comparable grid resolution in relevant areas.

Another effect of adaption on the unstructured approach is the adjustment of the first wall distance to reach a $y^+ \sim 1$ on the whole surface. This leads to a good resolution of the sub-laminar boundary layer without slowing down iterative convergence by highly stretched cells in areas of $y^+ \ll 1$. On the other hand the first wall distance in the structured grid generation is adjusted to reach a $y^+ \geq 1$ and therefore the resolution of the boundary layer is not as good as in the unstructured grid in all areas.

The discussion of grid resolution should also cover confluent wakes behind the components of the high lift configuration, because they play an important role in flow separation. In the structured grid generation the grid lines from trailing edges continues to the farfield and resolve the confluent wakes well, Fig. 12.8b. In the unstructured grid generation the prism layers cover only the vicinity of the elements, the wakes in the basic grid are not resolved properly by tetrahedra, Fig. 12.8a.

An adaptation to pressure loss improves the resolution of the wakes, but requires some adaptations, because an adaptation in the flow field needs a sufficient gradient of a flow variable to initiate a refinement of the grid. To accelerate this, a feature of the grid generator CENTAUR is available to insert so called wake-panels in the wakes of elements to improve the grid resolution in these areas.

Summarized both grids have a comparable resolution on the surface and the differences between the structured and unstructured approach are partly due to different grid resolution in the flow field.

12.3.2 EUROLIFT F11 Configuration

Another contribution to the validation of MEGAFLOW software for high lift configurations stem from the European 5th framework project EUROLIFT [20]. The main focus of the EUROLIFT investigations lies on the assessment of the capability of state-of-the-art numerical methods to simulate maximum lift on transport aircraft high lift configurations.

Aircraft Configuration:

For EUROLIFT an Airbus-type wing/fuselage high lift model denoted as KH3Y or DLR-F11 was chosen. For the MEGAFLOW software validation the test case TC211 featuring the F11 configuration in take-off configuration for low Reynolds-numbers. As depicted in Fig. 12.9 it has an full-span flap and a full-span slat.

Grid Generation:

The aim of the presented subtask of the EUROLIFT project was to compare the block-structured and the unstructured approach under as identical conditions as possible [21]. This is achieved by computing polars of the F11 model running both approaches on a block-structured grid of DLR and a hybrid unstructured grid of the Swedish research establishment FOI, with using basically the same numerical and physical modeling on an identical CAD geometry. Both grids, depicted in Fig. 12.9, have a point number of $3.08 \cdot 10^6$ million due to project related reasons. The most pronounced difference between both grids is, that the block-structured grid allows a better stream wise resolution on the leading edges and normal resolution, as a higher cell anisotropy is feasible than that for the hybrid unstructured grid.

Numerical Results:

In Fig. 12.10a the resulting computed lift polars with Spalart-Allmaras turbulence model are shown, which compare well with each other and to the experimental data up to the maximum lift. Pressure distributions at mid span are shown in 12.10b. A good agreement to the experimental evidence is achieved with small deviation between both approaches.

Block-structured Chimera Technique:

A simplification in the structured grid generation process was achieved with the implementation of the Chimera-technique into the FLOWer code [21]. The idea of the Chimera-technique is to generate a grid around a complicated configuration by decomposing the configuration into simple components and generating a body conforming grid for each component without having to take care about the other components, Fig. 12.11. The main requirement is that the grids around the components overlap each other. All component grids are placed inside a simple background grid, which covers the whole computational domain.

FLOWer has a full Chimera functionality [22]. The method is not restricted to a certain grid hierarchy. Instead, each grid may interpolate data from any other grid. In order to establish intergrid communication for a target point at a chimera boundary, first the overlying grids are scanned for a source cell enclosing the target point. In order to mark points being inside a solid body, a simple auxiliary grid must be provide d by the user, which encloses the solid body. All points of the grid inside the auxiliary grid are excluded from the flow calculation.

The demonstration of the chimera technique for complex 3D high lift applications for the example of F11 configuration was one milestone of the MEGAFLOW project. In Fig. 12.11a the component grids around the fuselage, slat, main wing and flap are depicted with overall $3.9 \cdot 10^6$ points in 333 blocks. This component grids are build independently, which simplifies the structured grid generation considerably. The Chimera processes and the generation of the cartesian background grids are widely automated.

In Fig. 12.12 the results from measurements, a purely block-structured and a structured chimera simulation with the FLOWer code are depicted. In the linear range of the $C_L(\alpha)$-curve a good agreement between simulation an experiment was reached. Further on the total lift agrees quite well with the baseline FLOWer computations for the three considered incidences.

12.4 Conclusion

The numerical flow simulation of Reynolds averaged Navier-Stokes equations with the block-structured FLOWer-code and the unstructured TAU-code shows a good prediction of the aerodynamic parameters in comparison to wind tunnel experiments up to the maximum lift.

For the ALVAST-configuration the lift coefficient of the unstructured approach is slightly higher than the experimental value, while the lift coefficient of the structured approach has variations because of early flow separations and departs from the experiment. As discussed, several sources of uncertainty might be responsible for the mis-prediction like the different turbulence models, the CAD-geometry and its simplifications compared to the wind tunnel model and the possibly too low grid resolution. As shown with a hybrid unstructured grid based on an improved CAD-geometry and grid resolution the results are considerably improved compared to wind tunnel experiments.

For the F11 configuration with a block-structured and a hybrid unstructured grid of the same grid resolution and turbulence model both approaches show similar results. With the structured chimera technique the grid generation process is clearly simplified. The results are again in good agreement with results from the purely block-structured approach.

The assessment of the computations carried out using the structured FLOWer code within the MEGAFLOW project resulted in the conclusion, that the configuration complexity for this approach is more or less limited to wing/fuselage configurations. Although no principal complexity limit exists, the effort to set-up a baseline grid and the block topology in conjunction with the limited adaptation capability supported the usage of the unstructured approach, which is represented by the TAU code in the MEGAFLOW project. It has the potential for far more complex configurations. Moreover, even if compared to the structured approach extended by Chimera technique or non-matching boundaries, the grid adaptation of the unstructured data concept offers the capability to adjust the grid resolution quite flexible in the relevant areas of the configuration. Especially for high lift flows these areas change their locations considerably depending on the incidence. Further on the short turn-around time and accurate grid generation enables the handling of complex industrial high lift configurations in a industrial production process.

12.5 Current/Future Work

Currently and in the near future the most of the work associated with high lift aerodynamics at DLR is devoted to an improved physical understanding. As an example a detailed analysis of complex 3D-flow features (e.g. slat-horn vortex, Fig. 12.1b) on high lift configurations and this influence on maximum lift using the TAU code is presented in [23]. The area of engine airframe integration for high lift configurations is discussed in [24]. As an example the ALVAST high lift configuration with an UHBR-engine is depicted in Fig. 12.1a. The prediction of the aerodynamic performance depending on changes in the geometry or the simulation of industrial high lift configurations with all features (e.g. flap-track fairings, winglets, spoilers) is another field of future activity at DLR.

References

1. Rogers, S. E., Wiltberger, N. L., Kwak, D.: "Efficient Simulation of Incompressible Viscous Flow over Multi-Element Airfoils." AGARD-CP-515, pp. 7-1 - 7-9, 1993.
2. Larsson, T.: "Separated And High-Lift Flows Over Single And Multi-Element Airfoils." ICAS-94-5.7.3, pp. 2505 - 2518, 1994.
3. Lindblad, I. A. A., de Cock, K. M. J.: "CFD Prediction of Maximum Lift of a 2D High Lift Configuration." AIAA-paper 99-3180, 1999.
4. Rudnik, R.; Ronzheimer, A.; Raddatz, J.: "Numerical Flow Simulation for a Wing/Fuselage Transport Configuration with Deployed High Lift system" in Notes on Numerical Fluid Mechanics, Vol. 72, pp. 363-370, Vieweg-Verlag, Braunschweig/Wiesbaden, 1999.
5. Rudnik, R.; Melber, S.; Ronzheimer, A.; Brodersen, O.: "Three-Dimensional Navier-Stokes Simulations for Transport Aircraft High Lift Configurations." Journal of Aircraft, Vol. 38, pp.895-903, 2001.
6. Melber, S.; Rudnik, R.; Ronzheimer, A.: "Structured and Unstructured Numerical Simulation in High Lift Aerodynamics." Workshop on EU-Research on Aerodynamic Engine / Aircraft Integration for Transport Aircraft, 26-27 September 2000, DLR Braunschweig, 2000, pp. 13-1 - 13-10.
7. Melber, S.: "3D RANS Simulations for High Lift Analysis of Transport Aircraft Configurations." Notes on numerical fluid mechanics, Volume 77, Springer Verlag, Berlin, Heidelberg, New York, 2002, pp 27-34.
8. Rogers, S.E.; Roth, K.; Cao, V.Hoa; Slotnick, J.P.; Whilock, M.; Nash, S.M.; Baker, M.D.: "Computation of viscous Flow for a Boeing 777 Aircraft in Landing Conf." AIAA paper 2000-4221, 2000.

9. Mavriplis, D.J.: "Parallel Performance Investigations of an Unstructured Mesh Navier-Stokes Solver" ICASE Report No. 2000-13, March 2000.
10. Raddatz, J., Fassbender, J.K.: "Block Structured Navier-Stokes Solver FLOWer." this volume.
11. Wilcox, D.C.: "Reassessment of the Scale Determining Equation for Advanced Turbulence Models." AIAA Journal, Vol. 26, No. 11, pp. 1299-1310, 1988.
12. Rudnik, R.: "Evaluation of the Performance of Two-Equation Turbulence Models for Airfoil Flows." DLR FB 97-49, 1997.
13. Spalart, P. R., Allmaras, S.R.: "A One-Equation Turbulence Model for Aerodynamic Flows." AIAA-paper 92-0439, 1992.
14. Gerhold, T.: "Overview of the Hybrid RANS Code TAU" this volume.
15. Rudnik, R.; Melber, S.; Ronzheimer, A.; Brodersen, O.: "Aspects of 3D RANS Simulations for Transport Aircraft High Lift Configurations." AIAA paper 2000-4326, 2000.
16. Kiock, R.: "The ALVAST Model of DLR." DLR IB 129-96/22, 1996.
17. Brodersen, O., Hepperle, M., Ronzheimer, A., Rossow, C.-C., Schöning, B.: "The Parametric Grid Generation System MEGACADS." Proc. of the 5th Intern. Conference on Numerical Grid Generation in Computational field Simulations 1996, Mississippi, Ed.: Soni, B.K., Thompson, J.F., Hauser, J., Eisemann, P., pp. 353-362, 1996.
18. Kallinderis, Y.: "Hybrid Grids and Their Applications." Handbook of Grid Generation, CRC Press, Boca Raton / London / New York / Washington, D.C., pp. 25-1 - 25-18, 1999.
19. Puffert-Meissner, W.: "ALVAST Half-Model wind-tunnel Investigations and Comparison with Full-Span Model Results." DLR IB 129-96/20, 1996.
20. Thiede, P.: "EUROLIFT - Advanced High Lift Aerodynamics for Transport Aircraft." AIR & SPACE EUROPE, VOL. 3, No 3 / 4, 2001.
21. Rudnik, R.; Heinrich, R.; Eisfeld, B.; Scharz, Th.: "DLR Contributions to Code Validation Activities within the European High Lift Project EUROLIFT." 13th AG STAB/DGLR Symposium München, 12.-14. November 2002.
22. T. Schwarz: "Development of a Wall Treatment for Navier-Stokes Computations using the Overset-Grid Technique." 26th European Rotorcraft Forum, The Hague, The Netherlands, 26-29 September 2000.
23. Melber, S.; Wild, J.; Rudnik, R.: "Numerical High Lift Research - NHLRes. Annual Review 2001." High Performance Computing in Science und Engineering '02, Springer-Verlag Berling, Heidelberg, New York, 2002, pp. 406 - 421.
24. Melber, S.: "3D RANS-Simulations for High-Lift Transport Aircraft Configurations with Engines" DLR IB 124-2002/27, 2002.

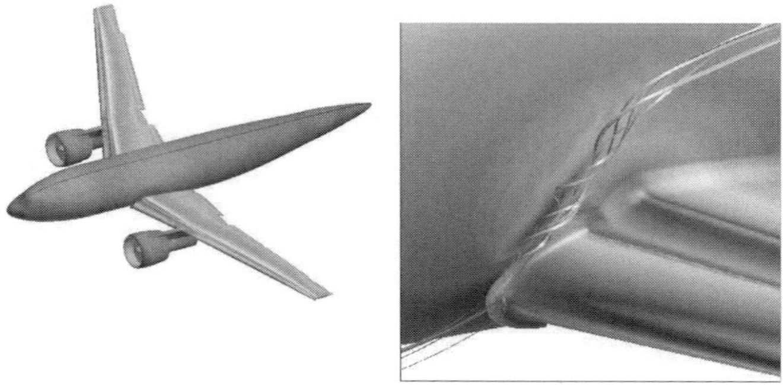

(a) (b)

Fig. 12.1. a) ALVAST high lift configuration with UHBR-engine, pressure distribution on the surface, b) Slat-horn in the area of wing-fuselage junction with slat-horn vortex.

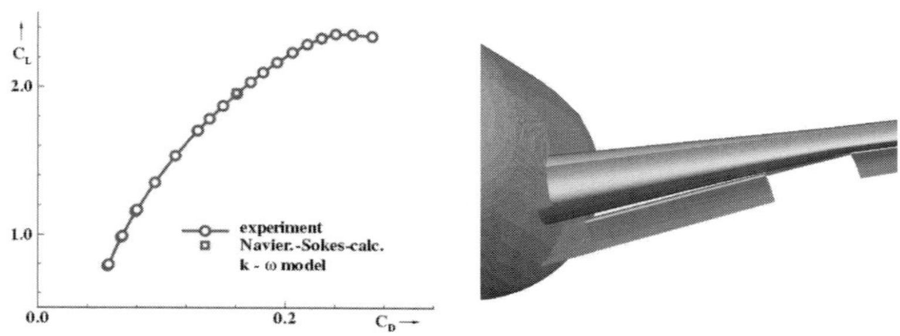

(a) (b)

Fig. 12.2. a) Polar of ALVAST high-lift configuration, experiments and FLOWer-calculation, b) ALVAST high-lift configuration, slat and flap gaps at the fuselage intersection.

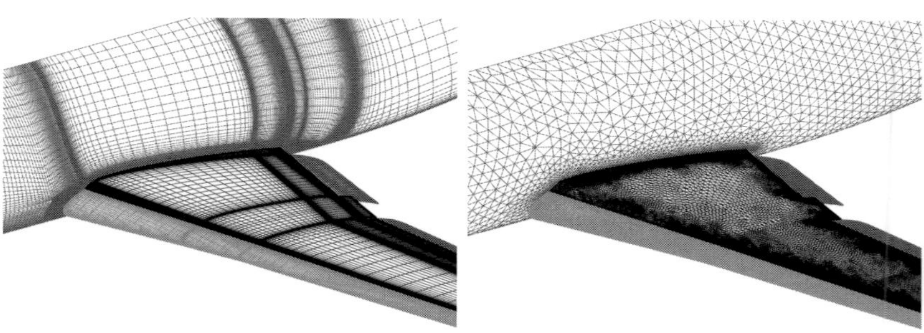

(a) (b)

Fig. 12.3. ALVAST high-lift configuration: a) block-structured surface grid, b) hybrid unstructured surface grid.

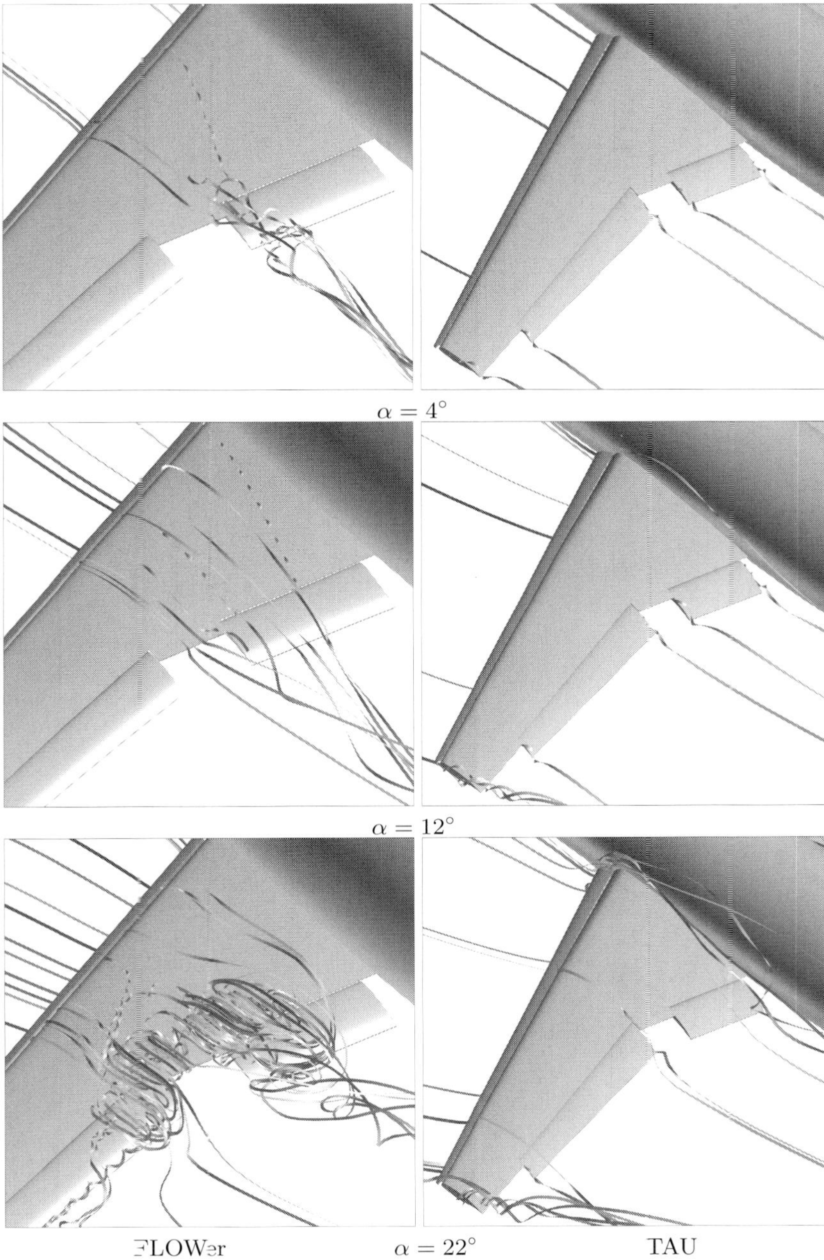

Fig. 12.4. ALVAST high-lift configuration, streamlines of the surface computed with FLOWer and TAU.

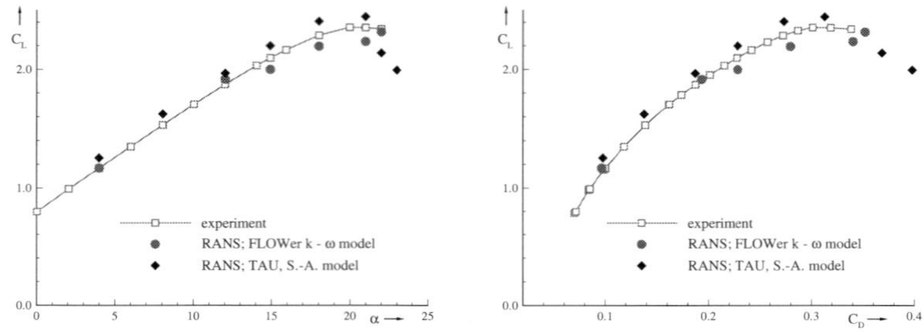

(a) (b)

Fig. 12.5. ALVAST high-lift configuration, $M = 0.22$, $Re = 2.0 \cdot 10^6$: a) lift-polar, b) drag-polar.

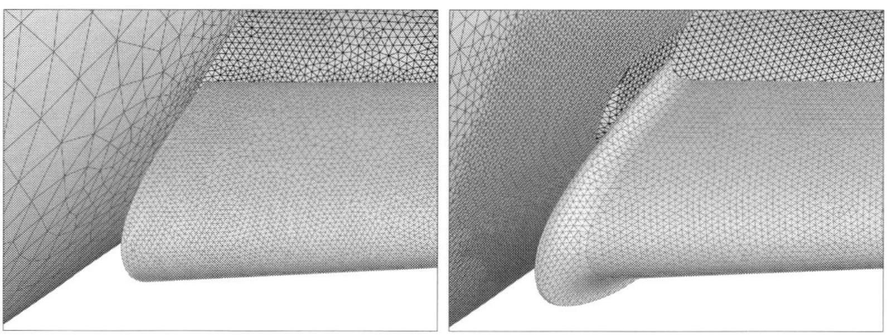

(a) (b)

Fig. 12.6. ALVAST high-lift configuration a) with basic and b) improved geometry.

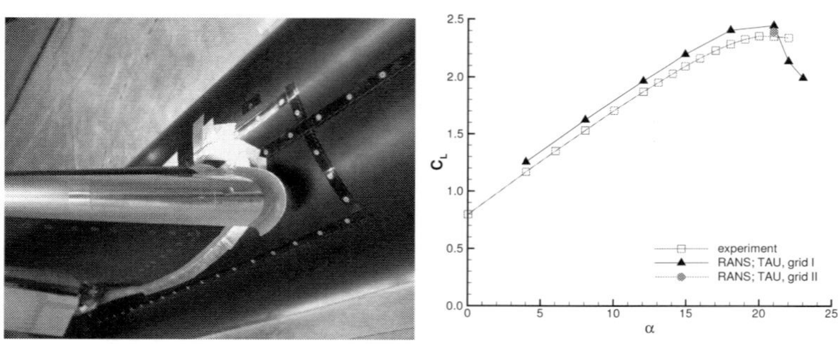

(a) (b)

Fig. 12.7. a) ALVAST high-lift configuration wind tunnel model, b) lift-polar for basic (grid I) and improved geometry (grid II).

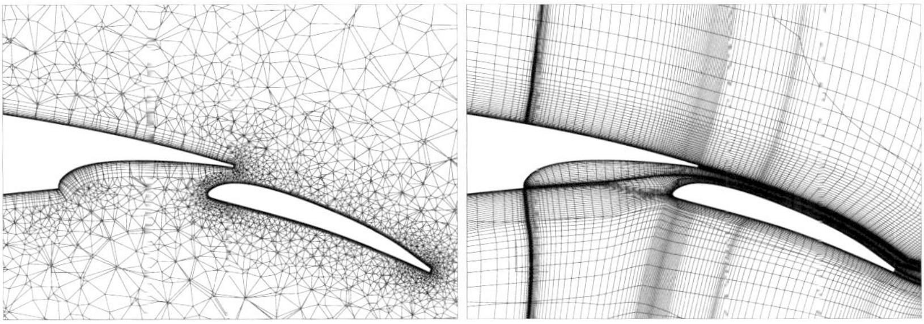

(a) (b)

Fig. 12.8. Cut through a) unstructured and b) structured grid at main-wing trailing edge.

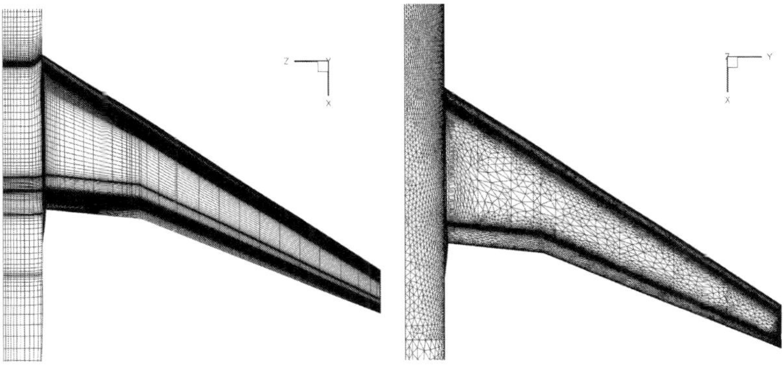

(a) (b)

Fig. 12.9. a) Block-structured and b) hybrid unstructured surface grids for F11 configuration.

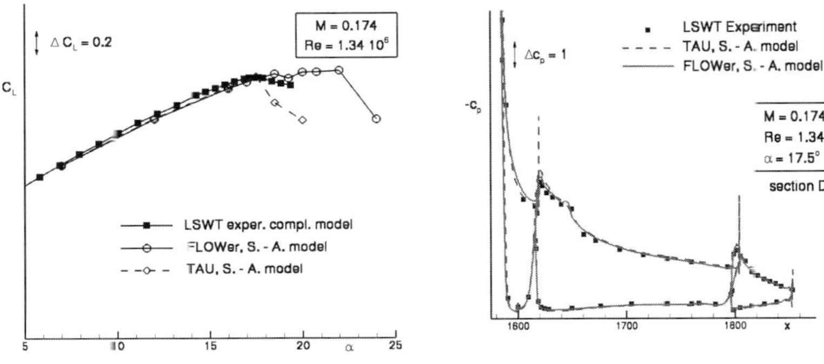

(a) (b)

Fig. 12.10. Lift polar for the F11 configuration a) with FLOWer and TAU, b) Pressure distributions at 68% half span, TC211.

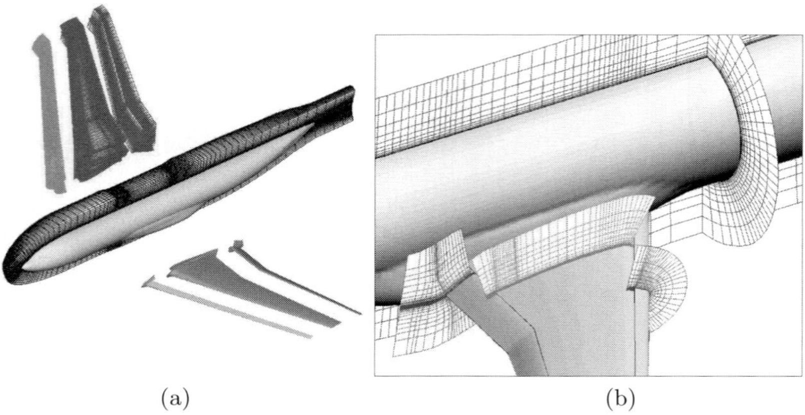

Fig. 12.11. F11 configuration with structured chimera grid a) components of the grid, b) detail view of the component grids with cutting-planes.

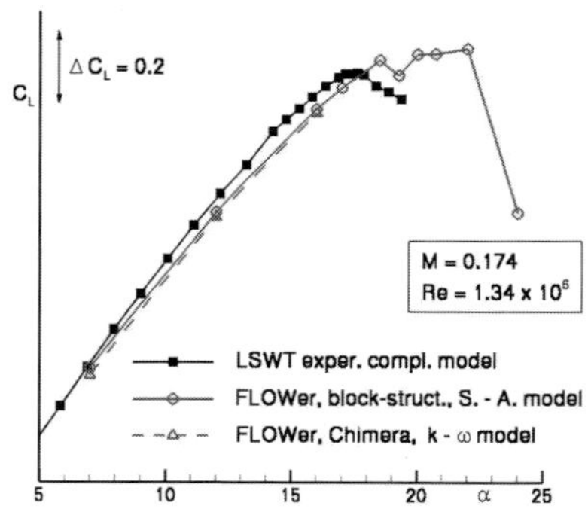

Fig. 12.12. Lift polars for the F11 configuration with FLOWER block-structured-chimera-grid and experiment.

Part V

Shape Optimization

13

The Continuous Adjoint Approach in Aerodynamic Shape Optimization

N.R. Gauger and J. Brezillon

DLR, Institute of Aerodynamics and Flow Technology,
Lilienthalplatz 7, D-38108 Braunschweig

Summary. Detailed numerical shape optimization will play a strategic role for future aircraft design. It offers the possibility of designing or improving aircraft components with respect to a pre-specified figure of merit subject to geometrical and physical constraints. However, the extremely high computational expense of straightforward methodologies currently in use prohibits the application of numerical optimization for industry relevant problems. Optimization methods based on the calculation of the derivatives of the cost function with respect to the design variables suffer from the high computational costs if many design variables are used. However, these gradients can be efficiently obtained by solution of the continuous adjoint flow equations.

13.1 Introduction

For an efficient aerodynamic optimization system, the development of adjoint solvers and adjoint optimization strategies is essential and an important part of the MEGAFLOW II project [1, 2].

The starting point for this development was the acquisition of the Jameson code flo87s [3, 4, 5]. This code is able to handle wing configurations in both inverse and drag optimization modes. The flo87s code is a complete optimization system which includes an Euler as well as an adjoint Euler solver, a mesh generator based on shared parabolic mappings and an optimization handler. The solvers as well as the mesh generator are only able to handle/generate (structured) one-block meshes. For the parameterization of the shape only the grid points of the surface of the wing are used.

Instead of extending the flo87s code, like Jameson did [6, 7], to handle more complex applications, DLR decided to implement an adjoint solver within the block-structured parallel Navier-Stokes solver FLOWer. FLOWer has been developed within the 1st phase of the MEGAFLOW project [8] and it is intensively used by DLR, aircraft industry and universities. The philosophy was that after the implementation of the adjoint solver all features of the

FLOWer code are also available for the adjoint mode [9, 10] (e.g. the capability to handle arbitrary multi-block topologies).

In this paper we give firstly the theoretical background of the continuous adjoint approach exemplarily for the 2D Euler equations and secondly we discuss its implementation in FLOWer and its validation for the aerodynamic cost functions drag, lift and pitching moment. The test case for the validation of the FLOWer ADJOINT code is the staggered, transonic RAE2822 airfoil, as a quasi-2D test case, because FLOWer ADJOINT is designed for 3D applications. Furthermore, a surface formulation for the adjoint gradients of drag, lift and pitching moment, which has been derived in [10], and its validation is presented for the transonic RAE2822 airfoil as well as the ONERA M6 wing as a first 3D example. Finally, the extension of FLOWer ADJOINT to viscous flows is presented and a first verification example for the viscous FLOWer ADJOINT is given.

13.1.1 Nomenclature

$(x, y) \in \mathbb{R}^2$	cartesian coordinates	M_∞	Mach number
$(\xi, \eta) \in [0, 1]^2$	body fitted coordinates	$)_\infty$	... at free stream
$D \subset \mathbb{R}^2$	flow field domain	γ	ratio of specific heats
$\partial D = B \cup C$	flow field boundary	C_{ref}	cord length
$B = \{(\xi, 1)\}$	farfield	C_p	pressure coefficient
$C = \{(\xi, 0)\}$	solid wall	C_D	drag coefficient
$\vec{n} = \begin{pmatrix} n_x \\ n_y \end{pmatrix} \perp D$	outward pointing normal unit vector	C_L	lift coefficient
α	angle of attack	C_m	pitching moment coefficient
ρ	density	(x_m, y_m)	pitching moment's reference point
$\vec{v} = \begin{pmatrix} u \\ v \end{pmatrix}$	velocity	I	cost function
p	pressure	$-d(I)$	adjoint boundary condition's RHS on C
E	specific total energy	$X \in \mathbb{R}^n$	vector of design variables
H	total enthalpy		

13.2 Gradient-based Aerodynamic Shape Optimization

Using genetic optimization strategies, the number of calls of flow calculations is very high. This is the reason why genetic algorithms are often too costly in detailed aerodynamic shape optimization and one has to choose deterministic gradient-based optimization strategies. For detailed aerodynamic shape optimization one needs the Navier-Stokes equations to control the cost function

I , like drag, lift or pitching moment, for physically meaningful results. The initial approach is to control the cost function by the Euler equations, as a logical step in the implementation phase.

Just for convenience reasons, the following analysis is restricted to the 2D Euler equations.

Let $X \in \mathbb{R}^n$ denote the vector of design variables (e.g the control points of a B-spline parameterization). Then X determines the airfoil $C(X)$ and its physics $w(X)$, where $w = \begin{pmatrix} \rho \\ \rho u \\ \rho v \\ \rho E \end{pmatrix}$ is the vector of the conserved variables. w is assumed to be the solution of the quasi-unsteady Euler equations

$$\frac{\partial w}{\partial t} + \frac{\partial f}{\partial x} + \frac{\partial g}{\partial y} = 0 \qquad \text{in } D, \tag{13.1}$$

where $\vec{n}^\top \vec{v} = 0$ on $C = C(X)$, with $f = \begin{pmatrix} \rho u \\ \rho u^2 + p \\ \rho uv \\ \rho uH \end{pmatrix}$ and $g = \begin{pmatrix} \rho v \\ \rho vu \\ \rho v^2 + p \\ \rho vH \end{pmatrix}$.

On the farfield free stream conditions are assumed. For a perfect gas

$$p = (\gamma - 1)\rho(E - \frac{1}{2}(u^2 + v^2)) \tag{13.2}$$

holds for the pressure, and finally C_p, C_D, C_L and C_m are defined as

$$C_p := \frac{2(p - p_\infty)}{\gamma M_\infty^2 p_\infty} , \tag{13.3}$$

$$C_D := \frac{1}{C_{\text{ref}}} \int_C C_p(n_x \cos\alpha + n_y \sin\alpha) dl , \tag{13.4}$$

$$C_L := \frac{1}{C_{\text{ref}}} \int_C C_p(n_y \cos\alpha - n_x \sin\alpha) dl , \tag{13.5}$$

$$C_m := \frac{1}{C_{\text{ref}}^2} \int_C C_p(n_y(x - x_m) - n_x(y - y_m)) \, dl . \tag{13.6}$$

If the geometry is now perturbed from $C(X)$ to $C(X + \delta X)$, then via the solution of

$$\begin{aligned} &\frac{\partial(w + \delta w)}{\partial t} + \frac{\partial(f + \delta f)}{\partial x} + \frac{\partial(g + \delta g)}{\partial y} = 0 \\ \Leftrightarrow \quad &\frac{\partial(\delta w)}{\partial t} + \frac{\partial(\delta f)}{\partial x} + \frac{\partial(\delta g)}{\partial y} = 0 \qquad \text{in } D, \end{aligned} \tag{13.7}$$

where

$$\vec{n}^\top \vec{v} = 0 \quad \text{on} \quad C = C(X + \delta X) , \tag{13.8}$$

the associated variation of pressure is as follows

$$\delta C_p = \frac{2\delta p}{\gamma M_\infty^2 p_\infty} \approx \frac{2(p(X+\delta X) - p(X))}{\gamma M_\infty^2 p_\infty} . \tag{13.9}$$

Finally via

$$\delta n_x \approx n_x(X+\delta X) - n_x(X) \tag{13.10}$$

and

$$\delta n_y \approx n_y(X+\delta X) - n_y(X) , \tag{13.11}$$

the variations of C_D, C_L and C_m are obtained as

$$\begin{aligned} \delta C_D = & \frac{2}{\gamma M_\infty^2 p_\infty C_{\text{ref}}} \int_C \delta p (n_x \cos\alpha + n_y \sin\alpha) dl \\ & + \frac{1}{C_{\text{ref}}} \int_C C_p (\delta n_x \cos\alpha + \delta n_y \sin\alpha) dl , \end{aligned} \tag{13.12}$$

$$\begin{aligned} \delta C_L = & \frac{2}{\gamma M_\infty^2 p_\infty C_{\text{ref}}} \int_C \delta p (n_y \cos\alpha - n_x \sin\alpha) dl \\ & + \frac{1}{C_{\text{ref}}} \int_C C_p (\delta n_y \cos\alpha - \delta n_x \sin\alpha) dl , \end{aligned} \tag{13.13}$$

$$\begin{aligned} \delta C_m = & \frac{2}{\gamma M_\infty^2 p_\infty C_{\text{ref}}^2} \int_C \delta p (n_y (x - x_m) - n_x (y - y_m)) \, dl \\ & + \frac{1}{C_{\text{ref}}^2} \int_C C_p \delta (n_y (x - x_m) - n_x (y - y_m)) \, dl . \end{aligned} \tag{13.14}$$

Proceeding as described above for the n perturbations $\delta_i X$ in each of the n components of the design vector X, the gradient of the cost function I (e.g. drag, lift or pitching moment coefficients) is obtained as $\nabla_X I = (\delta_i I)_{i=1,\ldots,n}$ after $n+1$ flow calculations. The easiest gradient-based optimization strategy is the steepest descent method. There recursively a line search in the direction $-\nabla_{X^{(k)}} I$, starting from the point $X^{(k)}$, leads to an optimal geometry

$$X^{(k+1)} = X^{(k)} - \varepsilon^{(k)} \nabla_{X^{(k)}} I \tag{13.15}$$

with respect to the cost function I in that direction. This is repeated until the norm of the gradient of the cost function becomes zero.

But one can see that the numerical costs, for the determination of the gradient of the cost function, are directly proportional to the number of design variables. This finite differences or brute force approach becomes more and more inefficient if the number of design variables increases.

13.3 Continuous Adjoint Formulation

In order to determine the gradient of the cost function independently of the design variables with respect to the numerical costs one should switch to the following continuous adjoint formulation. Let $\psi = \begin{pmatrix} \psi_1 \\ \psi_2 \\ \psi_3 \\ \psi_4 \end{pmatrix}$ denote the vector of the adjoint variables. Instead of solving $n+1$ times the quasi-unsteady Euler equations to get the gradient, the Euler equations are solved just once in order to get the transposed Jacobians $\left(\frac{\partial f}{\partial w}\right)^\top, \left(\frac{\partial g}{\partial w}\right)^\top$ and then the quasi-unsteady continuous adjoint Euler equations

$$-\frac{\partial \psi}{\partial t} - \left(\frac{\partial f}{\partial w}\right)^\top \frac{\partial \psi}{\partial x} - \left(\frac{\partial g}{\partial w}\right)^\top \frac{\partial \psi}{\partial y} = 0 \qquad \text{in } D, \tag{13.16}$$

where

$$n_x \psi_2 + n_y \psi_3 = -d(I) \quad \text{on} \quad C = C(X)\ , \tag{13.17}$$

and

$$\delta x_\xi, \ldots, \delta y_\eta = 0,\ \delta w = 0 \quad \text{on} \quad B = B(X)\ , \tag{13.18}$$

are also solved just once.

The right hand side $-d(I)$ of the wall boundary condition of the quasi-unsteady adjoint Euler equations is dependent on the cost function I. The adjoint farfield boundary condition describes just that the geometrical position of the farfield is fixed and free stream conditions apply there.

Finally the components of the gradient $\nabla_X I = (\delta_i I)_{i=1,\ldots,n}$ can now be determined via an integration just over the adjoint solution and the metric sensitivities $\delta x_\xi, \ldots, \delta y_\eta$ and

$$\begin{aligned} \delta I = &- \int_C p(-\psi_2 \delta y_\xi + \psi_3 \delta x_\xi)\ dl + K(I) \\ &- \int_D \psi_\xi^\top (\delta y_\eta f - \delta x_\eta g) + \psi_\eta^\top (-\delta y_\xi f + \delta x_\xi g)\ dA \end{aligned} \tag{13.19}$$

is obtained, where $K(I)$ is again a term dependent on the cost function I.

For the gradient of the drag, the following right hand side adjoint boundary on C is used

$$d(C_D) = \frac{2}{\gamma M_\infty^2 p_\infty C_{\text{ref}}} (n_x \cos\alpha + n_y \sin\alpha) \tag{13.20}$$

and to get the corresponding gradient, $K(I)$ is

$$K(C_D) = \frac{1}{C_{\text{ref}}} \int_C C_p(\delta n_x \cos\alpha + \delta n_y \sin\alpha) dl\ , \tag{13.21}$$

for the gradient of the lift

$$d(C_L) = \frac{2}{\gamma M_\infty^2 p_\infty C_{\text{ref}}}(n_y \cos\alpha - n_x \sin\alpha) \tag{13.22}$$

and

$$K(C_L) = \frac{1}{C_{\text{ref}}} \int_C C_p(\delta n_y \cos\alpha - \delta n_x \sin\alpha) dl \tag{13.23}$$

are used, and for the gradient of the pitching moment

$$d(C_m) = \frac{2}{\gamma M_\infty^2 p_\infty C_{\text{ref}}^2}(n_y(x - x_m) - n_x(y - y_m)) \tag{13.24}$$

and

$$K(C_m) = \frac{1}{C_{\text{ref}}^2} \int_C C_p \delta(n_y(x - x_m) - n_x(y - y_m))\, dl \tag{13.25}$$

are used. For more details see [10].

13.4 Implementation of the Adjoint Solver

Comparing the quasi-unsteady adjoint Euler equations (13.16) with the non-conservative form of the quasi-unsteady Euler equations

$$\frac{\partial w}{\partial t} + \frac{\partial f}{\partial w}\frac{\partial w}{\partial x} + \frac{\partial g}{\partial w}\frac{\partial w}{\partial y} = 0 \qquad \text{in } D, \tag{13.26}$$

one can see that they have the same form and it is obvious that one can use the same solver structure for both of them. Both equations can be solved in the quasi-unsteady formulation by an explicit central finite volume scheme blending artificial dissipation of 1. and 3. order. This type of scheme is implemented in DLR's flow solver FLOWer [1, 2, 8] for the cell centered metric.

Consequently, the procedure is to implement additionally adjoint flux and boundary treatment in FLOWer for 3D inviscid flows first and later to extend it for 3D viscous flows. The Jacobians and their transpose have the same spectral radii so that the filtering and the calculation of the time steps remain the same. The details on the spatial discretization of the adjoint equations within FLOWer can be found in [9, 10].

Of course one can also solve the adjoint equations by direct numerical inversion because they are linear, but the cost of the associated matrix inversion becomes prohibitive as the number of cells increases.

For an efficient calculation of the adjoint gradients, under the use of equation (13.19), the so called 'grid moving technique' based on J. Reuther [6] is implemented [9, 10, 11] in order to reduce the number of remeshings.

13.5 Validation of the Continuous Adjoint Approach

A first test case for the validation of the adjoint Euler solver FLOWer is the staggered, transonic RAE 2822 airfoil, as a quasi-2D test case, because FLOWer ADJOINT is designed for 3D applications. A C-mesh (see figure 1a) with 192x32x16 grid points is generated by the shared parabolic mappings routine of Jameson's flo87s code. The grid is transformed to the format of FLOWer and split along the cord into two blocks, because in addition we want to check the multi-block capability of FLOWer ADJOINT. The flight conditions are Mach number $M_\infty = 0.73$ and an angle of attack of $\alpha = 2°$. That FLOWer ADJOINT works very well on multi-block meshes can be seen in the convergence history in figure 1b by the comparison of the residuals by runs on one- and two-block meshes. After a proper convergence of the Euler solution FLOWer ADJOINT writes out a restart file and automatically restarts in the adjoint mode, where it converges again in comparable manner. Looking at the comparison between FLOWer ADJOINT and flo87s of the contours of the adjoint fields for the first adjoint variable ψ_1. As we can see in figure 2a the comparison shows a very favorable conformity.

The procedure to validate the adjoint based design method is to perform a check of the gradients of the drag, lift and pitching moment it produces with those obtained by finite differences. Here the design variables are 50 control points of a B-spline parameterization. The finite difference gradients are provided by forward differencing. Thus 51 Euler solutions are required to get the complete gradients. In order to be accurate enough, a convergence of 10 orders of magnitude is used.

For the adjoint approach just one Euler main and three Euler adjoint solutions (an adjoint solution per each boundary condition, each right hand side $-d(I)$) are necessary. The adjoint gradients are obtained after a convergence of 10 orders of magnitude for the flow solution and just 3 orders for each adjoint flow solution. This comparably low convergence of the adjoint solutions is sufficient enough for getting accurate gradients in the sense that any further increase of the adjoint flow convergence has no more influence (see also [12]).

Figures 2b, 3a and 3b show the components of the gradients of the drag, lift, and pitching moment, respectively, according to the design variables. The 50 design variables span from the leading edge to the trailing edge along first the upper side and then the lower side. All three figures show a very good conformity of the finite differences and adjoint gradients.

As expected, a lot of CPU time is saved with the adjoint approach. While with finite differences one needs 510 minutes on a NEC-SX5, only 40 minutes are required with the adjoint approach in order to obtain all three gradients.

13.6 Surface Formulation of the Adjoint Gradients

Following the ideas of M. Giles (see [13]) the volume integrals (13.19)

$$\delta I = -\int_C p(-\psi_2 \delta y_\xi + \psi_3 \delta x_\xi)\, dl + K(I)$$
$$- \int_D \psi_\xi^\top (\delta y_\eta f - \delta x_\eta g) + \psi_\eta^\top (-\delta y_\xi f + \delta x_\xi g)\, dA$$

which lead to the adjoint gradients, can be expressed by the surface formulation

$$\delta I = -\int_C w_H^\top \psi (\delta n_x u + \delta n_y v)\, dl + K(I) \tag{13.27}$$

where $w_H = \begin{pmatrix} \rho \\ \rho u \\ \rho v \\ \rho H \end{pmatrix}$. For details on the derivation of this surface formulation see [10].

Previously, one had to use a grid moving technique (see [6]) in order to evaluate the volume integrals (13.19). Thinking of complex 3D applications, equations (13.19) would become more complicated and the grid moving technique would lead to problems, especially in the case of multi-block meshes. But now, with the surface formulation, one gets rid of these problems.

13.7 Validation of the Surface Formulation of the Adjoint Gradients

In the previous section 13.5, the adjoint gradients have been determined via the volume formulations (13.19).

Figures 4a, 4b and 5 show the components of the adjoint gradients of the drag, lift and pitching moment, respectively, for the same RAE2822 case, obtained by the volume formulations (13.19) and the surface formulations (13.27). The figures show a very good conformity of the two different adjoint gradient's formulations.

The surface formulation (13.27) is also used for the determination of the adjoint gradients of drag, lift and pitching moment for the ONERA M6 wing. The flight conditions are Mach number $M_\infty = 0.84$ and an angle of attack of $\alpha = 3.1294°$ to get a target lift of $C_L^t = 0.3$. Figure 6a shows the contour plot of the first adjoint flow variable for this flight condition and the cost function of drag reduction. The 32 design variables are the spanwise twists at 32 wing sections. The mesh for this case is with 129x33x49 grid points as coarse as possible. The volume formulation (13.19) is not implemented for 3D applications, due to reasons explained in the section above. Therefore, the validation of the implementation of the surface formulation (13.27) for 3D applications is to be done by comparisons with finite differences' gradients.

For this coarse mesh it is not easy to tune the finite differences. This can be seen in figures 6b, 7a and 7b, where the finite differences' values show strong oscillations at the outer wing parts. These oscillations are surely non physical but come from the error introduced by the finite difference approximation of the gradient and the coarse mesh. Nevertheless, the conformity of the adjoint and finite differences' gradients is quite proper, especially in the center of the wing.

The last validation case shows that the adjoint approach is not only much more efficient but also more accurate than the finite difference approach.

13.8 Extension of the Adjoint Solver to Navier-Stokes

The extension of FLOWer ADJOINT to viscous adjoint flux and boundary treatment has been done in the framework of the European project AEROSHAPE [14].

DLR and SAAB have agreed to cooperate within AEROSHAPE, where they shared the work on the implementation. This was possible due to the similar structure of DLR's and SAAB's flow solvers [15].

The derivation of the viscous adjoint flux terms can be found in [15]. The viscous adjoint flux is added to the convective adjoint Euler fluxes while the turbulence is assumed to be frozen. The adjoint far field condition remains the same like in the adjoint Euler case while the viscous adjoint wall condition changes, e.g. in the case of drag reduction, to

$$\psi_2 \equiv -\cos\alpha \ , \tag{13.28}$$

$$\psi_3 \equiv -\sin\alpha \ , \tag{13.29}$$

$$\vec{n} \cdot \nabla\psi_5 \equiv 0 \ . \tag{13.30}$$

The viscous FLOWer ADJOINT has been successfully verified against hand calculations for an academic test example of drag reduction of a flat plate. The viscous adjoint solver has also been applied to the drag reduction problem of the transonic RAE2822 airfoil (Baldwin Lomax turbulence model, $Re = 6.5 \cdot 10^6$, $M_\infty = 0.73$, $\alpha = 2°$). Figure 8b shows the convergence history of FLOWer in MAIN mode as well as in ADJOINT mode for single grid calculations. Finally, figure 8a shows the viscous adjoint flow field for the first component of the adjoint vector.

13.9 Conclusion

The principles underlying the adjoint approach, its implementation and validation have been presented. It has been shown that a lot of CPU time is saved using adjoint compared to brute force finite differences approaches for determining the cost functions' gradients per each optimization stage. Furthermore, the accuracy of the gradients provided by the continuous adjoint approach was demonstrated by direct comparisons with finite differences' gradients for the adjoint volume as well as surface formulations of the cost functions' gradients.

The adjoint method is an effective and promising way for industry relevant 3D aerodynamic optimization - especially for detailed optimization. This is demonstrated in [11] and in particular in the next paper [16], where the integration of the adjoint approach in the optimization framework Synaps Pointer Pro [17] is described and its application for multi-constrained as well as multi-point optimization problems for 2D and 3D applications is presented.

13.10 Figures

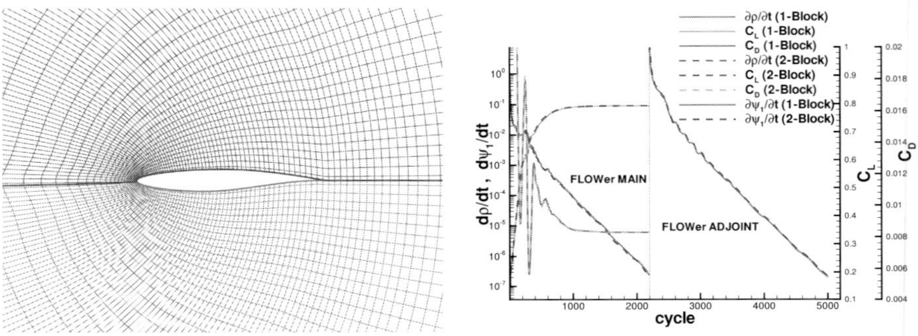

Fig. 13.1. a) Staggered 2-block C-mesh with 192x32x16 grid points for the RAE2822 airfoil and **b)** Convergence history of FLOWer in MAIN as well as ADJOINT mode for drag reduction ($M_\infty = 0.73$, $\alpha = 2°$).

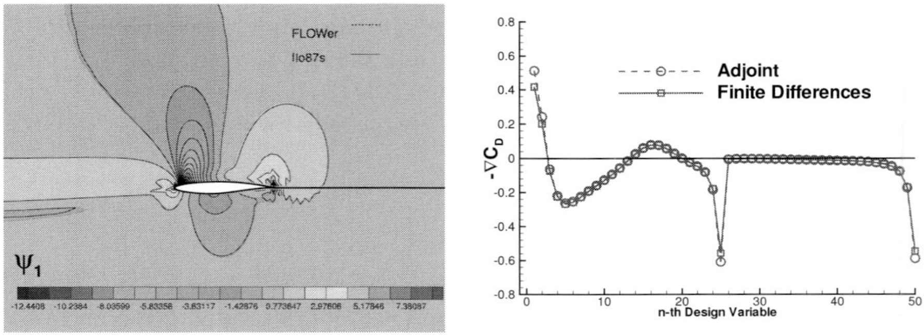

Fig. 13.2. a) Contour plot of the first adjoint flow variable calculated with FLOWer as well as flo87s and **b)** Gradient of the drag for the RAE2822 airfoil (50 B-spline parameters - $M_\infty = 0.73$, $\alpha = 2°$) calculated with FLOWer.

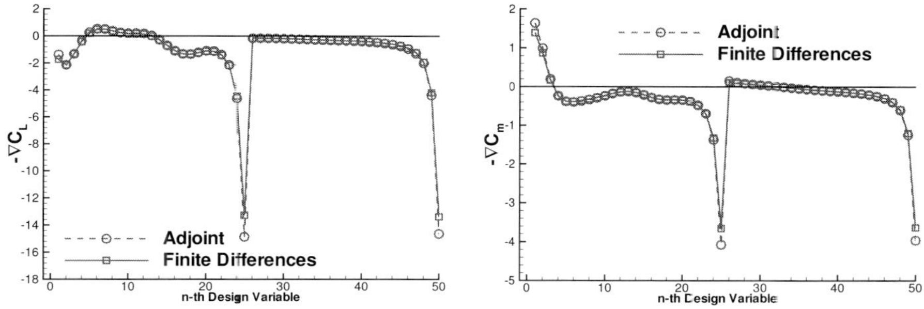

Fig. 13.3. a) Gradient of the lift and **b)** Gradient of the pitching moment for the RAE2822 airfoil (50 B-spline parameters - $M_\infty = 0.73$, $\alpha = 2°$) calculated with FLOWer.

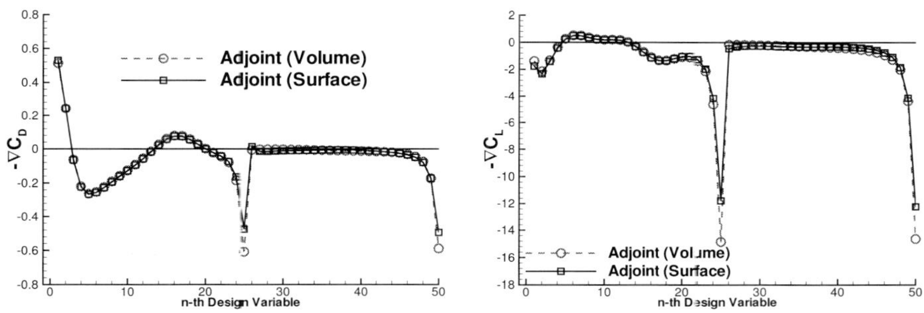

Fig. 13.4. a) Gradient of the drag and **b)** Gradient of the lift for the RAE2822 airfoil (50 B-spline parameters - $M_\infty = 0.73$, $\alpha = 2°$) calculated with FLOWer.

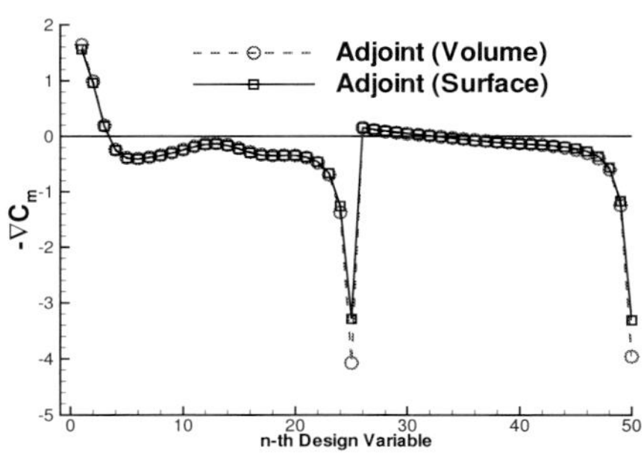

Fig. 13.5. Gradient of the pitching moment for the RAE2822 airfoil (50 B-spline parameters - $M_\infty = 0.73$, $\alpha = 2°$) calculated with FLOWer.

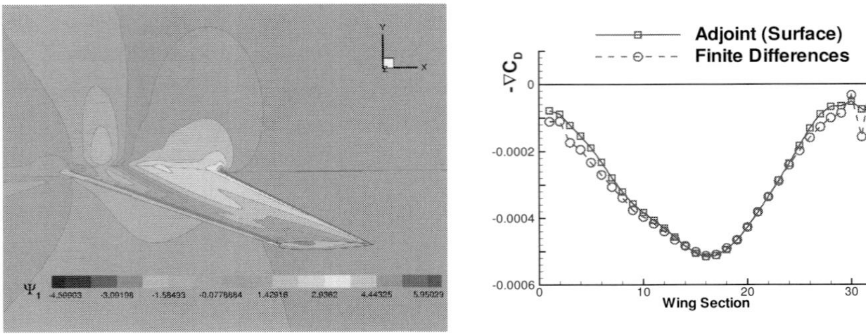

Fig. 13.6. a) Contour plot of the first adjoint flow variable and **b)** Gradient of the drag for the ONERA M6 wing (spanwise twist distribution at 32 wing sections - $M_\infty = 0.84$, $\alpha = 3.1294°$, $C_L^t = 0.3$) calculated with FLOWer.

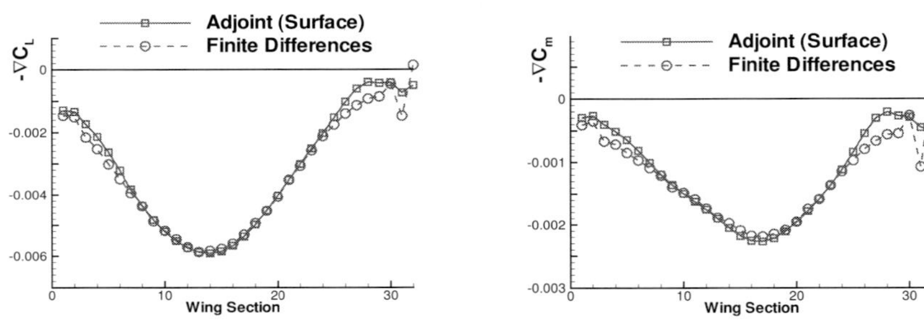

Fig. 13.7. a) Gradient of the lift and **b)** Gradient of the pitching moment for the ONERA M6 wing (spanwise twist distribution at 32 wing sections - $M_\infty = 0.84$, $\alpha = 3.1294°$, $C_L^t = 0.3$) calculated with FLOWer.

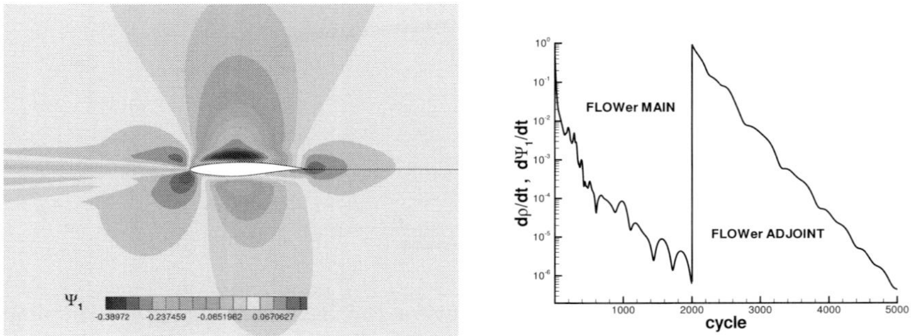

Fig. 13.8. a) Contour plot of the first adjoint flow variable and **b)** Convergence history of FLOWer in MAIN as well as ADJOINT mode for drag reduction (Baldwin Lomax turbulence model, $Re = 6.5 \cdot 10^6$, $M_\infty = 0.73$, $\alpha = 2°$).

References

1. Kroll, N., Rossow, C.C., Becker, K. and Thiele, F., "The MEGAFLOW project", Aerosp. Sci. Technol., Vol. 4, pp. 223-237, 2000.
2. Kroll, N., Rossow, C.C., Schwamborn, D., Becker, K. and Heller, G., "MEGA-FLOW — A Numerical Flow Simulation Tool for Transport Aircraft Design", ICAS 2002-1.10.5, 23rd International Congress of Aeronautical Sciences, Toronto, 2002.
3. Jameson, A., "Aerodynamic design via control theory", Journal of Scientific Computing, Vol. 3, pp. 233-260,1988.
4. Jameson, A., "Computational aerodynamics for aircraft design", Science, Vol. 245, pp. 361-371, 1989.
5. Jameson, A., "Optimum aerodynamic design via boundary control", AGARD-R-803, pp. 3.1-3.33, 1994.
6. Reuther, J., Jameson, A. et al, "Aerodynamic shape optimization of complex aircraft configurations via an adjoint formulation", AIAA 96-0094, 1996.
7. Jameson, A., Martinelli, L. and Pierce, N.A., "Optimum aerodynamic design using the Navier-Stokes equations", Theoret. Comput. Fluid Dynamics, Vol. 10, pp. 213-237, Springer, 1998.
8. Kroll, N., Rossow, C.C., Becker, K. and Thiele, F., "MEGAFLOW - a numerical flow simulation system", 21st ICAS Symposium, paper 98-2.7.4, Melbourne, Australia, 1998.
9. Gauger, N., "Aerodynamic shape optimization using the adjoint Euler equations", Proceedings of the GAMM Workshop 'Discrete Modelling and Discrete Algorithms in Continuum Mechanics', pp. 87-96, Logos Verlag Berlin, 2001.
10. Gauger, N., "Das Adjungiertenverfahren in der aerodynamischen Formoptimierung", to appear as DLR-Report No. DLR-FB–2003-05 (ISSN 1434-8454), 2003.
11. Gauger, N. and Brezillon, J., "Aerodynamic Shape Optimization Using Adjoint Method", Journal of Aero. Soc. of India, Vol. 54, No. 3, 2002.
12. Nadarajah, S. and Jameson, A., "Studies of the continuous and discrete adjoint approaches to viscous automatic aerodynamic shape optimization", AIAA 2001-2530, 2001.
13. Giles, M.B., "Adjoint equations in CFD: duality, boundary conditions and solution behaviour", AIAA 97-1850, 1997.
14. Selmin, V., "Multi-point aerodynamic shape optimization: The AEROSHAPE project", Proceedings of ECCOMAS, Barcelona, Spain, 2000.
15. Weinerfelt, P., Gauger, N., Quagliarella, D., Soemarwoto, B. et al, "Sensitivity computations based on continuous equations", Progress in Aerodynamic Shape Optimisation, Chap. 3, to appear in Notes on Numerical Fluid Mechanics.
16. Brezillon, J., "Application of the Adjoint Technique with the Optimization Framework Synaps Pointer Pro", Present Notes on Numerical Fluid Mechanics.
17. Frommann, O., "Conflicting criteria handling in multiobjective optimization using the principles of fuzzy logic", AIAA-98-2730, 1998.

14

Application of the Adjoint Technique with the Optimization Framework Synaps Pointer Pro

Joël Brezillon

DLR, Institute of Aerodynamics and Flow Technology, Lilienthalplatz 7, D-38108 Braunschweig

Summary. The present paper aims at describing the potential of the adjoint technique for aerodynamic shape optimization. After a brief description of the aerodynamic optimization process developed at DLR, specific requirements for an optimization framework combined with the adjoint technique are introduced. The drag reduction by constant lift and pitching moment for the RAE 2822 airfoil in transonic flow is then presented as validation case. An extension to multi-point optimization demonstrates the capability of the methodology to solve more complex problems. At the end, the body optimization and the wing optimization of a supersonic commercial aircraft confirm the flexibility of the framework and the efficiency of the adjoint technique.

14.1 Introduction

The numerical simulation in aerodynamics is now mature enough to play an important role in the design process of new aircrafts. A detailed aerodynamic shape optimization can therefore be done using accurate flow-solver based on the Navier-Stokes or at least the Euler equations. However flow simulations require an appropriate environment or framework in order to be integrated in an optimization loop. In other words, this framework is to connect all elementary processes, necessary to associate a given geometry to an aerodynamic state, and to carry this information over to the optimizer. In order to be able to handle a broad range of applications, this optimization framework should be flexible enough to allow the integration of new modules. Because detailed aerodynamic shape optimizations suffer from high computational costs, an efficient strategy has to be developed. The adjoint approach is seen nowadays as a promising alternative and an adjoint solver has been accordingly developed and validated during the MEGAFLOW II project [4, 5]. The present paper assesses the potential of the adjoint technique for aerodynamic shape optimizations.

14.2 The Optimization Framework

14.2.1 Requirements for the Aerodynamic Optimization

During the 1st phase of the MEGAFLOW project, the optimization system MEPO has been developed by DLR in cooperation with the Technical University of Braunschweig and Airbus Bremen [1]. This system proved its high efficiency in solving aerodynamic problems, but the structure was not flexible enough to be able to easily integrate new modules that were to be developed during the 2nd phase of the MEGAFLOW project. Thus a collaboration with the Synaps Ingenieur-Gesellschaft mbH was initiated in order to assess the capabilities of the SynapsPointer Pro optimization framework [3]. Flexibility has been checked against integration of modules developed in MEGAFLOW projects and specifically required for an aerodynamic optimization:

- geometry deformation according to B-splines, Hicks-Henne bump functions or a more advanced strategy such as the free form deformation;
- automatic mesh generation: structured meshes can be generated by the DLR tool MegaCads, developed and improved during MEGAFLOW projects, and unstructured meshes by Centaur. An other strategy based on mesh deformation has also been developed and used for optimizations;
- evaluation of the aerodynamic state extracted from the flow computation and based either on the structured code FLOWer or on the unstructured code Tau.

The SynapsPointer Pro optimization framework proves to allow an easy connection of all these building-blocks and manages automatically the connection to the optimizer: the aerodynamic optimization chain can then be built up and be summarized in Fig. 14.1. A snapshot of the framework showing the equivalent chain is depicted in Fig. 14.2. The framework allows also the integration of user-supplied optimizers and indeed several gradient-based strategies - like constrained steepest descent or quasi-Newton trust region - have been developed and integrated during the project. This framework offers the possibility to run the aerodynamic chain on a wide range of platforms: workstations (SGI, IBM,...), supercomputer (NEC) or even PC clusters. Compared to the previous optimization system MEPO, the SynapsPointer Pro optimization framework has proven to be more flexible and offers more possibilities. Thus it has been retained for further developments.

14.2.2 Specific Requirement for the Adjoint Technique

Let us now consider an unconstrained problem. With a gradient-based strategy, the approach for solving such a problem consists first in determining the optimal or negative gradient direction in the design space, that leads to the most promising decrease of the cost function. Then, a local minimum is searched along this optimal direction. These two steps are repeated until no more reduction of the cost function can be achieved.

The optimizer classically manages the evaluation of the optimal direction by finite differences: evaluations of the variation of the cost function in each direction of the design space. In an n-dimensional design space, all variations are obtained after n+1 evaluations of the flow. The evaluation of a single cost function can take a lot of time and, in the frame of a detailed aerodynamic optimization, this brute force strategy is particularly costly when the number of design variables is large, furthermore without any guaranty on the accuracy. Efficiency and accuracy can be improved in defining and solving an adjoint problem, in which case the evaluation of the optimal direction is now independent of the number of design variables. The development of an adjoint solver for aerodynamic problems and how to compute the optimal direction have been important parts of the MEGAFLOW II project [4, 5].

Using this attractive technique now requires to provide the optimizer not only with the cost function but also with the optimal direction, which is a vector of dimension n. This new functionality has been implemented in the framework and checked by Synaps GmbH in close collaboration with DLR. All applications presented below have been obtained using the adjoint technique and the improved SynapsPointer Pro system.

14.3 Validation

In the paper on the adjoint strategy in this book [5], the authors aim at showing how to get the gradient and at validating it on different cases. We now concentrate on how using it in an optimization process.

14.3.1 Test Case Definition

The optimization problem of the drag reduction by constant lift and pitching moment at a fixed angle of attack is chosen for the validation of the optimization system. The flight conditions are given by the Mach number $M_\infty = 0.73$ and the angle of attack fixed to $2\,°$. The initial geometry is the RAE 2822 airfoil. Before showing the results, the aerodynamic chain is described and the way how aerodynamic constraints have been handled is presented.

14.3.2 Description of the Aerodynamic Chain

For the parameterization, the airfoil is decomposed into thickness and camber distributions, but only the camberline is modified. This procedure allows the thickness of the airfoil to remain unchanged during the optimization process. Results first obtained with a B-spline parameterization were not smooth and satisfying enough. Therefore the parameterization has been switched to the basis of Hicks-Henne functions [6], which are "sine bumps" in form. A number of 20 Hicks-Henne functions have been assigned to the camberline.

During the whole optimization process, meshes for the modified geometries were generated by MegaCads. These are all of the same topology (a staggered C-Mesh around the airfoil) and of the same size (321x57x7).

The flow solution is computed using the FLOWer code in Euler mode. Therefore the drag improvement consists only of the wave drag decrease. The flow solution is sufficiently converged with typically 10 orders of magnitude on RMS residuals which is easily obtained by means of the multigrid option available in FLOWer. This level of convergence ensures that the noise in the cost function lies well below the level of realizable improvements. With adjoint solutions, only 3 orders of magnitude are enough to get accurate gradients. Both the flow and the adjoint flow computations are done on a NEC-SX5.

14.3.3 Treatment of the Aerodynamic Constraints

At this point, it is assumed that all three gradients of the drag, lift and pitching moment are accurately determined with the adjoint approach, as described in [5].

As mentioned in the previous part, for optimization problems without constraints, the optimal direction along which a local minimum is searched corresponds to the direction pointed to by the negative gradient of the cost function. However, for constrained problems, this optimal direction can lead to a violation of the constraints. One way to solve constrained problems consists then in finding a "feasible" direction along which the cost function decreases while constraints are kept constant. In other words, along this "feasible" direction, the constraints should have no variation. As an example, let us consider a single constraint termed C and depending on its position in the n-dimensional design space. It is mathematically well known that the variation of C along a particular direction in the design space can be approximated, at first order accuracy, by the scalar product of the gradient of C with the vector pointing into the specified direction. Therefore, a direction normal to the gradient of C yields a null scalar product and no deviation of C, at least in a neighborhood where the first order accuracy is predominant. Furthermore, finding such direction orthogonal to the gradient has more than one solution and defines in fact an hyperplane of dimension n-1: any vector included in the subspace orthogonal to the gradient of C is a candidate for no-variation of C.

Decreasing the drag while maintaining the lift constant can be reached in projecting the negative drag-gradient onto the hyperplane normal to the lift-gradient. This approach is simple to implement and can be generalized to several constraints using a Schmidt orthogonalization. The resulting direction leads to a decrease of the cost function while keeping the constraints constant, at least if the variation is small enough not to introduce second order effects.

As a conclusion, the treatment of the constraints can be based on an optimal direction that depends now not only on the gradient of the cost function but also on the gradients of the constraints themselves.

14.3.4 Results

Figure 14.3a shows the evolution of the drag, lift and pitching moment over several optimization stages. Upon convergence, a decrease of more than 60% of the cost function (drag) is achieved while both the constraints (lift and pitching moment) remain unchanged. As expected, the strong shock appearing on the RAE 2822 airfoil is eliminated at the end of the design process as showed in Fig. 14.3b. The pressure distribution is also characterized by an increase of the suction peak and an increase of the rear loading in order to keep the lift constant. An analysis of the optimal geometry evidences a lower camber in the front part and a higher one in the rear part.

Due to the magnitude of the changes in the cost function at the beginning and to the non-linearity of the optimization problem, small deviations in the constraints do occur. Therefore a correction step is included at the fourth stage, which introduces an increase of the cost function as can be seen in Fig. 14.3a. The optimization procedure is then continued until convergence.

This first result is meant as a validation of the complete optimization strategy: from how to handle constraints down to the correct integration of modules in the SynapsPointer Pro framework.

14.4 Multi-Point Optimization

14.4.1 Introduction

The design problem is now the minimisation of a linear combination of the drag at two flow conditions for the RAE 2822 airfoil under aerodynamic and geometric constraints. The objective function reads:

$$Obj = \sum_{i=1}^{2} W_i C_D(\alpha_i, M_i) \tag{14.1}$$

where C_D is the drag coefficient as a function of the angle of incidence α and the Mach number M, and W_i are the weighting coefficients.

The two design conditions are :

1. $\alpha_1 = 2.8°$, $M_1 = 0.734$, $Re_1 = 6.5 \cdot 10^6$, $W_1 = 2$
2. $\alpha_2 = 2.8°$, $M_2 = 0.754$, $Re_2 = 6.2 \cdot 10^6$, $W_2 = 1$

The aerodynamic constraints on both points read:

- the lift is not allowed to decrease and the angle of attack is fixed

- only a variation of ±2% is allowed on the pitching moment.

Four additional geometrical constraints are set:

- the maximum thickness is frozen
- the thickness at 5% chord should not be lower than 96% of the initial one,
- the leading edge radius should be equal to or higher than 90% of the initial leading edge radius,
- the trailing edge angle should not be lower than 80% of that of the initial geometry.

14.4.2 Strategy

The parameterization is the same as described in the previous part: the RAE 2822 geometry is split into its thickness and camberline descriptions and in order to meet automatically the geometrical constraints, only the camberline is deformed using 20 Hicks-Henne bump functions. The structured mesh is provided by the MegaCads grid generator.

As for the aerodynamic flow, the Reynolds-averaged Navier-Stokes equations are solved using FLOWer. Turbulence effects are taken into account with the cost-efficient algebraic Baldwin-Lomax model rather than with a more accurate two-equation transport model. In order to compute the required gradients, the inviscid adjoint approach is used and this original approach corresponds to a "hybrid optimization". It is important to point out that the resulting gradients only take into account the inviscid effects and neglect the viscous ones.

14.4.3 Results

The complete optimization took less than 1 day and at the end of the optimization process, both design points exhibit a significant drag reduction. Aerodynamic coefficients for the baseline and the optimized geometries are combined in table 14.1: a decrease of about 60 drag counts was obtained with the same lift coefficient for the first design point, and 85 drag counts with a slight increase of the drag for the second. The constraints on the pitching moment are also fulfilled.

Table 14.1. 2D Multi-point aerodynamic optimization (where ^{b} refers to the baseline geometry)

Pt	$\alpha(^\circ)$	M	W	$C_D^b \times 10^{-4}$	$C_D \times 10^{-4}$	ΔC_D	ΔC_L	ΔC_m
1	2.8	0.734	2	197.1	135.5	-31.2 %	0 %	-1.6 %
2	2.8	0.754	1	300.8	215.0	-27.4 %	+2.7 %	-2.0 %

The optimized geometry and camberline are compared to the original ones in Figs. 14.4a and 14.4b respectively: the optimized shape is characterised by a decrease of the camberline in the front part and an increase in the rear part. This leads to a decrease of the shock strength, in particular for the first design point (figure 14.5a) that was defined as the major flight condition. An increased suction peak can also be observed on both design points (Figs. 14.5a and 14.5b).

14.5 Optimization of the Supersonic Commercial Aircraft

14.5.1 Introduction

Let us consider a supersonic commercial aircraft of 252 seats capacity (Fig. 14.6), designed for 5500 nautical miles range with a supersonic cruise flight at Mach number $M_\infty = 2.0$ [7, 8]. The aim of the present work is the aerodynamic improvement via the redesign of the body and the wing.

The initial geometry is based on a simplified supersonic transport aircraft, defined for the AEROSHAPE project [9]. These simplifications were introduced in order to ease CFD grid generations and analysis work. A minimum allowable value of the fuselage radius and a minimum wing thickness law were set in order to prevent unrealistic aircraft.

Two drag optimizations have been performed, both at Mach number $M_\infty = 2.0$ and at fixed lift coefficient of $C_L = 0.12$ for the complete aircraft. The first optimization concerns the improvement of the body while the wing geometry has been kept constant. In the second case the wing has been optimized without the body.

The aerodynamic solutions have been provided by FLOWer, running in Euler mode. The target lift mode available in FLOWer has been activated in order to reach the desired lift coefficient by adjusting automatically the angle of attack. The gradient has been provided by the continuous adjoint approach developed in FLOWer and by the surface formulation based on SAAB formulation [2].

Because the lift constraint is automatically handled by the FLOWer code via the angle of attack, these problems have been treated as unconstraint optimizations by using the conjugate gradient strategy. The aerodynamic chain has been integrated in the SynapsPointer Pro framework and has run on PC clusters.

14.5.2 Body Optimization

The fuselage contraction has been parameterized with 10 design variables which change the stream wise law of the body radius. After deformations, the sections have become of elliptical form. The body centerline has been kept unchanged during the optimization. The cross-section between wing and body has been recalculated at each optimization iteration by MegaCads.

A new mesh has been created for each geometry with MegaCads: it has supplied a grid of the same topology which consists of a structured multi-block type with about 196.000 cells. This grid has been designed for multigrid computations.

Figure 14.7a shows the evolution of the drag and the lift coefficients during the optimization. About 7 optimization stages have been required to get the optimum, which represent about 40 aerodynamic computations and 6 adjoint flow evaluations. The optimal configuration has 3.5 less drag counts than the baseline. It can be seen that FLOWer has kept the lift constant during the complete optimization.

The body radius law in stream wise direction is presented in Fig. 14.7b: while sections located close to the wing apex are expanded, with a maximum at x=30m, sections close to the wing trailing edge are contracted up to the minimum allowed radius. Such deformation is not surprising if one look at the evolution of the area rule [10] along the stream wise direction in Fig. 14.7c.

14.5.3 Wing Optimization

The wing has been optimized without taking into account the body. The Mach number has been identical with the previous case but the lift has been set to $C_L = 0.1207$ in order to compensate the influence of the body. This lift value has been obtained by setting the angle of incidence obtained for the complete aircraft at $C_L = 0.12$ to the baseline wing.

Because the use of an adjoint strategy allows to evaluate the gradient independently of the number of design variables, the wing has been parameterized without any a priori: 122 design variables have been chosen to change the twist, the thickness and the camber line at specific wing sections (Fig. 14.8a). The resulting wing is constructed by linear lofting of modified wing sections.

The thickness deformation has been based on B-splines which set free the range and the chord wise position of the maximum thickness, the leading edge radius and the trailing edge angle at 8 wing sections. The positions of these sections are chosen according to the span wise distribution of the geometrical constraints on maximum thickness. The camber line has been modified by adding 10 Hicks-Henne functions at 8 wing sections. The twist distribution has been described by a Bezier curve defined by 10 nodes. The center of rotation for the twist has been set at the leading edge of the wing.

An initial mesh has been performed by MegaCads and has consisted of a CH mono-block of about 40.000 grid points. During the optimization process the mesh has been deformed by a function which propagates smoothly the deviation from the wing to the farfield. In the same way as for the body optimization the aerodynamic solutions have been computed by FLOWer, running in Euler mode and in the target lift mode in order to maintained the desired lift. The gradients are computed by the continuous adjoint.

Figure 14.8b shows the evolution of the drag and lift coefficients during the optimization. The optimal configuration has been obtained after 6 optimization stages, which represent about 39 aerodynamic computations and 5 adjoint flow evaluations. The optimized wing has 12.5 less drag counts than the baseline and FLOWer was able to maintain the lift constant.

A classical optimization based on finite differences to evaluate the gradients would require about 600 aerodynamic evaluations more compared to the adjoint approach, which represents a factor up to 14 in terms of CPU time: the benefit of the adjoint approach increases accordingly to the number of design variables.

The geometry of the wing is plotted at 6 different span wise sections — Fig. 14.9. One can clearly see two different types of optimized airfoils:

- for sections close to the root, the airfoils are mainly characterized by a round leading edge, with a shift of the maximum thickness location to the leading edge
- for sections located close to the wing tip, the radius of the leading edge becomes smaller and the maximum thickness location is shifted to the trailing edge.

The pressure distributions are also plotted for the 6 corresponding wing sections in Fig. 14.9: the pressure peak is reduced all over the wing more particularly for wing sections located close to the root.

14.5.4 Optimized Wing Combined with Optimized Body

In the present section, the optimized wing is combined with the optimized body. The aerodynamic performances on the resulting geometry have been compared to the baseline geometry in order to check the meaningful of the optimizations.

The mesh has been generated by MegaCads based on a new script file and the resulting mesh is of mono-block type with 230.000 cells.

Table 14.2 summarizes the aerodynamic performances for the baseline configuration and for the combination of the optimized wing with the optimized body: a decrease of 13.2 drag counts has been obtained at the same lift coefficient which leads to an increase of the lift over drag ratio of 15%. While the angle of incidence of the combined geometry is 1 degree lower, the pitching moment has been slightly increased which could be penalized for the trim of the aircraft.

Table 14.2. Baseline configuration compared to the combination of the optimized body with the optimized wing.

Geometry	$\alpha(^\circ)$	$C_D \times 10^{-4}$	Δ	C_L	C_m	C_L/C_D	Δ
Baseline	3.2	99.7	-	0.12	0.0363	12.0	-
Combined	2.2	86.6	-13.2 %	0.12	0.0379	13.9	+15 %

It can be noticed that the drag decrease obtained with the combined geometry is lower than the sum of drag improvement obtained separately with the optimized body and optimized wing. This can be explained by the aerodynamic interactions between the wing and the body which were not taken into account during the wing optimization. However the comparison of the Mach number distribution over the wing in Fig. 14.10 confirms the aerodynamic improvement of the combined geometry and therefore validates the optimization strategy described in this paper.

14.6 Conclusion

During the MEGAFLOW II project, a new optimization framework has been tested and its superiority in terms of flexibility has made it the main optimization framework used at the DLR Institute of Aerodynamics and Flow Technology. This framework was successfully coupled with the adjoint technique, also developed within this project, and a step forward in terms of efficiency has been achieved. This system has first been validated for the drag reduction by constant lift and pitching moment of the RAE 2822 airfoil in transonic flow. The resulting geometry is a shockfree airfoil fulfilling all constraints. The extension of this configuration to a multi-point optimization confirms the capability of the framework to treat more complex configurations. The efficiency of the adjoint approach has also been demonstrated on the body optimization and wing optimization of a supersonic transport aircraft where a lot of CPU time has been saved compared to brute force approaches.

The combination of the SynapsPointer Pro framework with the adjoint approach allows to perform aerodynamic shape optimizations in a faster way and make them attractive in an industrial context.

References

1. Axmann J.K., Hadenfeld M. and Frommann O.: "Parallel Numerical Airplane Wing Design". New Results in Numerical and Experimental Fluid Mechanics: Contributions to the 10th AG STAB/DGLR Symposium", Vieweg Verlag, 1997.
2. Enoksson O. and Weinerfelt P.: "Numerical Methods for Aerodynamic Optimisation". 8th International Symposium on Comp. Fluid Dynamics, Sep. 5-10 1999, Bremen, Germany.
3. Frommann O.: "SynapsPointer Pro V2.50", Synaps Ingenieur-Gesellschaft mbH, Bremen, Germany 2002. AIAA-98-2730, 1998.
4. Gauger N. R.: "Aerodynamic Shape Optimization using Adjoint Euler Equations". Proceeding of the GAMM Workshop 'Discrete Modelling and Discrete Algorithms in Continuum Mechanics', pp 87-96, Logos Verlag Berlin, 2001
5. Gauger N. R. and Brezillon J: "The Continuous Adjoint Approach in Aerodynamic Shape Optimization". Present Notes on Numerical Fluid Mechanics

6. Hicks R.M. and Henne P.A.: "Wing design by numerical optimization". Journal of Aircraft, Vol. 15, pp. 407-412, 1978.
7. Lovell D.A.: "Aerodynamic Research to Support a Second Generation Supersonic Transport Aircraft - the EUROSUP Project". Eccomas 98, 1998
8. Lovell D.A.: "European Research of Wave and Lift Dependant Drag for Supersonic Transport Aircraft". AIAA Paper No. 99-3100, 1999
9. Selmin, V.: "Multi-point aerodynamic shape optimization: The AEROSHAPE project". Proceedings of ECCOMAS, Barcelona, Spain, 2000.
10. Whitcomb R.T., Sevier J.R.: "A Supersonic Area Rule and an Application to the Design of a Wing-Body combination with high Lift-Drag Ratios". NASA TR R-72, 1960

Figures

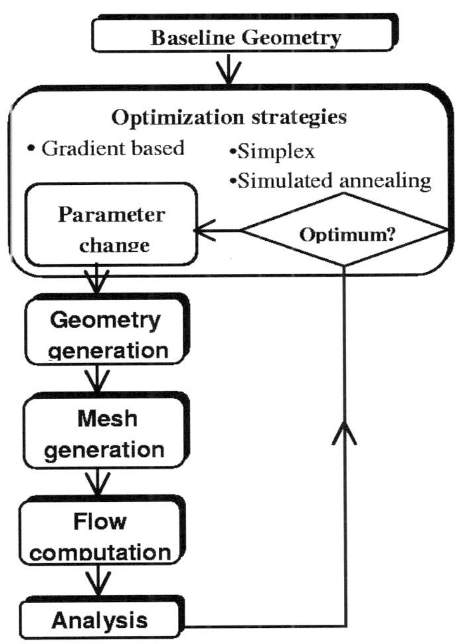

Fig. 14.1. Structure of the aerodynamic optimization loop

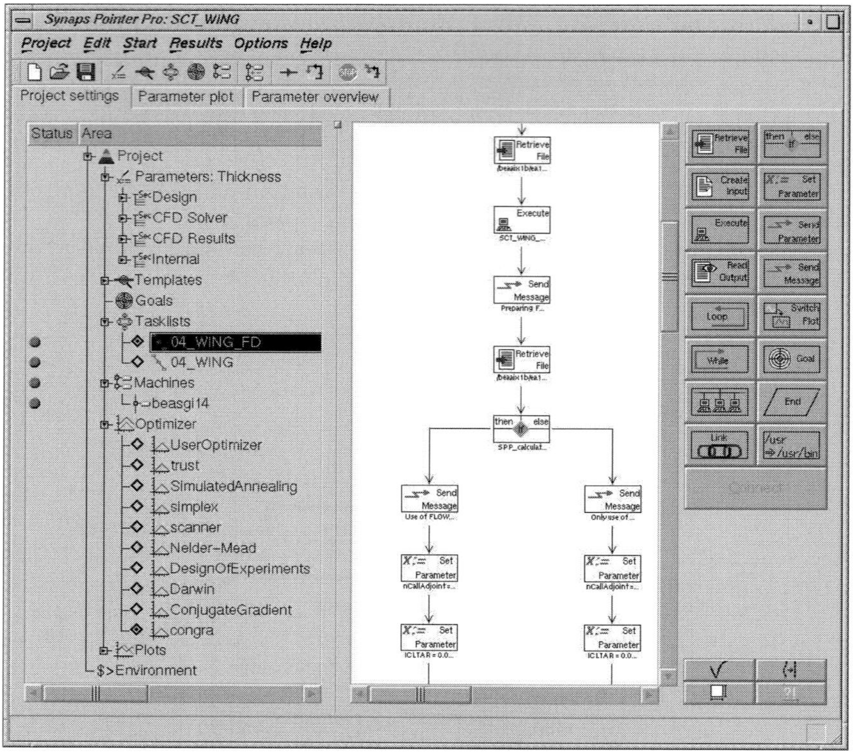

Fig. 14.2. Snapshot of the Synaps Pointer Pro optimization framework

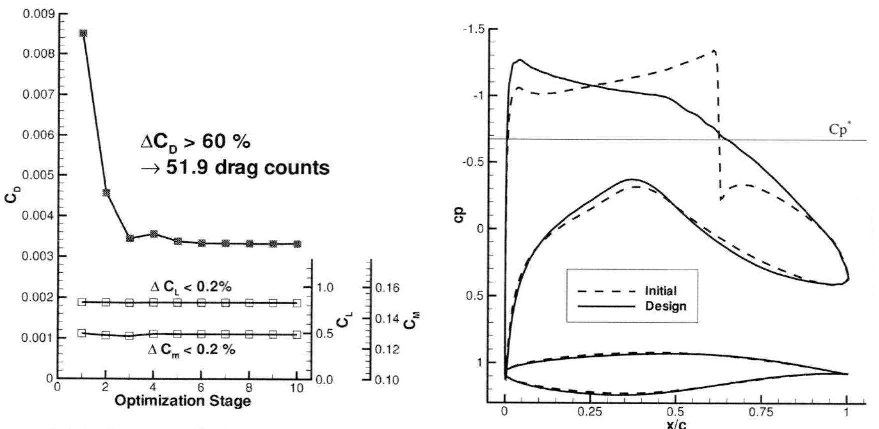

a) Evolution of the aerodynamic coefficients during the optimization.

b) Initial and final pressure distribution.

Fig. 14.3. Drag reduction of the RAE 2822 airfoil at fixed lift, pitching moment and angle of attack ($M_\infty = 0.73, \alpha = 2°$)

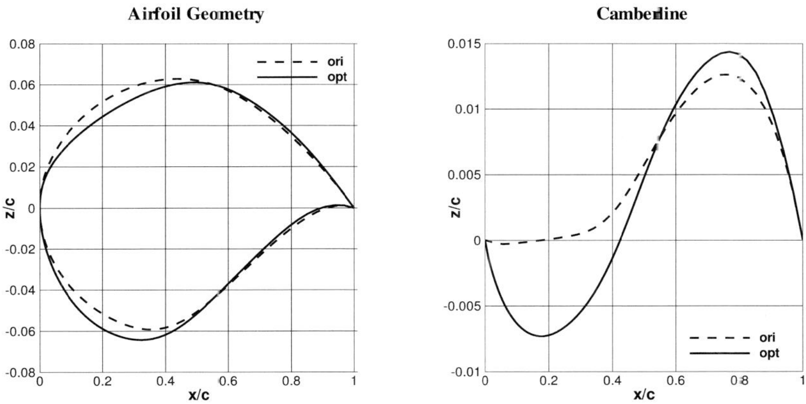

a) Baseline and optimized geometry. b) Baseline and optimized camberline.

Fig. 14.4. Multi-point drag optimization of the RAE 2822 airfoil

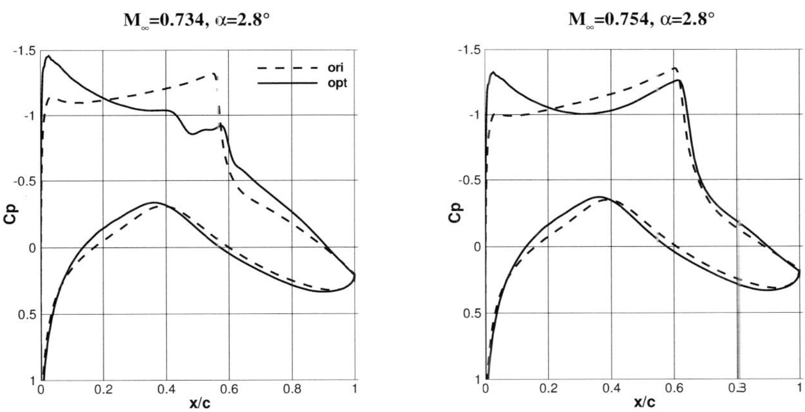

a) Pressure distribution for the first design point. b) Pressure distribution for the second design point.

Fig. 14.5. Pressure distribution obtained for the multi-point optimization

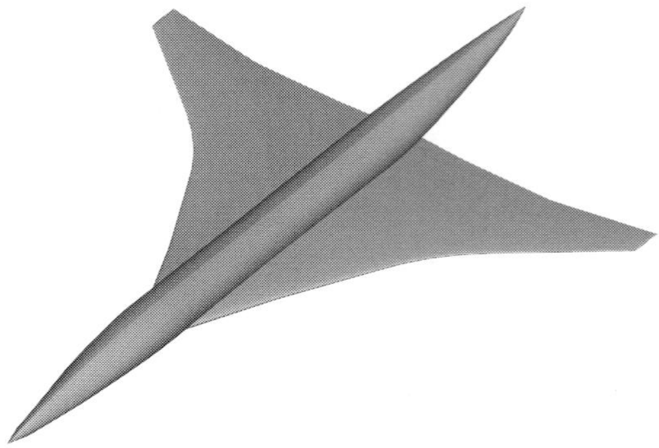

Fig. 14.6. Geometry of the Supersonic Transport Aircraft

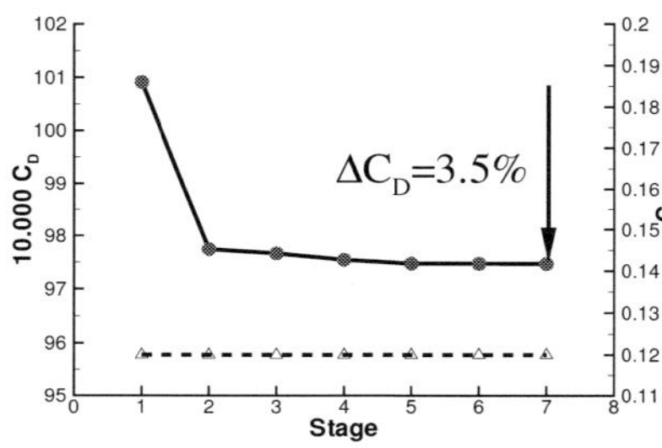

a) Evolution of the drag and lift during the optimization.

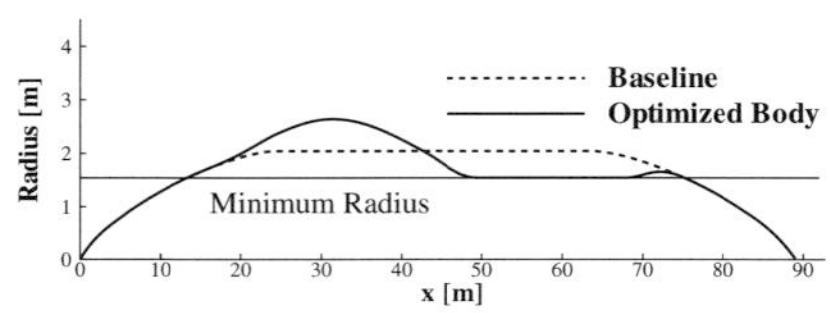

b) Baseline and optimized body radius.

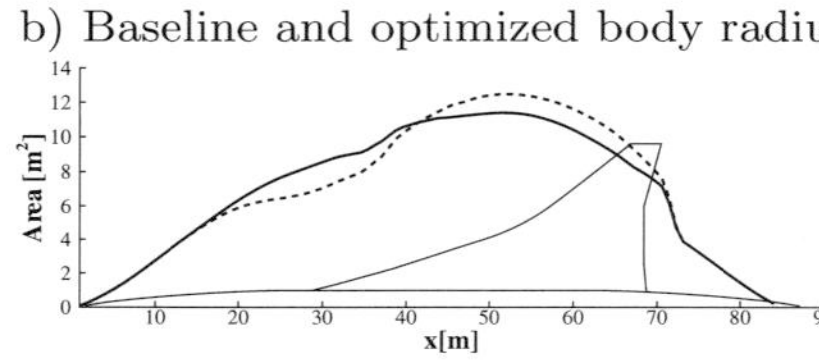

c) Area rule for the baseline and optimized geometry.

Fig. 14.7. Body optimization of the supersonic commercial aircraft

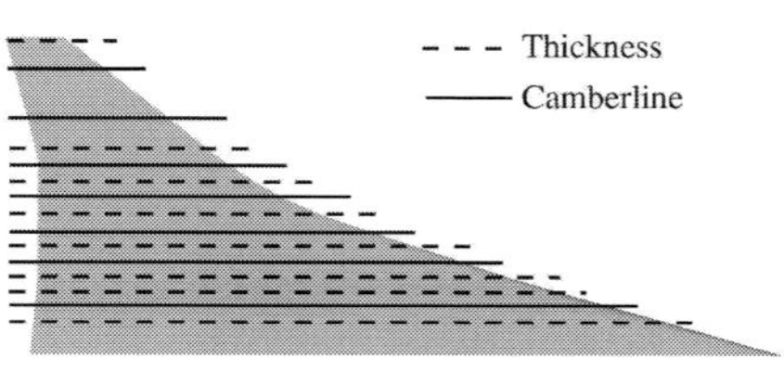

a) Parameterized section of the wing.

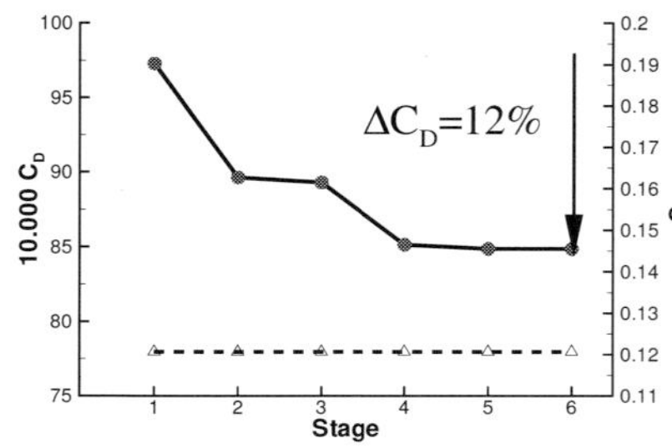

b) Evolution of the drag and lift during the optimization.

Fig. 14.8. Wing optimization of the supersonic commercial aircraft

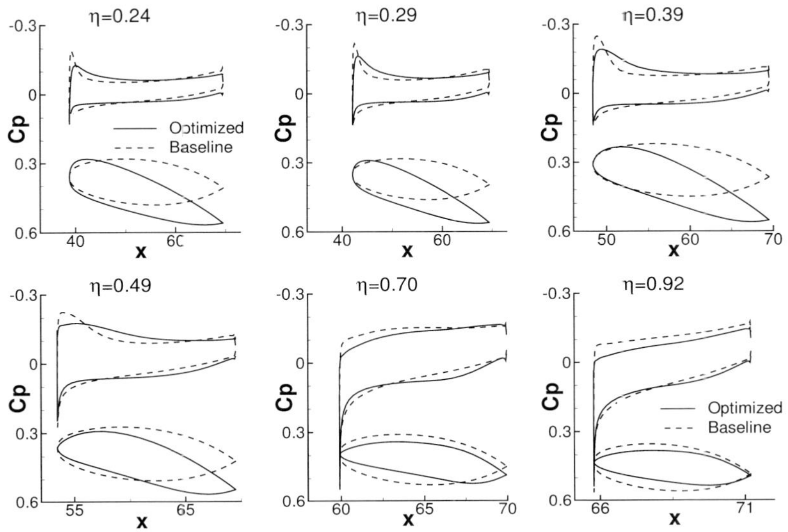

Fig. 14.9. Pressure distribution at 6 sections for the baseline and optimized wing

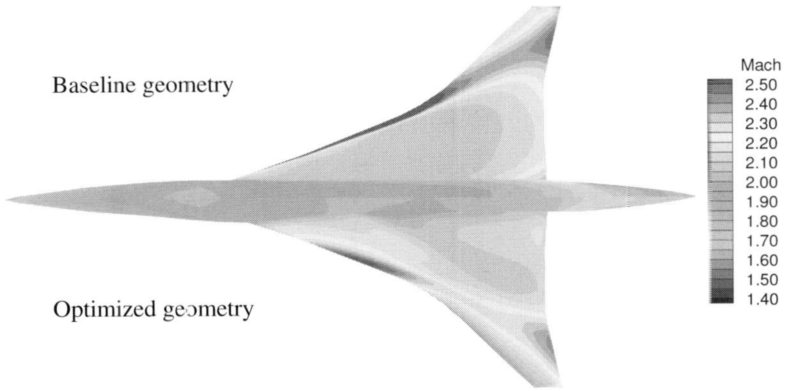

Fig. 14.10. Mach number distribution over the baseline geometry and over the optimized body combined with the optimized wing

15

Shape Parametrization Using Freeform Deformation

Arno Ronzheimer

DLR Braunschweig, Institute of Aerodynamics and Flow Technology, Transport Aircraft, Lilienthalplatz 7, 38108 Braunschweig, Germany

Summary. Shape parametrization has been identified as an import issue in aerodynamic design optimisation based on high-fidelity CFD-methods. For given shapes, which are available as CAD-models, post-parameterization method, based on freeform deformation, has been established to simplify and to automate the generation of geometrical variants to be used for CFD analyses. To create the necessary deformation lattices, structured grid generation techniques of a grid generation system, developed at DLR, are utilized. As this grid generation system has the salient feature to store and to replay a sequence of processes with different parameter settings, modifications of shapes, given by polygonal curves and surfaces, can be performed instantly.
The present freeform deformation method has reached a state, where it can be integrated into design loops to handle a variety of shape optimisation tasks. In two examples the applicability of the method for aerodynamic wing design and detailed design of a wing tip is demonstrated.

15.1 Introduction

In the beginning of the era of computational fluid dynamics (CFD), shape modelling plays only a minor role, since development was mainly focused on the improvement of CFD solvers concerning accuracy, efficiency and robustness. As generally flow codes prevailed with algorithms relied on body-fitted grids, geometry modelling was only an unavoidable subtask of former grid generation codes, running in batch mode. These grid generation codes were able to produce grids for specific tasks, as for example a grid around the DLR F6 wing-fuselage-engine-pylon configuration [9], on key press and permits slight geometrical variations to study the influence of different nacelle types and positions.

However, as configurational changes became the objective of certain studies, the use of task-specific grid generation codes was very limited, due to a hard-wired binding of geometry modelling and block topology of the structured grid. To over-come this inflexibility, the development of the grid generation system MegaCads [1] became a major part of the MEGAFLOW project [8]. The intention behind the MegaCads system was, to provide a user with an interactive tool to model a geometry with a certain complexity, based on cross-sectional polygons, which may have come out of a preliminary aircraft design method, and to generate a structured volume grid around this geometry. For this purpose several CAD-functionalities have been implemented to create curves and surfaces, to translate rotate and intersect curves and surfaces, and to project points onto curves and surfaces. For grid generation customary techniques are available to distribute points along curves, to create surface and volume grids using bilinear and trilinear interpolation techniques. Beside the prescribed functionalities, MegaCads has been provided with a scripting technique, which stores the complete sequence of construction steps, and enables to replay the parts of this sequence in batch mode with new defined parameters to generate a similar geometry with an appendant grid. This parametric concept has been successfully applied to optimise the settings of slat and flap of multi-element airfoils in high-lift aerodynamics [13].

In the process of industrial aircraft development the use of the MegaCads grid generation system in CFD-loops was primarily seen in the conceptional and preliminary design phase, where a CAD-representation of the aircraft geometry is still pending, and geometrical complexity is still moderate. But CFD methods in computational fluid dynamics have been continuously improved and particularly un-structured CFD-methods have gained a degree of automation and reliability to perform routinely flow analyses on geometries, which only can be handled by powerful CAD-systems, having nearly any detail of a final aircraft. Presently CAD-models of the outer shape of an aircraft may contain several hundred panels to be used for grid generation. If all panels form a watertight shape, unstructured grid generation methods are able to perform the grid generation task highly reliable and automated on a CAD-representation of the outer aircraft shape [2].

Currently, at first CAD-based automated grid generation has lead to the desired acceptance of CFD methods in aircraft industry and has enabled the integration of CFD-tools into computer aided engineering (CAE) processes. However, to improve an aircraft design in a late phase, especially when aerodynamic interference effects occur, which may degrade the promised performance of the aircraft, the turnaround time for further design studies is mainly governed by geometrical complexity and the effort to construct alternate geometry models within the CAD-System. Ideally a parametric CAD-system has been used, which will now allow to generate geometrical variants of the initial geometry on key-stroke. But to improve a design to meet fluid dynamic demands a parametric CAD-model may not always provide the appropriate parameters to vary. Since fluid flow is very sensitive to curvature and particularly to changes in curvature, it is more desirable to have additional variation capabilities as those, which are provided and constituted by a parametric CAD-model.

The same uncertainty persists if aerodynamic shape optimization is targeted, where an appropriate geometry parametrisation represents the base of the optimisation loop. To eliminate this lag of automation in geometry generation for redesign and shape optimisation, freeform deformation techniques, which are extensively used in the area of soft object animation, have been adapted, similar to the approach described in [10]. However, to gain a higher flexibility in applying freeform deformation to a great variety of forthcoming tasks, the freeform deformation method, described in section 15.2, has been implemented into the grid generation system MegaCads [1]. MegaCads provides import and export of geometry data and the present functionalities are used to create the set-up for a specific deformation task.

15.2 Principle of Freeform Deformation

The classical method of freeform deformation has been proposed by Sederberg and Parry [12] in 1986. Extensions to this method have been made by Griessmair and Purgathofer [7] in 1989 and by Coquillart [3] in 1990.

The first step of freeform deformation is to map a given object, defined by polygonal curves and surfaces with vertices $\vec{P}_{Object}$ into a parametric volume $\vec{P}(u,v,w)$. In the present implementation a trivariate B-spline volume, given by

$$\vec{P}(u,v,w) = \sum_{i=0}^{n_u}\sum_{j=0}^{n_v}\sum_{k=0}^{n_w} N_{i,m_u}(u)N_{j,m_v}(v)N_{k,m_w}(w)\vec{Q}_{i,j,k} \tag{15.1}$$

with B-spline basis functions $N_{i,m_u}(u)$, $N_{j,m_v}(v)$, $N_{k,m_w}(w)$ of degree m_u, m_v, m_w, defined on uniform knot vectors $u_i \in \{0,\ldots,0,\ u_{m_u+1},\ldots,u_{n_u},\ 1,\ldots,1\}$, $v_i \in \{0,\ldots,0,\ v_{m_v+1},\ldots,v_{n_v},\ 1,\ldots,1\}$, $w_i \in \{0,\ldots,0,\ w_{m_w-1},\ldots,w_{n_w},\ 1,\ldots,1\}$ is used. The basis functions are calculated recursively:

$$\begin{aligned} N_{i,1}(u) &= \begin{cases} 1 \text{ if } & u_i \le u < u_{i+1} \\ 0 & \text{otherwise} \end{cases} \\ N_{i,m_u}(u) &= \frac{u-u_i}{u_{i+m_u-1}-u_i} N_{i,m_u-1}(u) + \frac{u_{i+m_u}-u}{u_{i+m_u}-u_{i+1}} N_{i+1,m_u-1}(u)\,. \end{aligned} \tag{15.2}$$

The definitions of $N_{j,m_v}(v)$ and $N_{k,m_w}(w)$ are similar.

The control points $\vec{Q}_{i,j,k}$ define the initial deformation lattice, that encloses the initial object, as shown in Figure 15.1. Each vertex point $\vec{P}_{Object}$ of the initial object is mapped into the B-spline volume by solving for the parametric co-ordinates (u_0, v_0, w_0) of $\vec{P}_{Object}$ from eq. (15.1), using a Newton method. In a second step the deformation of the initial object is achieved by re-mapping the previously calculated parametric co-ordinates (u_0, v_0, w_0) to yield new vertex points:

$$\vec{P}_{NewObject}(u,v,w) = \sum_{i=0}^{n_u}\sum_{j=0}^{n_v}\sum_{k=0}^{n_w} N_{i,m_u}(u_0)N_{j,m_v}(v_0)N_{k,m_w}(w_0)\vec{R}_{i,j,k} \tag{15.3}$$

Eq. (15.3) is mainly identical to eq. (15.1), but with a different control point lattice $\vec{R}_{i,j,k}$, which determines finally the deformation of the object.

15.3 Control of Deformation

As the present method of freeform deformation is an indirect method, where the surrounding space is deformed, rather than the object itself, proper deformation lattices have to be provided, to achieve a desired geometrical variation of an object. For this purpose quite simply the structured grid generation methodology of MegaCads has been adducted and the described deformation method has been implemented into the MegaCads system, using the present structures and functionalities to construct adequate deformation lattices.

Besides the prescribed functionalities of MegaCads the underlying replay concept allows to repeat a complete construction sequence in batch mode with altered parameters. These parameters are used among other things to scale the length of curves and lines, which are used to construct the wire frame of the volume grid to be used for freeform deformation. As MegaCads supports arithmetic and the definition of dependencies between parameters, finally a control of the deformation incorporating certain geometrical constraints is enabled.

In Figure 15.2 freeform deformation is exemplarily applied to the geometry of the ONERA M6-wing [11], which is represented by a structured surface grid. For this purpose at first a structured grid of size $6 \times 2 \times 1$ cells has been spanned between two corresponding rectangular 2-dimensional grid sections at root and tip. This grid represents the initial lattice, which is used for the mapping step. The second grid, used for the re-mapping step, has been generated in restart mode, but now with altered grid sections at root and tip. In Figure 15.3 initial 2-dimensional section and altered section at the root are shown together with those parameters that control the vertical shifting of grid points at upper and lower boundary. These parameters have been defined during interactive construction of the initial lattice and are now used to generate the final lattice and to perform the ultimate deformation in restart mode.

The manual effort to construct the prescribed deformation mechanism for the M6-wing using MegaCads is conceivably small and is already done in a few hours. Beyond this small effort, a geometrical variant by editing one or more parameters, which are stored in the replay script, can be generated in seconds.

15.4 Application of Freeform Deformation

The basic question is now, if the freeform deformation method, which at first is used to animate imaginary creatures, is applicable in aerodynamic design. In the above deformation example a variation of a single parameter curls up a curved bump on the wing surface, what in matter of an improvement of the aerodynamic performance appears to be devious. However, the present example of the M6-wing deformation has been used as a test case for aerodynamic shape optimisation.

For this purpose the batch version MegaBatch of MegaCads has been integrated into the Synaps Pointer optimisation framework [4], which manages the complete optimisation procedure, (Figure 15.4). MegaBatch performs the deformation of the structured surface grid of the ONERA M6-wing by replaying the script, which has been stored during interactive work by MegaCads. For the generation of the structured surface grid an available code, written years ago for validation purposes, has been split into a surface grid generation part and a volume grid generation part. The surface grid generator is used to generate a single grid around the M6-wing geometry. The volume grid generator produces O-O type volume grids and has become part of the optimisation loop. Furthermore for the solution of the Euler equations the DLR FLOWer code [8] has been integrated. The optimisation algorithm is based on a conjugated gradient method and parameters from the optimisation algorithm are routed to the replay file of MegaCads via a template.

The objective of the optimisation is to minimise the inviscid drag coefficient c_D, which is extracted from the output of the flow solver and passed to the optimisation algorithm. Although there was an adjoint method available [5], gradients have been computed explicitly during optimisation. To avoid expensive computations, parameters at the tip of the wing $p_1, \ldots, p_{11}$ have been linked to corresponding parameters of the root section. Additionally a parameter p_{12}, which determines a twist of the wing, has also been incorporated into the deformation lattice.

The resulting surface grid of the optimised wing and the appropriate parameters, which have been obtained after 85 optimisation loops, are shown in Figure 15.5. As all flow solutions have been computed for a target lift coefficient $c_L = 0.3$ and an onflow Mach number $M_\infty = 0.84$ the resulting surface is cambered near the trailing edge. Along the leading edge the thickness is decreased. In Figure 15.6 lines of constant Mach number on the upper surface and pressure distributions at a span-wise cut are compared for initial and optimised wing. The typical λ-type shock on the upper surface of the basic M6-wing geometry has vanished and the level of suction pressure has been reduced on the upper surface after optimisation. More significant is the fact, that the inviscid drag has been reduced more than 40%. But despite the fact, that parameters have been linked to preserve thickness, from Figure 15.6 a small reduction can be observed.

To overcome this problem of vanishing thickness an alternate relationship between parameters has been investigated (Figure 15.7). In this case thickness has been moved from the upper surface to the lower surface by the optimizer, resulting in a fewer drag reduction of 25%, as shown in Figure 15.8.

15.5 Parametrization of CAD-based Shapes

In the above example freeform deformation has been applied successfully to parameterise an elementary geometry. However, due to the simplicity of the freeform deformation method, there should be no obstacle to adopt freeform deformation for more complex design problems. As there is only one restriction, which limits the application of freeform deformation to points of polygonal curves and polygonal surfaces, either knot-points or collocation points or discretized representations of spline curves and spline surfaces can be treated.

Ideally the tool for grid set-up of the CentaurSoft hybrid grid generation system [2], which is used for repair of CAD-data and definition of boundary conditions, provides a suitable re-discretisation of CAD-data, which are already cleaned and are representing a watertight domain for grid generation. Therefore an interface for importing and exporting CentaurSoft geometry data has been integrated into MegaCads to enable freeform deformation of CAD-data. As far as the topology of the CAD-model is preserved, whereas the geometry may have changed by freeform deformation, the exported geometry data can be processed immediately by the CentaurSoft intrinsic grid generation, which runs automatically in batch mode. In combination with the unstructured flow-solver TAU [6] the main prerequisites are given for the set-up of shape optimisation loops (Figure 15.9), but now based on the unstructured approach to handle more complex aircraft models.

A possible application of freeform deformation is demonstrated for a CAD-model of a transport aircraft in cruise configuration (Figure 15.10). In this case freeform deformation is applied to a special detail, namely the wing tip. In Figure 15.11 shaded surface grids of possible geometrical variants are shown, which have been generated upon deformed CENTAURSoft geometry data. All deformation lattices have been defined using the same MegaCads replay file, but with different parameter settings.

15.6 Conclusion

The freeform deformation method provides a great potential in aerodynamic shape design. However, the flexibility and applicability of this method to parameterise a given geometry retrospectively is at first enabled by a combination with the structured grid generation system MegaCads. MegaCads manages import and export of geometrical data, supports the construction of necessary deformation lattices and provides the control of deformation through the underlying replay concept. In all cases shown, the present functionalities have been applied, which are normally used to generate structured grids for CFD analyses.

The presented examples have demonstrated, that routinely applications of the freeform deformation method to a variety of aerodynamic design problems are established. Preferably the present method would be placed in front of the essential grid generation, instead of applying freeform deformation to surface grids, what had to be done in the M6-wing optimization example, as there was no usable surface geometry available. The main advantages of this procedure are the availability of a suitable geometry for grid adaptation and at last, if an optimized shape has been found, a re-feeding to a usable CAD-model may be performed more simplified. However the effort to extend this deformation method to unstructured surface data, which may be used immediately for volume grid generation, is estimated to be very small, since the topology of the data is simply copied, whereas coordinates are changed by the present freeform deformation method. Furthermore this would open the opportunity to handle FEM-model data in common with CFD-shape data, what will be a key prerequisite to perform multidisciplinary optimizations.

References

1. O. Brodersen, A. Ronzheimer, R. Ziegler, T. Kunert, J. Wild, M. Hepperle: "Aerodynamic Applications using MegaCads", 6th Intern. Conf. on Numerical Grid Generation, Ed.: M.Cross et.al., London, 1998, pp. 793-802.
2. CentaurSoft: http://www.centaursoft.com.
3. S. Coquillart: "Extended Free-Form Deformation: A Sculpting Tool for 3D Geometric Modeling", Proceedings of SIGGRAPH '90, Dallas, Texas, Aug. 6-10, 1990.
4. O. Frommann: "Conflicting Criteria Handling in Multiobjective Optimization Using the Principles of Fuzzy-Logic", AIAA-98-2370, 1998.
5. N. Gauger, J. Brezillon: "Aerodynamic Shape Optimization Using Adjoint Method", Journal of Aero. Soc. of India, Vol. 54, No. 3, 2002.
6. T. Gerhold, J. Evans: "Efficient computation of 3D-flows for complex configurations with the DLR-TAU code using automatic adaption", New Results in Numerical and Experimental Fluid Mechanics II, Vieweg Notes on Numerical Fluid Mechanics, Vol. 72, 1998, pp. 178-185.
7. J. Griessmair, W. Purgathofer: "Deformation of Solids with Trivariate B-Splines", EUROGRAPHICS '89, Elsevier Science Publishers (North Holland), 1989, pp. 137-148.
8. N. Kroll, C.-C. Rossow, K. Becker, F. Thiele: "MEGAFLOW — A Numerical Flow Simulation System", 21st ICAS Congress, Melbourne, paper 98-2.7.4, 1998.
9. C.-C. Rossow, A. Ronzheimer: "Investigations of Interference Phenomena of Modern Wing-Mounted High-Bypass-Ratio Engines by the Solution of the Euler-Equations", AGARD Conference Proceedings 498, Aerodynamic Engine/Airframe Integration for High Performance Aircraft and Missiles.
10. J. A. Samareh: "A Novel Shape P Approach ", NASA TM-1999-209116, NASA Langley Research Center, Hampton, VA 23681.
11. V. Schmitt, F. Charpin: "Pressure Distributions on the ONERA-M6-Wing at transonic Mach Numbers", AGARD Advisory Report No. 138, May 1979.
12. T. W. Sederberg, S. R. Parry: "Freeform Deformation of Solid Geometric Models", Proceedings of SIGGRAPH' 86, Dallas, Texas, Aug. 18-22, 1986.
13. J. Wild: "Numerische Optimierung von zweidimensionalen Hochauftriebskonfigurationen durch Lösung der Navier-Stokes-Gleichungen", DLR Forschungsbericht 2001-11, ISSN 1434-8454.

15.7 Figures

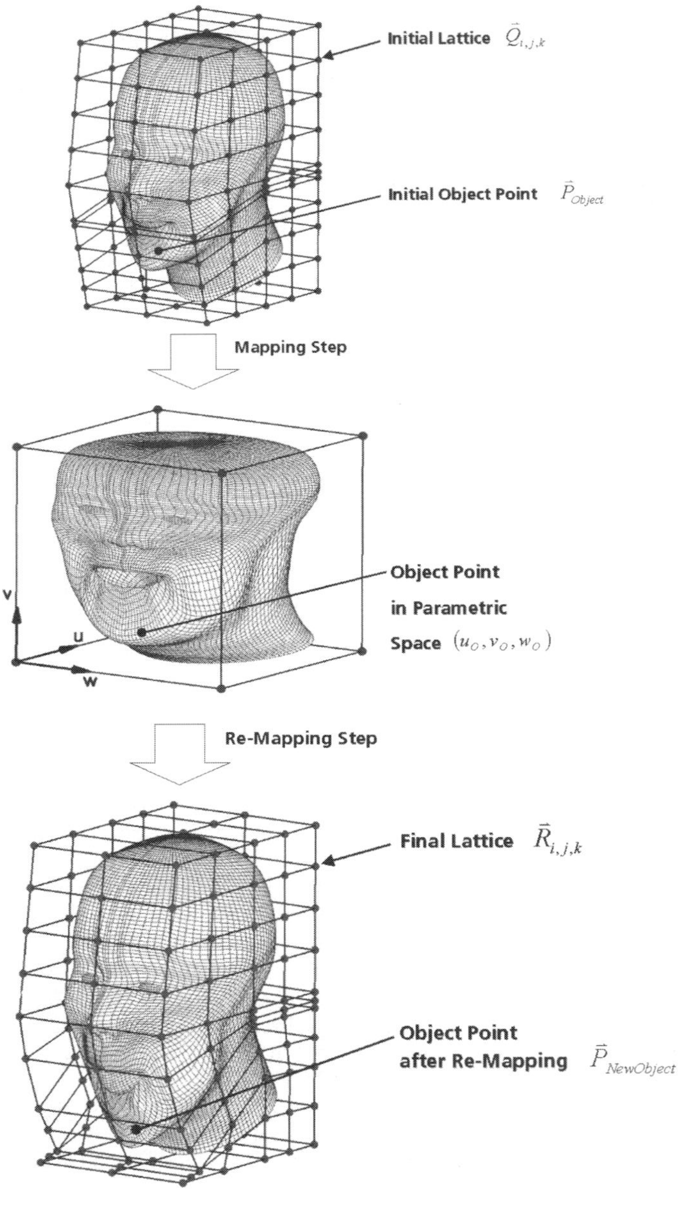

Fig. 15.1. Principle steps of freeform deformation

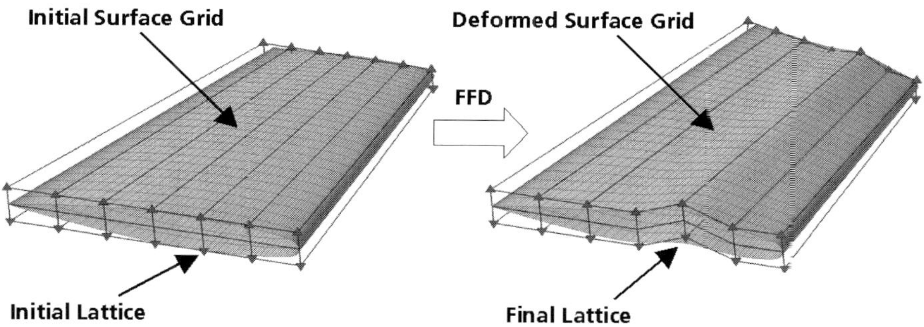

Fig. 15.2. Deformation lattices for ONERA M6-wing

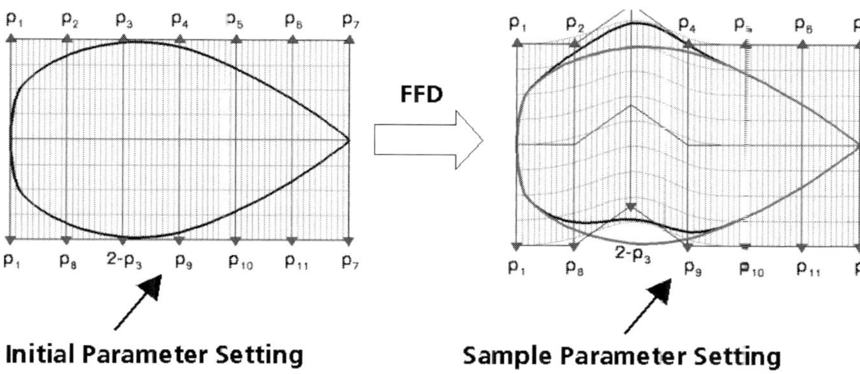

Fig. 15.3. Deformation of 2D root section with parameter adjustments

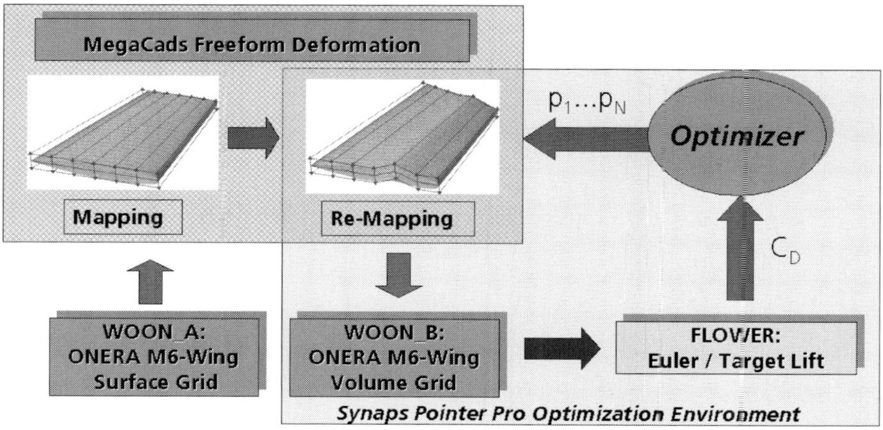

Fig. 15.4. Structure of optimisation loop

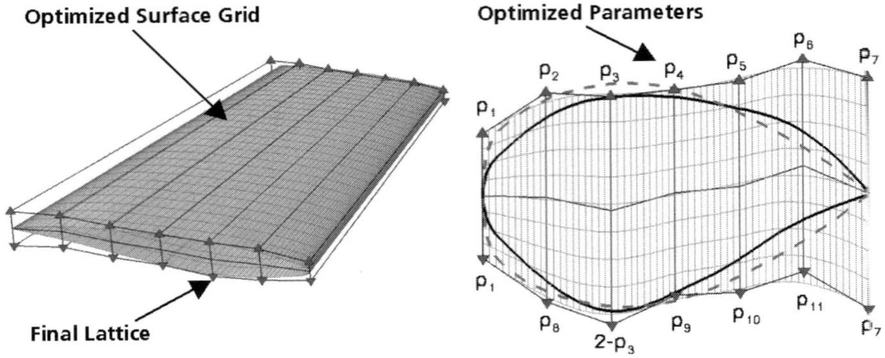

Fig. 15.5. Optimised M6-wing geometry

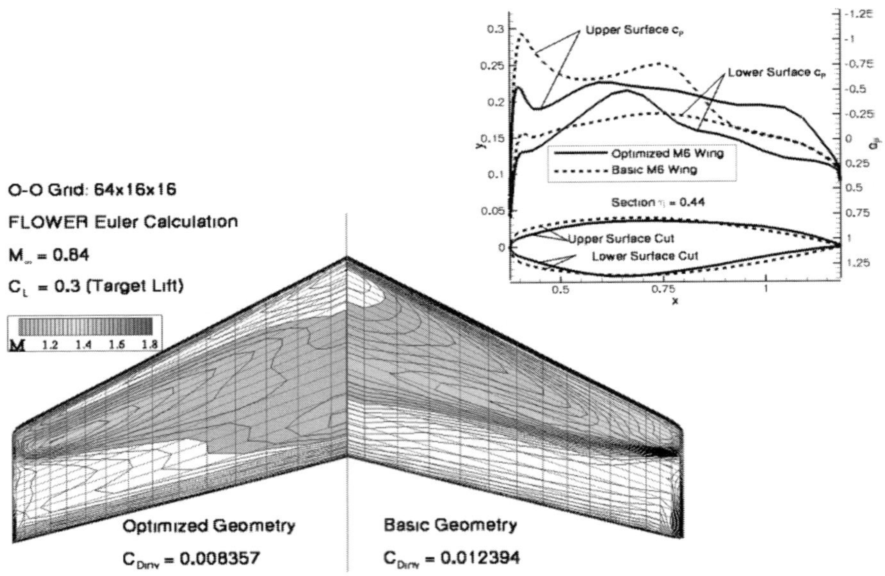

Fig. 15.6. CFD results of basic and optimised M6-wing geometry

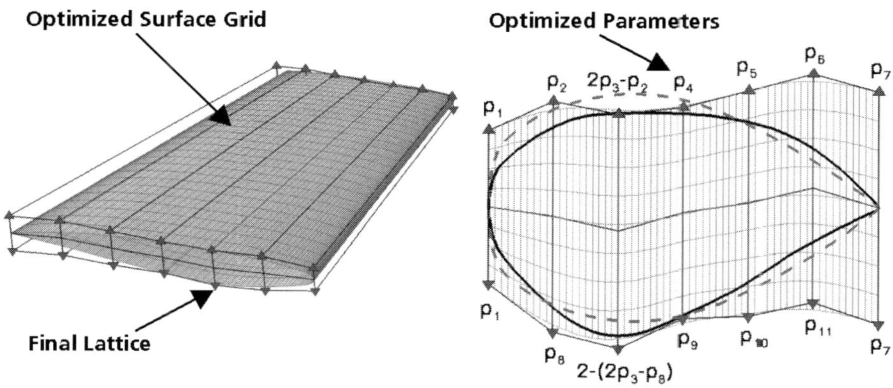

Fig. 15.7. Optimised M6-wing geometry with modified parameter relationship

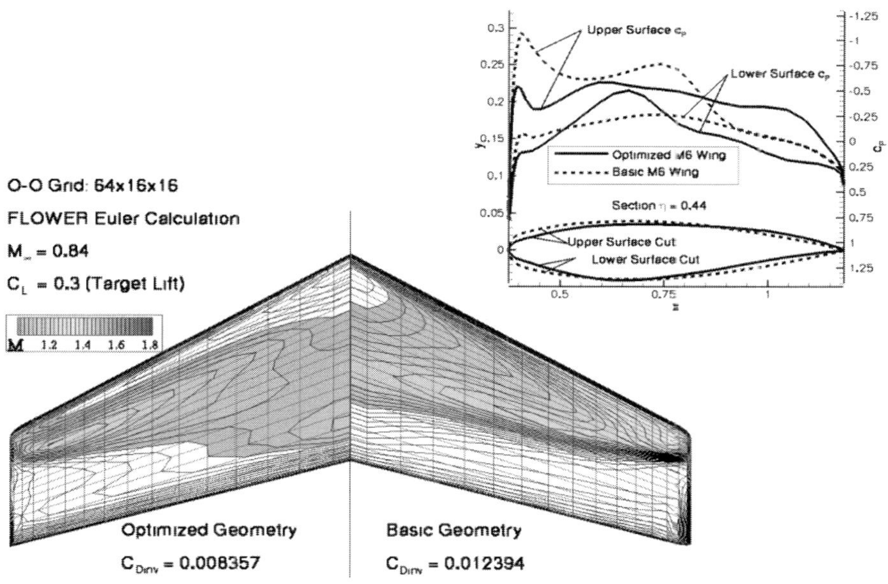

Fig. 15.8. CFD results of basic and optimised M6-wing geometry with modified parameter relationship

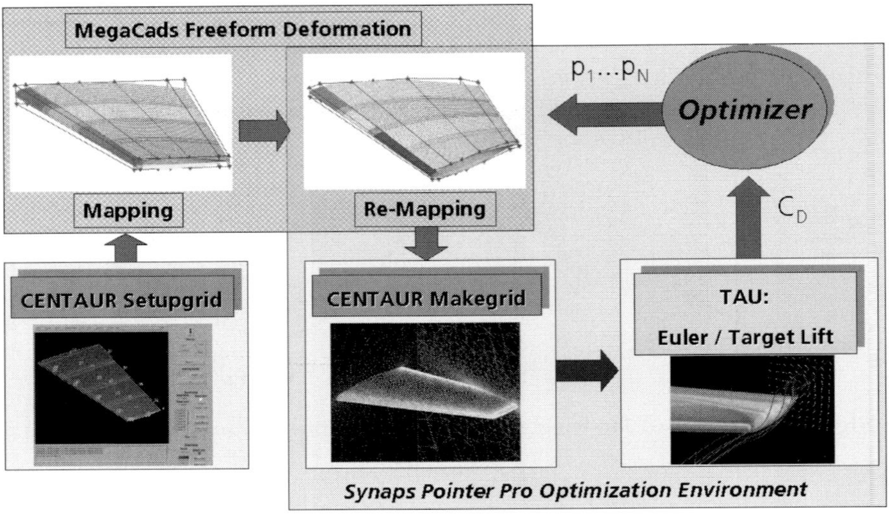

Fig. 15.9. Structure of optimisation loop in case of CAD-based geometries

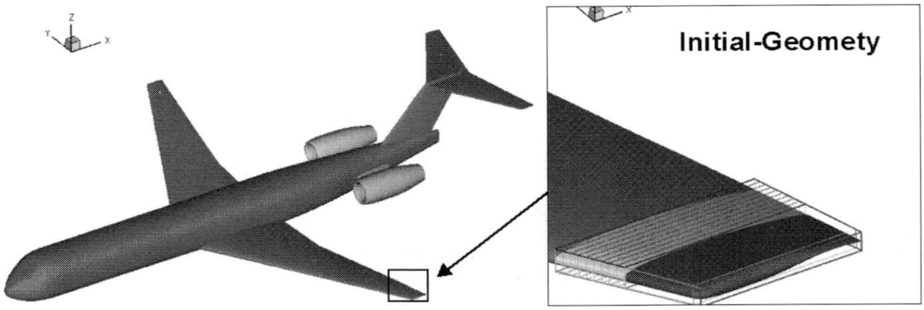

Fig. 15.10. CAD geometry with wing tip prepared for deformation

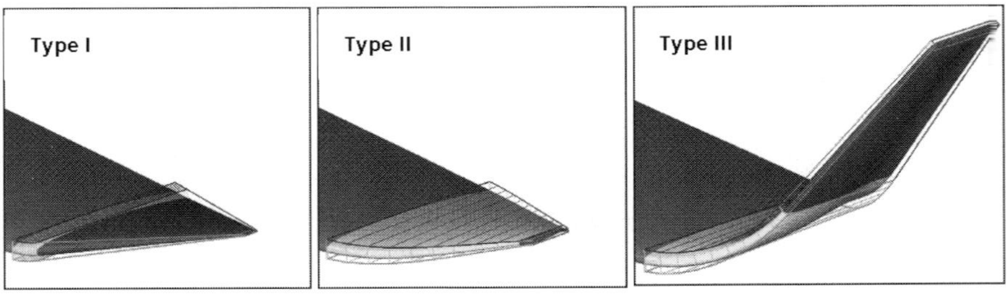

Fig. 15.11. Geometrical variations of wing tip

Part VI

Contributions of Universities

16

Advanced Turbulence Modelling in Aerodynamic Flow Solvers

Martin Franke, Thomas Rung and Frank Thiele

TU Berlin, Hermann-Föttinger-Institute of Fluid Mechanics,
Müller-Breslau-Str. 8, 10623 Berlin, Germany
franke@pi.tu-berlin.de

Summary. In the aerodynamic industrial design process, the use of numerical simulation is of ever increasing importance. In order to adequately capture flow features such as pressure-induced separation or shock-boundary-layer interaction, an appropriate representation of turbulence is needed. This contribution summarizes the efforts undertaken at TU Berlin to develop, implement and validate advanced linear and non-linear models in the aerodynamic flow solvers FLOWer and TAU in the framework of MEGAFLOW and related projects. The accuracy of the approaches is discussed on various cases and statements with respect to their computational performance are given. The results indicate that improved predictive accuracy can be obtained from advanced Eddy-Viscosity Models at a moderate computational surplus.

16.1 Introduction

Computational Fluid Dynamics (CFD) has become an integral part of the aircraft design process [1]. Over the past 15 years, research efforts concentrated on the accurate and reliable computation of aerodynamic flows with significant viscous effects. This includes high-lift flows featuring pressure-induced separation, viscous wake interaction and confluent boundary layers as well as cruise-flight conditions incorporating shock-boundary-layer interaction and shock-induced separation. As turbulent effects play a significant role in these flows, an adequate representation is crucial for successful aerodynamic performance prediction.

Owing to the currently prohibitive demand with respect to computer resources, any approach other than solving the Reynolds-Averaged Navier-Stokes (RANS) equations will not be feasible for practical problems in the near future [2]. A variety of RANS methods exist, ranging from simple algebraic models up to Reynolds-Stress Transport Models (RSTM). While the former have to be considered outdated, the use of the latter is still prohibitive in complex applications, mainly due to convergence and stability problems as well as their computational expenses. For that reason, transport-equation

Eddy-Viscosity Models (EVM) remain the backbone of turbulent viscous computations in the industrial design process.

This motivates the work on improved EVM, linear and non-linear, performed within the MEGAFLOW framework and subsequent projects. By devising, implementing and validating such approaches based on a physically sounder representation of turbulence, a more accurate simulation quality is aspired.

16.2 Turbulence Modelling

Simple, standard Boussinesq-viscosity models have proven insufficient to correctly predict complex flow situations. Their inability to render the fundamental physics of turbulence, such as curvature-driven shear-stress variation and secondary motion, is a well-known fact. Thus, attention is drawn to more reliable approaches towards the modelling of turbulence. Taking into account that RSTM have not yet reached the state of industrial maturity, and furthermore considering that no difference in priority between accuracy and efficiency is made in the design process, the attention is focussed on improved one- and two-equation practices. A variety of models was either developed or adopted from literature, implemented in the MEGAFLOW software and thouroughly validated. Six different approaches have been considered, and will be briefly described. While the first two models represent linear two-equation models which are expanded for an enhanced range of validity, the second two models are non-linear Explicit Algebraic Stress Models (EASM), which are characterized by a mathematically rigorous derivation from RSTM. Finally, two one-equation models are included. The classical Wilcox k-ω formulation [3] serves as a baseline reference.

The Menter SST (Shear Stress Transport) k-ω model [4] was designed to remedy two major flaws incorporated in the Wilcox approach, viz. the freestream dependency and the unsatisfactory predictive performance in adverse-pressure-gradient flows. The former is adressed by blending from k-ω in the inner region of the boundary layer to k-ε in the outer region and free shear flows (BSL model). The latter is tackled by sensitizing the eddy viscosity to the transport of the shear stress magnitude by incorporating the Bradshaw hypothesis (SST modification). The Menter model has gained significant popularity in the aeronautical community and can be regarded as one of the standard approaches today [5]. It can be used with either both BSL and SST (usually referred to as the full Menter SST model) or with SST only (labelled as Wilcox+SST).

The LLR (Local Linear Realizable) k-ω model [6] is a local linear two-parameter model derived from realizability and non-equilibrium turbulence constraints. The coefficients of the stress–strain relation and the turbulence-transport equations are all functions of the non-dimensional invariants of the mean strain and vorticity rates. The approach thus tries to accomplish consistent stress–strain distributions not only in plane shear flow, but also in more general flow situations.

The RQEVM (Realizable Quadratic Explicit Algebraic Stress Model) [7] stems from an explicit solution to the second-moment closure in the limit of equilibrium turbulence. This approach can be regarded as a generalized (non-linear) two–parameter model, which retains the predictive benefits of the second–moment closure methodology, while numerical advantages of the Boussinesq-viscosity concept are conserved. Additional key features of the modelling practice are topography-independent low-Re formulations, obedience of the realizability principle, consistency to the hydrodynamic stability theory and an approximately self-consistent representation of non-equilibrium turbulence. The current formulation is cast in terms of the Wilcox k-ω background model. Besides the full non-linear model, a linear truncation of the non-linear constitutive relation named LEA (Linearized Explicit Algebraic Stress Model) k-ω is available.

The EARSM (Explicit Algebraic Reynolds Stress Model) [8] by Wallin is another EASM derived similarly to the RQEVM. However, contrary to the latter, it is based on a fully self-consistent formulation. Following a suggestion by Wallin [9], the Kok k-ω approach [10] is used as the background model. Additionally, in an early stage of the project, a linear truncation of this model based on Wilcox k-ω, thus designated as L-EARSM + Wilcox k-ω, was investigated.

Due to their cost-effectiveness, one-equation models enjoy a wide popularity in practical application-oriented methods. Especially the Spalart-Allmaras (SA) model [11], which solves a transport equation for a modified eddy viscosity, has, besides the SST approach, become a standard model in aeronautical applications [5]. The model was assembled in a bottom-up manner, mainly by arguments of empiricism and dimensional analysis, also incorporating the Bradshaw hypothesis.

Finally, the SALSA (Strain-Adaptive Linear Spalart-Allmaras) model [12] is included, which is based on the original SA formulation. However, it offers an enhanced range of validity with respect to non-equilibrium flows. Unlike in standard one-equation approaches, which inherently contain the assumption of local equilibrium of production and destruction of turbulent energy, the reconstruction of the production-to-destruction ratio from mixing-length hypothesis elements via a sensitisation of the production term coefficient to variable strain rates allows for a more realistic representation of non-equilibrium states.

16.3 Computational Approach

The MEGAFLOW development includes two different Finite-Volume flow solvers, viz. the block-structured FLOWer and and the unstructured TAU code. They will be introduced in brief here, however, only the features relevant for this work will be given. FLOWer is is a density-based solver with different metric schemes on block-structured grids. For the purpose of the work reported, a cell-vertex scheme is used. Spatial discretization is based on central differences with added artificial dissipation, an explicit Runge-Kutta scheme being employed to integrate in time. The code is second-order accurate in space and time and uses multigrid acceleration in conjunction with a multiblock formulation. TAU has very similar features, however, as the solver is unstructured, a dual-mesh approach is used, rendering the solver independent of the element type used in the primary mesh. Both solvers run on various computer systems, including vector and massively parallel architectures.

In the framework of the MEGAFLOW project, six different linear models were implemented and validated in FLOWer at TU Berlin, viz. LLR k-ω, Menter SST k-ω, LEA k-ω, L-EARSM + Wilcox k-ω, SA and SALSA. All models were integrated in conjunction with low-Reynolds boundary conditions. More recently, under the umbrella of the ongoing EU 5th Framework Programme TAURUS, the two non-linear EASM, viz. RQEVM + Wilcox k-ω and EARSM + Kok k-ω, were implemented in the TAU solver. This work is included here, as it forms a natural continuation of the MEGAFLOW efforts.

16.4 Results and Discussion

The turbulence models described in Chap. 16.2 were thoroughly validated on a broad variety of high-lift as well as transonic testcases. Obviously, a complete description of these computations is beyond the scope of this paper. Thus, only a few selected cases using linear and non-linear two-equation models will be presented, ordered along increasing physical and geometric complexity. For a more detailed review, the reader is referred to [13, 14, 15, 16]. The three-dimensional testcases presented here are subject to certain restrictions, thus, no quantitative comparisons can be given here.

16.4.1 RAE 2822 Aerofoil

The RAE 2822 is a supercritical aerofoil [17], which has become a standard testcase for turbulence modelling validation. Two cases of transonic flows are considered, i.e. a subcritical one where little or no separation occurs downstream of the shock position ("Case 9", $Ma = 0.73$, $Re = 6.5 \cdot 10^6$, $\alpha = 2.8°$) and a supercritical one, where the shock-boundary-layer interaction induces massive separation ("Case 10", $Ma = 0.75$, $Re = 6.2 \cdot 10^6$, $\alpha = 2.8°$).

Unfortunately, the experiments suffer from wind tunnel influences, which cannot be quantified a posteriori. Several corrections have been suggested, however, as, none can be considered superior, the flow conditions used in [18] have been selected for comparability reasons. The computations presented here are performed with the TAU solver on a structured mesh (albeit computed in an unstructured manner) with $312 \cdot 65$ nodes generated by DLR. Transition is fixed at 3% chord length. Turbulence models included here are RQEVM + Wilcox k-ω and EARSM + Kok k-ω, Wilcox k-ω serves as baseline reference.

In Fig. 16.1, the pressure distribution is given for both cases. Turning the attention to the shock location, it is evident that Wilcox k-ω, while being able to quite accurately predict Case 9, computes the shock about 8% chord length too far downstream for Case 10. In contrast, both EASM results yield a much better shock location for Case 10, almost identical to the experiments. Looking at Case 9, it can be seen that, compared to Wilcox, the results of the EASM are only slightly worse, thus, the over-all gain in predictive accuracy is encouraging.

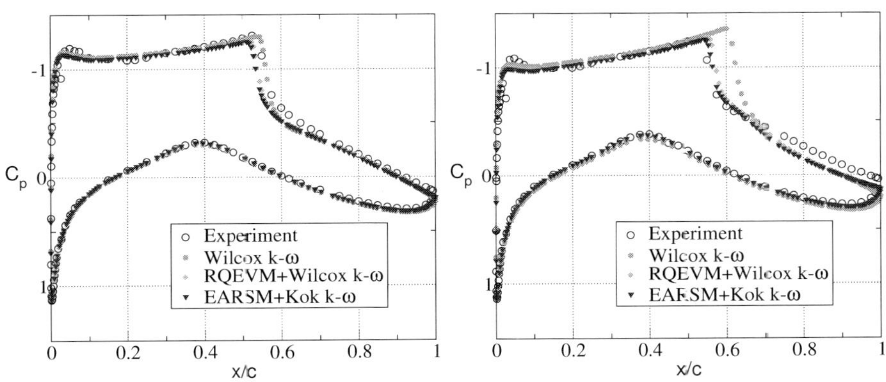

Fig. 16.1. RAE 2822: Pressure distribution for Case 9 (left) and Case 10 (right)

16.4.2 UHCA Wing-Body Configuration

Turning the attention to three-dimensional testcases, an Ultra-High Capacity Aircraft (UHCA) wing-body configuration is computed at $Ma = 0.85$, $Re = 2.8 \cdot 10^6$ and $\alpha = 2.66°$. This case is of particular interest since it represents a combination of high Mach numbers and a comparatively high angle-of-attack for transonic flows, yielding a shock-induced separation over large portions of the wing, see Fig. 16.2. A grid generated by Airbus Germany of 52 equal-sized blocks with $25 \cdot 41 \cdot 45$ mesh nodes and 38,642 surface points, totalling to approximately 2.4 million points, is employed. A $y^+ \approx 1$ is guaranteed over most of the wing surface. The transition location is prescribed according to given data. The computations are performed with FLOWer, LLR and LEA k-ω are compared to the baseline Wilcox model. The pressure distribution in selected wing sections are given in Fig. 16.3. While the pressure side including the rear loading is captured very well by all models, a clear improvement in the prediction of the shock location on the suction side from Wilcox over LEA to LLR k-ω is visible in the mid-span section. However, in the most outboard section, the shock appears to be somewhat smeared out, probably owing to an insufficient grid resolution in this region. Thus, it is not possible to make a clear judgement whether LLR or LEA k-ω yields a superior prediction here. Nonetheless, the enhanced predictive accuracy with respect to the Wilcox model is evident.

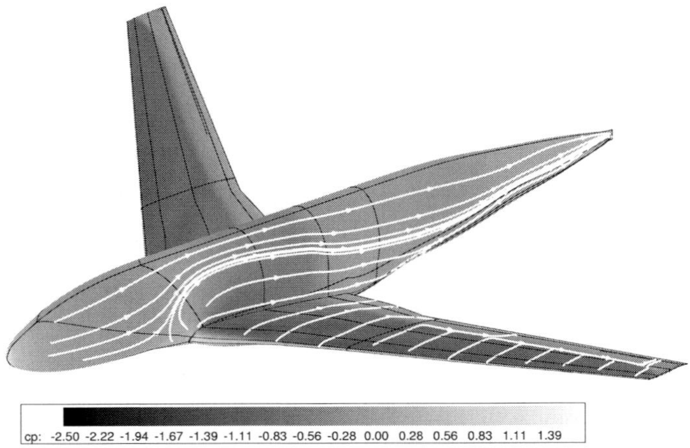

Fig. 16.2. UHCA: Surface pressure distribution and streamlines

16.4.3 Wing-Body Configuration at High Angle-of-Attack

The third testcase is a wing-body configuration (WBC) in the subsonic regime at a high angle-of-attack, where the flow on the upper surface of the wing is partially separated, see Fig. 16.4. The inflow conditions are set to $Ma = 0.2$, $Re = 2.7 \cdot 10^6$ and $\alpha = 10°$. This case is particularly interesting as a streakline picture of the wing is available. Computations are performed with both FLOWer and TAU. The structured grid used consists of 48 equal-sized blocks with $25 \cdot 45 \cdot 45$ points and 33,882 surface points, yielding approximately 2.4 million points. The hybrid grid also contains about 2.4 million points. Again, for both grids, which were supplied by Airbus Germany, a $y^+ \approx 1$ is guaranteed over most of the wing surface. Since no transition locations are given, all computations are performed in a fully turbulent manner. Results presented here include Wilcox and LLR k-ω with FLOWer as well as RQEVM + Wilcox k-ω with TAU.

Fig. 16.5 compares the computed surface pressure distribution and streamlines with an experimental oil flow picture. The measurements show a pocket of separated flow in the outer third of the wing. Both LLR and RQEVM predict the extent of the separation quite correctly, whereas Wilcox k-ω only displays a tiny separation near the trailing edge. However, not even the more advanced models deliver the correct flow topology, thus, a quantitative agreement cannot be expected, see Fig. 16.6. As the grid has to be considered as comparatively coarse for such computations and the effects of transition are neglected, no final conclusion on the accuracy of the turbulence models can be drawn here. Furthermore, it becomes evident that correct predictions are harder to achieve in high-lift flows as compared to the transonic regime, which mainly has to be attributed to the more complex transition patterns which can hardly be prescribed (as seen in the oil flow picture, Fig. 16.5) and the fact that larger laminar regions are encountered. It should be noted, though, that even here, the advanced models' predictions do outperform the standard $k-\omega$ approach.

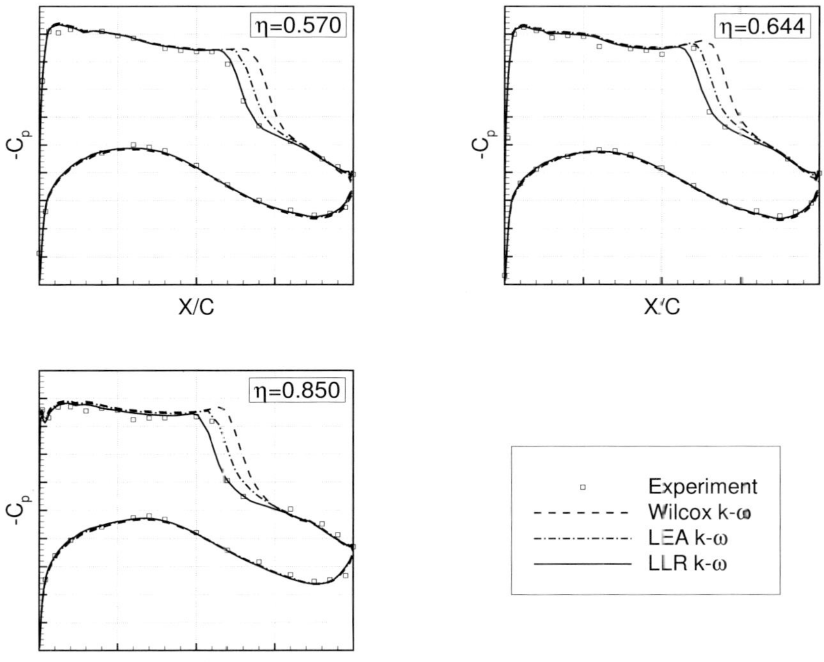

Fig. 16.3. UHCA: Pressure distribution in selected wing sections

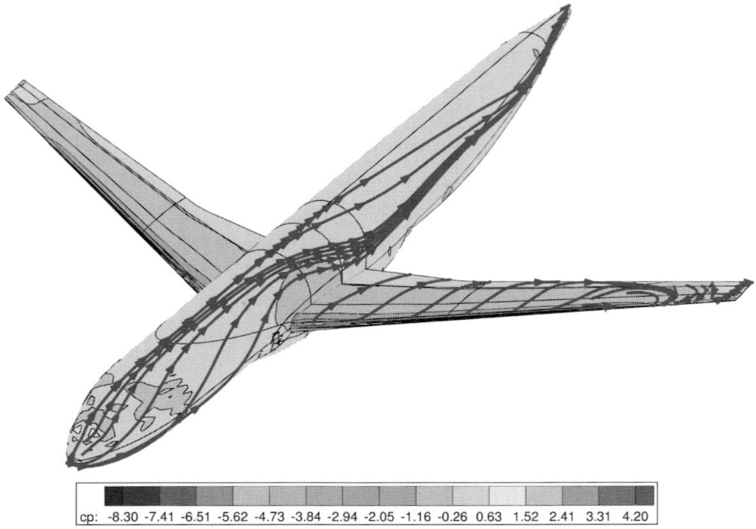

Fig. 16.4. WBC: Surface pressure distribution and streamlines

16.4.4 Wing-Body-Pylon-Nacelle Configuration

This geometry (WBPN), which has been a mandatory testcase in the European validation project AVTAC [19], is presented to demonstrate the applicability of the advanced turbulence models to configurations of practical interest. Results are reported for typical transonic conditions, the inflow is set to $Ma = 0.8$, $Re = 10.8 \cdot 10^6$ per meter and $\alpha = 2.2°$. A grid supplied by Airbus Germany consisting of approximately 5.3 million nodes and 78 blocks largely varying in size is employed. A $y^+ \approx 1$ can be kept over most parts of the geometry, only in the rear part of the fuselage higher y^+ have to be tolerated. The results are obtained with FLOWer using LEA k-ω, computations by Airbus Germany with FLOWer and Wilcox k-ω as well as L-EARSM + Wilcox k-ω from [19] are included for comparison. Owing to the rather complex geometry, no transition is set.

The surface pressure distribution and the streamlines, see Fig. 16.7, show a strong shock on the suction side of the wing, however, the flow does not separate. Furthermore, the influence of the pylon and the nacelle on the shock structure are clearly visible. Turning the attention to the pressure distribution in selected wing sections, Fig. 16.8, a generally good agreement with the experiment can be seen. LEA k-ω determines the shock slightly upstream of L-EARSM + Wilcox k-ω while Wilcox k-ω predicts a shock location down-

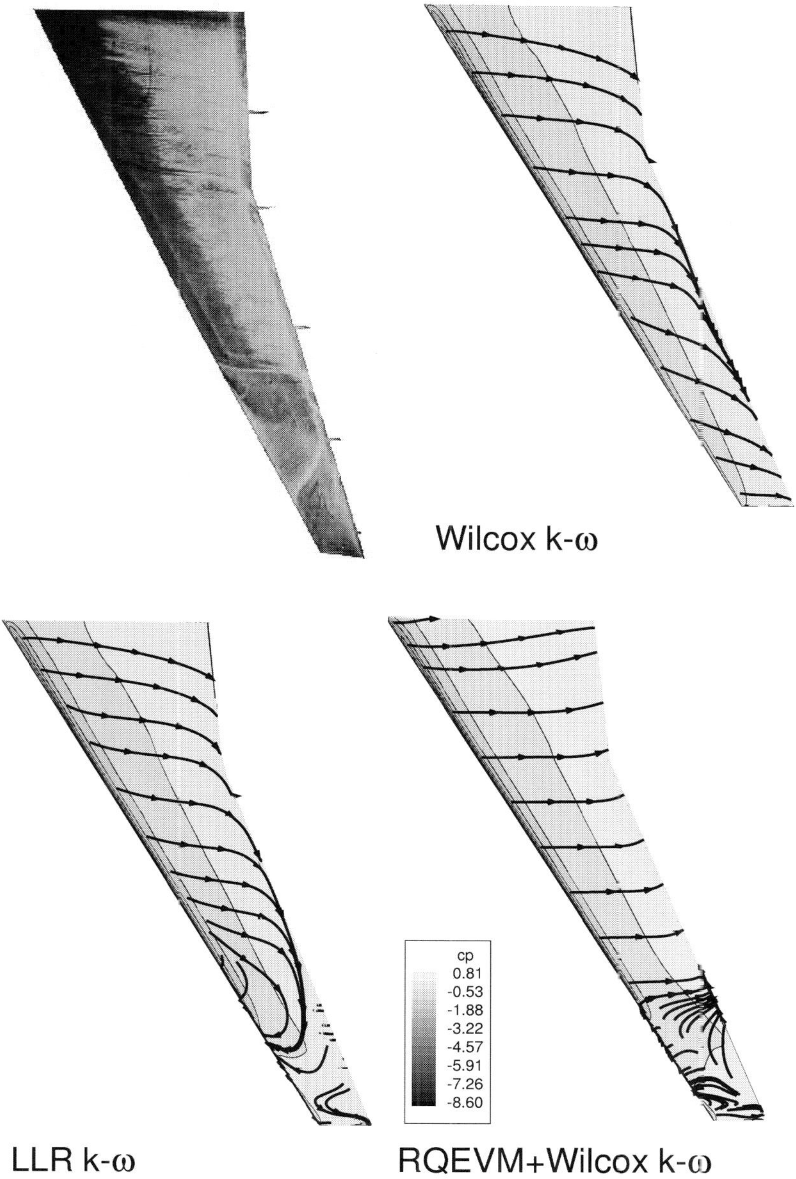

Fig. 16.5. WBC: Surface pressure distribution and streaklines on the wing, comparison with experimental data (top left)

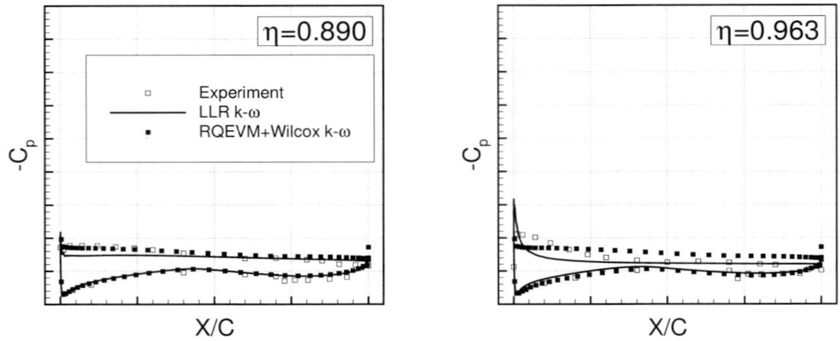

Fig. 16.6. WBC: Pressure distribution in selected wing sections

stream of the linearized EASM. However, the differences remain small, hinting at a weak shock-boundary-layer interaction. This is confirmed by the flow remaining attached over the whole suction side of the wing.

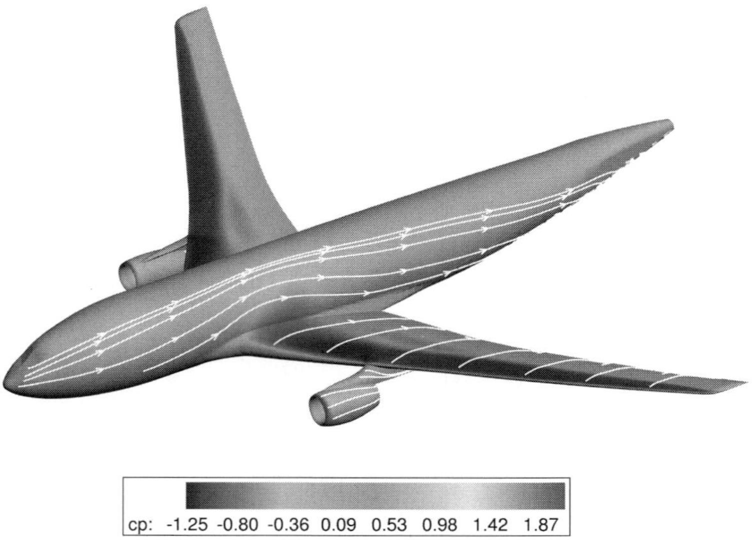

Fig. 16.7. WBPN: Surface pressure distribution and streamlines

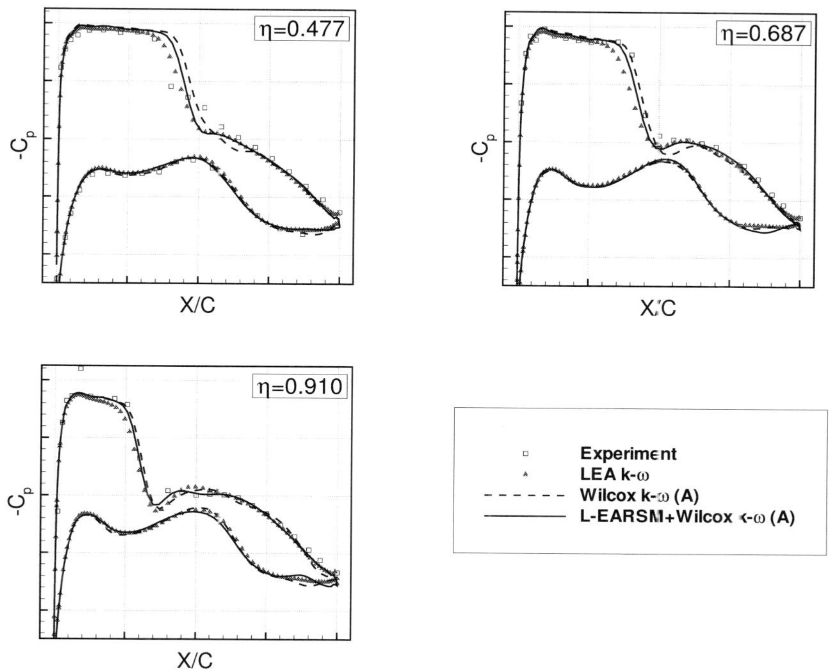

Fig. 16.8. WBPN: Pressure distribution in selected wing sections, comparison with data from Airbus Germany (A)

16.5 Performance Issues

In order to judge the industrial applicability of the turbulence models under consideration, their respective computational surplus over standard approaches has to be quantified, a few exemplary figures will be given here.

Table 16.1. WBC: Comparison of performance rates for the structured FLOWer code on a CRAY T3E-900

Block number and size	48 - 50625 Pts.		
Model	Wilcox	LEA	LLR
Performance per PE [MFLOPS]	53.9	55.6	58.0
Total performance [GFLOPS]	2.64	2.73	2.84

Looking at the linear models in the structured FLOWer code, it turns out that in terms of additional CPU costs, the LLR $k - \omega$ model requires approximately 25% more time per iteration than Wilcox $k - \omega$ and LEA $k - \omega$ about 17%. Additionally, it should be mentioned that SALSA only requires approximately 3% more CPU time than the original SA model [15]. For the two- as well as the one-equation models, the increase in memory requirements is not significant. The subsonic wing-body configuration was also investigated concerning the performance figures on a massively parallel architecture, viz. a CRAY T3E-900. Table 16.1 summarizes the achieved performance rates obtained with three different k-ω models. The figures also demonstrate that the more elaborate algorithms of the advanced models are in part counterbalanced by the higher performance rates achievable.

Turning the attention to the non-linear models in the unstructured TAU solver, performance measurements, as yet performed on the RAE aerofoil only, show that RQEVM costs about 6% and EARSM about 10% more than Wilcox k-ω. Expectedly, this is less than the values determined for comparable models in a structured solver, owing to the fact that the computational effort for turbulence modelling constitutes a smaller fraction of the total cost in an unstructured code.

Finally, it should be mentioned that the use of models independent of topographical information, i.e. local models turned out to be advantageous, since the computation of the wall distance can be very tedious in complex three-dimensional configurations.

16.6 Conclusion

Moving towards the routine use of Navier-Stokes methods in aerodynamic design, the adequate representation of turbulent effects constitutes an absolute necessity. Thus, the improved modelling of turbulence formed an integral part of the MEGAFLOW project throughout its course and is continued in subsequent projects such as the ongoing TAURUS programme. At the start of MEGAFLOW, algebraic models and the standard Wilcox k-ω two-equation model formed the backbone of applied viscous simulations. Owing to their unsatisfactory predictive accuracy, the development and transfer of more advanced approaches from academia to industry became an important issue in the project.

To this end, a variety of enhanced one- and two-equation models have been either taken from the literature or developed, integrated and validated in FLOWer and, more recently, also in the TAU code. Starting from one-equation and enhanced linear two-equation approaches, the focus gradually shifted towards the more general non-linear EASM, with their linear truncations forming an intermediate stepping stone.

The validation exercises performed, ranging from simple profiles to three-dimensional configurations featuring a high degree of physical and geometric complexity, demonstrate the capabilities of the integrated turbulence models. In general, the application of enhanced modelling concepts leads to a predictive accuracy that has improved significantly as compared to the standard approaches available at the start of MEGAFLOW. Table 16.2 gives a qualitative rating for high-lift and cruise-flight conditions, respectively. Furthermore, as the models presented imply only a tolerable computational overhead, their use in the industrial design process can be advocated, especially when their applicability in parallel algorithms is ensured. This will become of even bigger importance in the future, since flow solver parallelization (whether MPP or PVP) is the only feasible strategy to attack large-size problems.

Table 16.2. Qualitative rating of the EVM available in FLOWer and TAU (Predictive performance indicator: -: poor, 0: average, +: fair, ++: good)

class	model	high lift	cruise flight	remarks
one-eq.	SA	0	0	non-local
	SALSA	0	0	non-local
standard two-eq.	Wilcox	-	0	
advanced two-eq.	LLR	+	-	
	Wilcox+SST	+	-	non-local
	Menter SST	-	+	non-local
linearized EASM	LEA	0	+	
	L-EARSM	0	+	
non-linear EASM	RQEVM	0	++	
	EARSM	0	++	

Additionally, it should be noted that the robustness of the numerical algorithm can be significantly increased by means of physically consistent modelling, e.g. by enforcing the length-scale variable to obey an integral formulation of the Schwarz inequality resulting from realizability constraints. Such a limiter has been integrated in both FLOWer and TAU and proved very helpful.

However, a few further remarks seem to be in order here. While cruise-flight simulations can be computed quite accurately today, albeit not with the exactness needed for drag prediction, it has to be stated that in high-lift flows, quantitative agreements with experimental data are much harder to achieve than in transonic conditions. Nevertheless, even a qualitative improvement of the flow simulation accuracy is useful to the aerodynamic design. Enhanced turbulence modelling can help to propel the evaluation of such systems, especially if the general behaviour of a turbulence model has been determined on known cases. However, contrary to earlier aspirations, as no universal approach is available, it might prove necessary to use several computations with different models and the design engineer's expertise to judge a new configuration.

While turbulence modelling efforts in MEGAFLOW concentrated on steady simulations, future developments will also include unsteady flows. Suggestions of hybrid methods such as Detached-Eddy Simulation (DES) have shown promising results [20], however, whether DES will become a pillar of practical aerodynamic computations soon remains a controversial issue [2]. Additionally, measures dedicated to enhance efficiency and robustness, such as generalized boundary conditions [21] or interation schemes ensuring positivity [22] will be investigated. Finally, as also the results presented here indicate, the pressing need for adequate engineering methods for transition prediction should be mentioned. Thus, the effort towards accurate and reliable viscous flow prediction methods will continue.

Acknowledgements

This work was mainly sponsored by the German Ministry of Education and Research (BMBF) under the umbrella of the MEGAFLOW project and the European Commission in the 5th Framework Programme TAURUS. The three-dimensional FLOWer computations presented here were calculated on the CRAY T3E of ZIB Berlin, the TAU results were obtained on the SGI Origin 3800 at Poznan Supercomputing Center. Their respective support is gratefully acknowledged. Finally, the authors would like to thank DLR and Airbus Germany for continued cooperation and support.

References

1. J.B. Vos, A. Rizzi, D. Darracq, E.H. Hirschel: Navier-Stokes Solvers in European Aircraft Design. Progress in Aerospace Sciences **38** (8), 2002, pp. 601-697.
2. P.A. Durbin: A Perspective on Recent Developments in RANS Modeling. In: W. Rodi, N.Fueyo (Eds.): Engineering Turbulence Modelling and Experiments 5, Elsevier, Amsterdam, 2002, pp. 3-16.
3. D.C. Wilcox: Turbulence Modeling for CFD. DCW Industries, Inc., La Cañada, CA, USA, 1993.
4. F.R. Menter: Two-Equation Eddy-Viscosity Turbulence Models for Engineering Applications. AIAA Journal **32** (8), 1994, pp. 1598-1605.
5. P.R. Spalart: Trends in Turbulence Treatments. AIAA Paper 2000-2306, Fluids 2000, Denver, USA, 2000.
6. T. Rung, F. Thiele: Computational Modelling of Complex Boundary-Layer Flows. In: Proc. 9th Intl. Symposium on Transport Phenomena in Thermal-Fluid Engineering, Singapore, 1996.

7. T. Rung, H. Lübcke, M. Franke, L. Xue, F. Thiele, S. Fu: Assessment of Explicit Algebraic Stress Models in Transonic Flows. In: W. Rodi, D. Laurence (Eds.): Engineering Turbulence Modelling and Experiments 4, Elsevier, Amsterdam, 1999, pp. 659-668.
8. S. Wallin, A:V. Johannson: An Explicit Algebraic Reynolds Stress Model for Incompressible and Compressible Turbulent Flows. Journal of Fluid Mechanics **403**, 2000, pp. 89-132.
9. S. Wallin: An Efficient Explicit Algebraic Reynolds Stress k-ω Model (EARSM) for Aeronautical Applications. FFA TN 1999-71, 1999. In: S. Wallin: Engineering Turbulence Modelling for CFD with a Focus on Explicit Algebraic Reynolds Stress Models. Dissertation, KTH Stockholm, 2000.
10. J.C. Kok: Resolving the Dependence on Free-Stream Values for the k-ω Turbulence Model. AIAA Journal **38** (7), 2000, pp. 1292-1295.
11. P.R. Spalart, S.R. Allmaras: A One-Equation Turbulence Model for Aerodynamic Flows. AIAA-Paper 92-0439, 30th AIAA Aerospace Sciences Meeting, Reno, USA, 1992.
12. U. Bunge, T. Rung, M. Schatz, F. Thiele: Restatement of the Spalart-Allmaras Eddy-Viscosity Model in a Strain-Adaptive Formulation. To be published as a Technical Note in the AIAA Journal, 2003.
13. M. Franke, T. Rung, E. Elsholz, P. Aumann, F. Thiele: Numerical Simulation of Three-Dimensional Transonic Flows Using Advanced Turbulence-Transport Models. In: W. Nitsche, H.-J. Heinemann, R. Hilbig (Eds.): New Results in Numerical and Experimental Fluid Dynamics II, Notes on Numerical Fluid Mechanics, Vol. 72, Vieweg, Braunschweig, 1999, pp. 154-161.
14. E. Monsen, M. Franke, T. Rung, P. Aumann, A. Ronzheimer: Assessment of Advanced Transport-Equation Turbulence Models for Aircraft Aerodynamic Performance Prediction. AIAA Paper 99-3701, 30th AIAA Fluid Dynamics Conference, Norfolk, USA, 1999.
15. M. Franke, T. Rung, M. Schatz, F. Thiele: Numerical Simulation of High-Lift Flows Employing Improved Turbulence Modelling. ECCOMAS 2000, 11.-14.9.2000, Barcelona, Spain.
16. M. Franke: Untersuchung zum Potential höherwertiger Turbulenzmodelle für den aerodynamischen Entwurf. Dissertation, TU Berlin, 2003.
17. P.H. Cook, M.A. McDonald, M.C.P. Firmin: Aerofoil RAE 2822 — Pressure Distributions and Boundary Layer and Wake Measurements. In: AGARD AR-138, 1979.
18. R. Rudnik: Untersuchung der Leistungsfähigkeit von Zweigleichungs-Turbulenzmodellen bei Profilumströmungen. DLR Forschungsbericht 97-49, Cologne, 1997.
19. A. Gould, J.-C. Courty, M. Sillen, E. Elsholz, A. Abbas: The AVTAC Project — A Review of European Aerospace CFD. ECCOMAS 2000, 11.-14.9.2000, Barcelona, Spain.
20. K.D. Squires, J.R. Forsythe, S.A. Morton, W.Z. Strang, K.E. Wurtzler, R.F. Tomaro, M.J. Grismer, P.R. Spalart: Progress on Detached-Eddy Simulation of Massively Separated Flows. AIAA Paper 2002-1021, 40th AIAA Aerospace Sciences Meeting, Reno, USA, 2002.
21. H. Grotjans, F.R. Menter: Wall Functions for General Application CFD Codes. In: Papailiou K.D. et al.: Computational Fluid Dynamics '98, Proc. Fourth European Computational Fluid Dynamics Conference, Athens, Greece; Wiley, Chichester, 1998, pp. 1112-1117.

22. P. Eliasson, S. Wallin: A Robust and Positive Scheme for Viscous, Compressible Steady State Solutions with Two-Equation Turbulence Models. FFA TN 1999-81, 1999. In: S. Wallin: Engineering Turbulence Modelling for CFD with a Focus on Explicit Algebraic Reynolds Stress Models. Dissertation, KTH Stockholm, 2000.

17

Large-Eddy Simulation of Attached Airfoil Flow

Qinyin Zhang, Matthias Meinke, and Wolfgang Schröder

Aachen University, Institute of Aerodynamics, Wüllnerstr. zw. 5 u. 7, D–52062 Aachen, Germany,
office@aia.rwth-aachen.de

Summary. A Large-eddy simulation version of the FLOWer code is introduced to compute attached flow around a quasi three-dimensional airfoil at a Mach number $Ma = 0.088$, Reynolds number $Re = 8 \times 10^5$, and an angle of attack of $3.3°$. For the treatment of the subgrid-scale stresses the MILES approach is chosen. The visualization of instantaneous flow fields shows the typical flow features such as the streaky structures in the near-wall region to be well resolved and the overall agreement of the computational results with experimental data to be satisfactory.

17.1 Introduction

Advanced numerical simulation tools are of increasing importance for the aircraft industry especially when nonlinear flow phenomena and their impact on the overall flow field have to be evaluated. The most commonly used approach is based on the approximate solution of the Reynolds averaged Navier-Stokes (RANS) equations. However, there are still flow problems which are difficult to tackle using RANS methods, e.g., flows which encounter an adverse pressure gradient and exhibit separation. Large-eddy simulations (LES) have successfully been applied for such flow problems for academic configurations over the last decades. With rapidly growing performance of modern supercomputers LES appears as a potential tool to analyze some problems encountered in the design of the next generation aircraft. For instance, the prediction of the radiated noise from an high-lift multi-element airfoil in landing configuration is of industrial interest. In this case, LES can be used to compute the detailed turbulence structure of the near wall flow field to feed the acoustic source terms of some kind of acoustic analogy.

In an LES the large scales of the turbulent motion are directly resolved while the small scales are modeled. Since the small structures exhibit predominantly dissipative characters, LES in conjunction with relatively simple subgrid scale models is expected to be a general concept to determine a wide range of turbulent flows. In recent years, large-eddy simulations have been used to predict flows at turbine blades, trailing-edge flows, shock-boundary layer interactions, free shear flows and so forth. However, although an LES of the flow over a complete fighter aircraft has been published lately it is fair to say that an LES at a Reynolds number typically encountered in aeronautical engineering applications still poses quite a challenge.

The flow over an airfoil at high Reynolds numbers ($Re \approx 10^6$) falls into this category. This kind of flow configuration has been investigated within the framework of the LESFOIL project, a joint European research project, which dealt with the Aerospatial A-airfoil at a Reynolds number of $Re = 2.1 \times 10^6$ and an angle of attack $\alpha = 13.3°$ [3] [5] [7]. The experience gathered in the simulations of this flow configuration showed the resolution of the computational mesh to possess the greatest influence on the findings while the subgrid-scale (SGS) model appears to play a subordinate role. When the resolution requirements of an LES for a wall bounded flow problem are met, the substantial features of the airfoil flow can be quantitatively predicted [7].

In this study, the applicability of the LES concept to high Reynolds number airfoil flows and its capability to capture the most important flow phenomena are demonstrated for a $Ma = 0.088$ airfoil flow at a Reynolds number based on the chord length $Re = 8 \times 10^5$ for an angle of attack $\alpha = 3.3°$.

17.2 LES Method

The governing equations are the Navier-Stokes equations for an ideal gas filtered by a low-pass filter of width Δ, which corresponds to the local average in each cell volume

$$\frac{\partial \overline{Q}}{\partial t} + \frac{\partial \overline{E}_a}{\partial \xi} + \frac{\partial \overline{F}_a}{\partial \eta} + \frac{\partial \overline{G}_a}{\partial \zeta} = \frac{1}{Re_0}\left(\frac{\partial \overline{E}_d}{\partial \xi} + \frac{\partial \overline{F}_d}{\partial \eta} + \frac{\partial \overline{G}_d}{\partial \zeta}\right) . \qquad (17.1)$$

The simulation expects the resulting subgrid scale (SGS) stresses to be at least one order of magnitude larger than the truncation error. Although the analysis shows that the SGS stresses are of the order $\mathcal{O}(\Delta^2)$, it is generally accepted that a numerical method of at least second-order accuracy is sufficient to perform an LES. It is, however, crucial to minimize the amount of numerical dissipation of the scheme. Since the turbulent flow is characterized by strong interactions between various scales of motion, schemes with a large amount of artificial viscosity significantly impair the level of energy distribution governed by the small-scale structures and therefore distort the physical representation of the dynamics of small as well as large eddies. It has been shown that a mixed central-upwind AUSM (advective upstream splitting method) scheme with low numerical dissipation could remedy this problem [6] which is why a second-order accurate method is used in this study. The subgrid scale stresses are approximated using the MILES (monotone integrated large-eddy simulation) approach [1].

The applied AUSM scheme is a slight variation of the original method introduced by Liou and Steffen [4]. The convective terms

$$\frac{\boldsymbol{E}_a}{J} = \underbrace{\begin{pmatrix} \rho U \\ \rho U u \\ \rho U v \\ \rho U w \\ \rho U(e + p/\rho) \end{pmatrix}}_{\boldsymbol{E}_a^K} + \underbrace{\begin{pmatrix} 0 \\ p\xi_x \\ p\xi_y \\ p\xi_z \\ 0 \end{pmatrix}}_{\boldsymbol{E}_a^P} \tag{17.2}$$

are expressed by inserting the local speed of sound

$$\boldsymbol{E}_a^K \Big|_{i\pm\frac{1}{2},j,k} = \frac{1}{4} \begin{pmatrix} 2\,\mathcal{F}_\pm \\ \mathcal{F}_\pm(u_- + u_+) + |\mathcal{F}_\pm|\,(u_- - u_+) \\ \mathcal{F}_\pm(v_- + v_+) + |\mathcal{F}_\pm|\,(v_- - v_+) \\ \mathcal{F}_\pm(w_- + w_+) + |\mathcal{F}_\pm|\,(w_- - w_+) \\ \mathcal{F}_\pm(h_- + h_+) + |\mathcal{F}_\pm|\,(h_- - h_+) \end{pmatrix}_{i\pm\frac{1}{2},j,k} \tag{17.3}$$

$$\text{with } \mathcal{F}_\pm = \frac{1}{2}(\widehat{Ma}_- + \widehat{Ma}_+)(\widehat{\rho a}_- + \widehat{\rho a}_+) + \frac{1}{2}\left|\widehat{Ma}_- + \widehat{Ma}_+\right|(\widehat{\rho a}_- - \widehat{\rho a}_+)\,,$$
$$h = e + p/\rho \quad .$$

These expressions are determined on the cell interfaces to second-order accuracy via a quadratic MUSCL interpolation of the primitive flow variables. The pressure part of the convective term

$$\boldsymbol{E}_a^P = \begin{pmatrix} 0 \\ \xi_x(\widetilde{p}_+ + \widetilde{p}_-) \\ \xi_y(\widetilde{p}_+ + \widetilde{p}_-) \\ \xi_z(\widetilde{p}_+ + \widetilde{p}_-) \\ 0 \end{pmatrix} \tag{17.4}$$

contains an additional expression, which is scaled by a weighting parameter χ, that represents the rate of change of the pressure ratio with respect to the local Mach number

$$\widetilde{p}_\pm = \widehat{p}_\pm\left(\frac{1}{2} \pm \chi\,\widehat{Ma}_\pm\right) \quad . \tag{17.5}$$

The parameter χ determines the numerical dissipation of the scheme. To avoid, on the one hand, too large a numerical dissipation ($\chi = 0.5$) and on the other hand, oscillations that could lead to unstable solutions ($\chi = 0$) the parameter χ was chosen in the range $0 \le \chi \le 1/96$ [6].

The discretization of the friction and heat conduction expressions plays a less important role in turbulent flows than that of the nonlinear inviscid terms. These viscous terms are approximated by second-order accurate central differences. An explicit five-step Runge-Kutta scheme is used for the temporal integration using the coefficients $\alpha_l = (\frac{1}{4}, \frac{1}{6}, \frac{3}{8}, \frac{1}{2}, 1)$.

On the body surface the no-slip condition is imposed and an adiabatic wall is prescribed. At the far-field non-reflecting boundary conditions [9] are implemented. To minimize the numerical reflections, which are generated by a non-perfect pressure relaxation, a sponge-layer zone is added in which a smoothing term based on the deviation of the numerical and the analytical far-field solution is computed. In the spanwise direction periodic boundary conditions are specified.

In the following, we discuss first the results of the flow past a cylinder and subsequently, the flow field around an airfoil is considered.

17.3 Results and Discussion

17.3.1 Cylinder Flow

To validate the LES version of the FLOWer code the wake of the flow past a three-dimensional circular cylinder at $Re = 3900$ is computed. The mixed C- and O-type mesh consists of 8 blocks with 128 and 48 cells in the circumferential and spanwise direction, and the minimum normal space step is 1.3 percent of the diameter of the cylinder. There are 196 cells placed in the wake, which results in a total number of grid points of 1,386,308. These mesh parameters are summarized in Tab. 17.1.

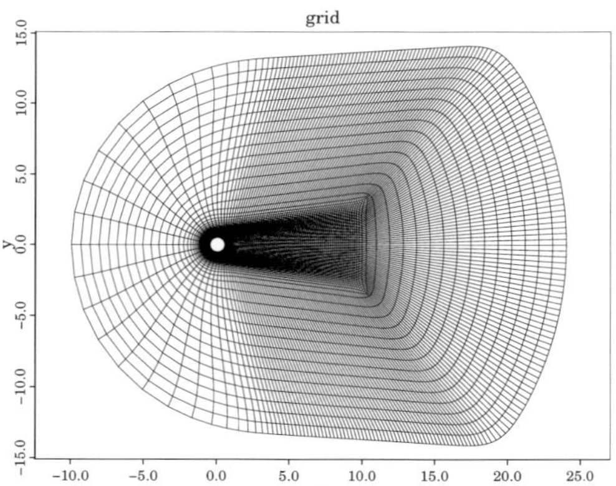

Fig. 17.1. Computational mesh for the turbulent cylinder wake flow at $Re = 3900$ (every second grid point is shown).

The numerical simulations are preformed using the FLOWer code, which contains the AUSMDV scheme for the Euler fluxes, and the new reformulated LES version, in which among other features the AUSM-central scheme has been implemented. The visualization of the flow field in Fig. 17.2 shows

Table 17.1. Computational region and number of grid points, Cylinder (Re=3900).

L_X/D	L_Y/D	L_Z/D	$N_x \times N_y$	N_z	Total no. of grid points
-10 ... 25	-15 ... 15	0 ... 3	28,292	49	1,386,308

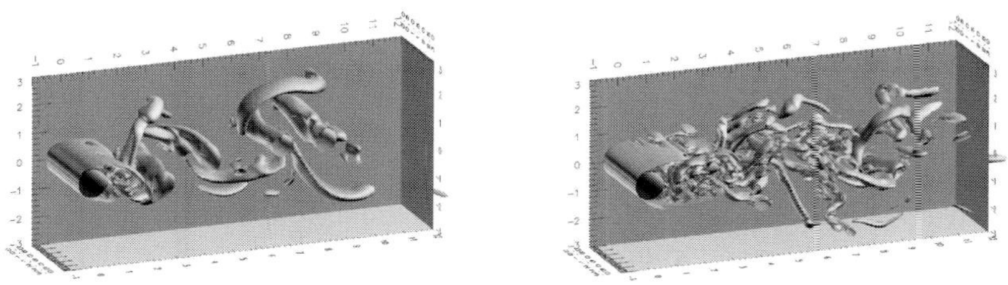

Fig. 17.2. Instantaneous vortex structures in the wake ($t = 650$), AUSMDV (left) and AUSM-central (right) based solutions.

the reduction of the numerical dissipation of the AUSM-central compared to the AUSMDV method. In the AUSMDV solution hardly any small turbulent structures exist, they are simply smeared out by the numerical dissipation (Fig. 17.2), whereas the AUSM-central finding in Fig. 17.2 evidences on exactly the same grid the intricacy of the vortical field, which consists of the main rollers and the streamwise vortices.

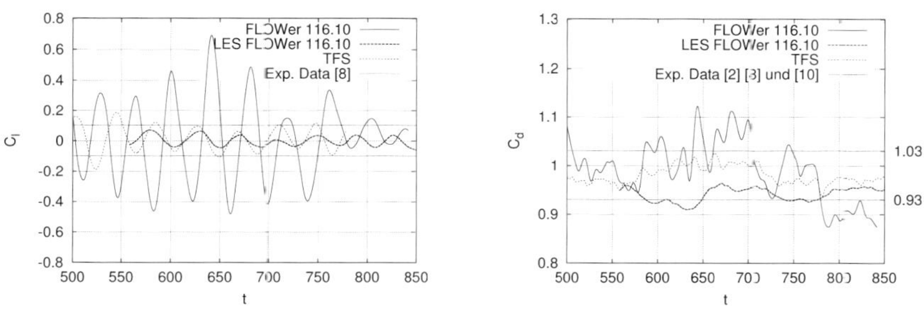

Fig. 17.3. Time history of the lift (left) and drag coefficients (right) of the cylinder.

Fig. 17.3 shows the time history of the aerodynamic coefficients. The label TFS denotes an algorithm, which has been developed at the Aerodynamisches Institut [6] and which is also based on the AUSM-central discretization. Note the pronounced amplitudes of the AUSMDV solution. These strong oscillations are caused by the separation of the unphysically coarse turbulent structures from the cylinder surface. With less numerical dissipation induced by the modified AUSM scheme, the vortex structures in the flow field behind the cylinder are finer, which is why the forces exerted on the cylinder are smaller. The time histories of the TFS and LES Flower solutions fall within the range of the experimental values for the lift and drag coefficient [2] [8] [10].

17.3.2 Airfoil Flow

To perform an LES for the flow over the quasi 3D-airfoil a block-structured mesh was generated whose inner block is a C-type and its outer block an O-type grid. Since the Reynolds number based on the chord length is $Re = 8 \times 10^5$ a spanwise extension of the mesh of 0.3 percent chord suffices. Experience has shown that an LES for the flow over an airfoil requires a resolution of the near wall cells in inner law scaling in the range of $\Delta x^+ \approx 100, \Delta y^+ \approx 2$ und $\Delta z^+ \approx 20$, where x, y, z denote the streamwise, normal, and spanwise coordinates, respectively. Fig. 17.4 evidences that

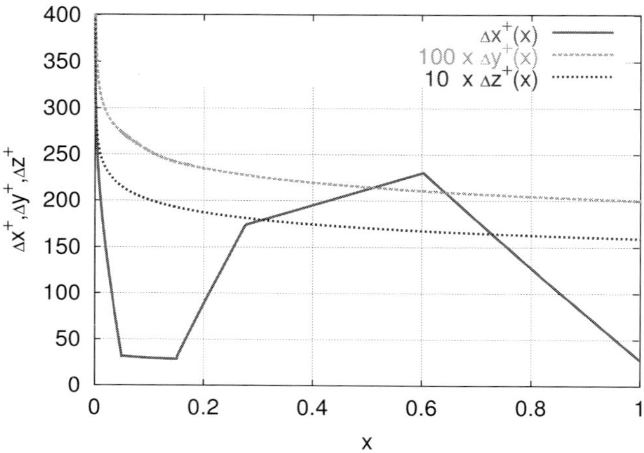

Fig. 17.4. Grid resolutions in inner law scaling in the streamwise, normal, and spanwise directions.

this requirement is approximately satisfied over the whole airfoil. The complete mesh contains 7,323,651 grid points, 1686 of which are distributed on the airfoil, 373 in the wake, 197 in the normal, and 17 in the spanwise direction (Fig. 17.5). To capture the laminar-turbulent transition, which is known from experiments to occur just downstream of the nose region, the mesh was clustered in the area $0 \leq x \leq 0.2$ (Figs. 17.4, 17.5). Furthermore, a high resolution was generated near the trailing edge to ensure that the vortical structures in the boundary layers on the upper and lower surface and in the wake are well captured, since these structures on the one hand, contain the history of the boundary layer and as such evidence the quality of the solution and on the other hand, can be understood as input data for, e.g., acoustic source terms to determine airframe noise.

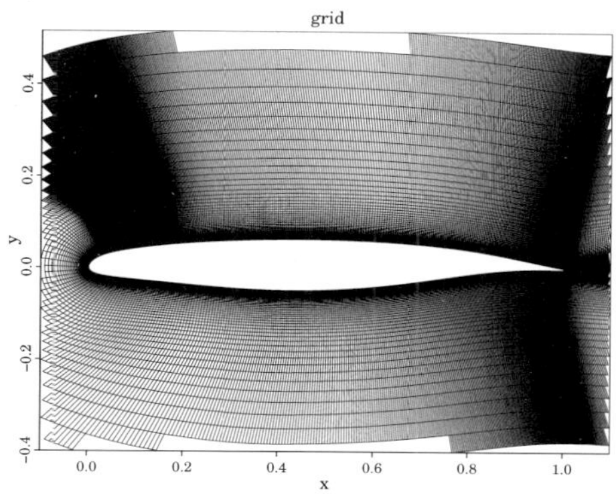

Fig. 17.5. Airfoil Mesh in the x-y plane (every second grid point is shown).

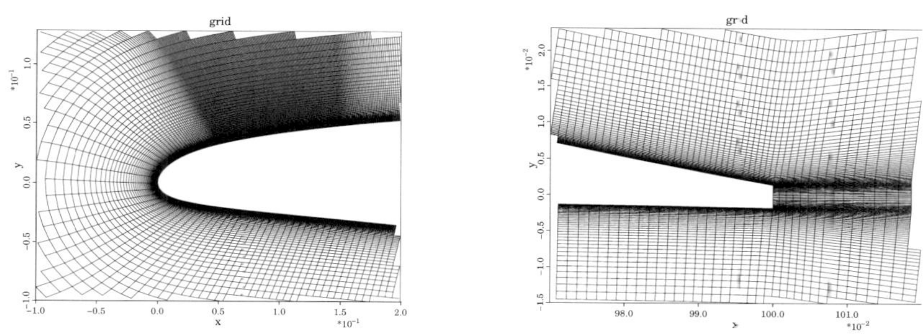

Fig. 17.6. Enlargement of the leading (left) and trailing edge region (right).

When large-eddy simulations over slender bodies at zero or small angle of attack are preformed it is one of the major difficulties to ensure the turbulent flow to stay turbulent, i.e., to avoid laminarization. Since this unphysical phenomenon is caused by too large a dissipation of the numerical scheme, this geometrically simple airfoil flow, which is turbulent on the upper surface over more than 90 percent of its chord length, is an excellent problem to show that the numerical method is well suited to compute attached wall-bounded turbulent flows with a large streamwise extension. The visualization of the flow field will show that the turbulent structures are well resolved by this simulation and that no relaminarization occurs neither on the upper nor on the lower surface.

For the $Ma = 0.088$ flow measurements of the skin-friction and the velocity distribution near the trailing edge have been performed by the Institute of Aerodynamics and Gasdynamics of the University of Stuttgart [11]. A comparison of the time averaged velocity profiles with the experimental data is given in Fig. 7(a). In the near wall region a satisfactory agreement is achieved whereas in the outer part of the boundary layer due to the reduced grid resolution quite a discrepancy of the numerical and experimental distributions occurs. The skin-friction coefficient distribution on the suction side in Fig. 7(b) evidences a little difference since the measurements do not capture the slight increase right at the trailing edge. In Fig. 17.9 the vortical structures in the aft region of the airfoil and in the wake are visualized. The streaky structures in the near wall region are elongated near the trailing edge and spanwise vortices form and grow in the wake region.

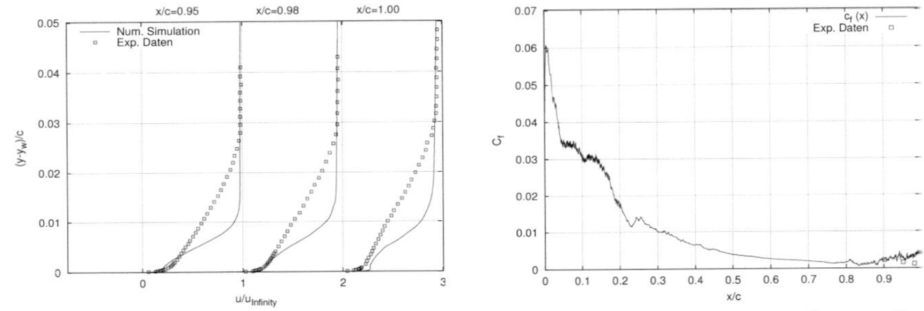

(a) Mean streamwise velocity profiles. (b) Mean skin-friction coefficient C_f

Fig. 17.7. Time averaged flow parameters

17.4 Conclusions and Future Work

In this work, a modified FLOWer version is worked out. A new AUSM formulation for the convective terms and new boundary conditions are implemented in the FLOWer code. A LES computation is performed for an airfoil flow at high Reynolds number. The overall agreement of the numerical results with the experimental data is reasonable. The visualization of the instantaneous flow field shows that the typical turbulent structures are well resolved by this simulation.

Next, the flow field with a strong adverse pressure gradient will be considered. That is, an LES of the flow over an NACA 4412 airfoil at high incidence with a massive separation area near the trailing edge will be performed.

Fig. 17.8. Coherent structures on the airfoil surface.

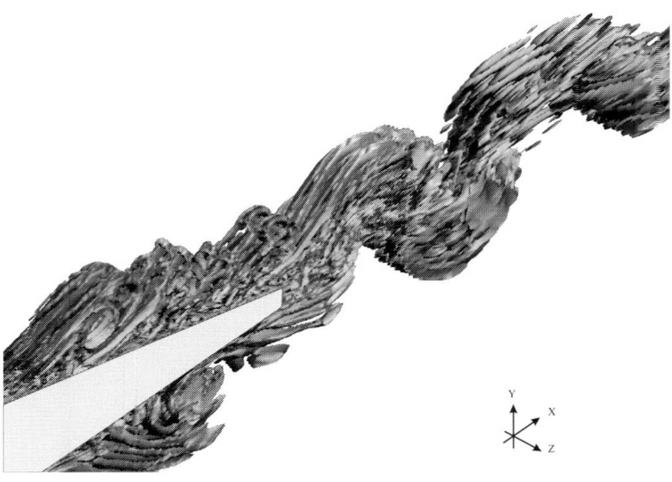

Fig. 17.9. Vortex structures in the wake.

References

1. J. P. Boris, F. F. Grinstein, E. S. Oran, R. L. Kolbe: "New Insights into Large Eddy Simulation". Fluid Dynamics Research. **10**, 1992, pp. 199-228.

2. G.S. Cardell: "Flow past a circular cylinder with a permeable splitter plate". PhD thesis, Graduate Aeronautical Lab., California Inst. of Technology, 1993
3. S. Dahlström, L. Davidson: "Large Eddy Simulation of the Flow Around an Airfoil". AIAA Pap. **2001-0425**, 2001.
4. M. S. Liou, Ch. J. Steffen Jr.: "A New Flux Splitting Scheme". J. Comp. Phys. **107**, 1993, pp. 23-39.
5. I. Mary, P. Sagaut: "Large Eddy Simulation of Flow Around an Airfoil Near Stall". AIAA J. **40(6)**, 2002, pp. 1139-1145.
6. M. Meinke, W. Schröder, E. Krause, Th. Rister: "A Comparison of Second- and Sixth-Order Methods for Large-Eddy Simulations". Computers and Fluids. **31**, 2002, pp. 695-718.
7. C.P. Mellen, J. Fröhlich, W. Rodi: "Lessons from the European LESFOIL project on LES of flow around an airfoil". AIAA Pap. **2002-0111**, 2002.
8. C. Norberg: "Effects of Reynolds number and low-intensity free stream turbulence on the flow around a circular cylinder". Publ. No. **87/2**, Dept. of Applied Thermoscience and Fluid Mech. Chalmers University of Technology, Gothenburg, Sweden, 1987.
9. T. J. Poinsot, S. K. Lele: "Boundary Conditions for Direct Simulations of Compressible Viscous Flows". J. Comp. Phys. **101**, 1992, pp. 104-129.
10. J. Son and T.J. Hanratty: "Velocity gradients at the wall for flow around a cylinder at Reynolds numbers from 5×10^3 to 10^5". J. Fluid Mech. **35**, 1969, pp. 353-368.
11. W. Würz, S. Guidati, S. Herr: "Aerodynamische Messungen im Laminarwindkanal im Rahmen des DFG-Forschungsprojektes SWING+ Testfall 1". Inst. für Aerodynamik und Gasdynamik, Universität Stuttgart, 2002.

18

Transition Prediction for 2D and 3D Flows using the TAU-Code and N-Factor Methods

C. Nebel, R. Radespiel, and R. Haas

Braunschweig Technical University, Institute of Fluid Mechanics, Bienroder Weg 3, 38106 Braunschweig, Germany

Summary. The 3D Navier-Stokes solver TAU is coupled with linear stability analysis methods in order to predict flows including transition due to Tollmien-Schlichting (TS) and crossflow (CF) instabilities. The new simulation capability is investigated for an airfoil and compared with data of 2D boundary layer methods that include transition prediction based on a well-known envelope method and with experiments. The results indicate the levels of grid and residual convergence needed for accurate transition prediction. First applications of transition prediction in 3D for a 1:6 prolate spheroid are discussed. It is shown that transition calculations for fully 3D flows are numerically feasible and yield physically reasonable results for moderate angles of attack.

18.1 Introduction

In many cases high-quality CFD simulations of viscous flows are only possible with detailed knowledge of the transition location. Current practise of most CFD calculations today is to apply procedures that run separately from the main CFD-tool or transition location is extracted from experiments. An important step towards the establishment of reliable and comprehensive design and analysis methods is the integration of transition prediction capabilities into the flow solver.

Unfortunately, transition processes are in no way predetermined in general flows around aerodynamic configurations. A numerically feasible transition prediction method is therefore based on the assumption of particular transition scenarios. A well-known transition scenario for 2D airfoils and wings with low sweep angles describes the break down of 2D waves in the boundary layer (Tollmien-Schlichting waves) as the result of large secondary instabilities within the flow field. Since the amplification of the primary waves takes a large portion of overall transition length before significant secondary waves and break down occur the linear behaviour of primary waves is quite a good measure of transition and this is the justification of the e^N-transition prediction approach. The situation is more complex for 3D flow fields with crossflow in the boundary layer. Strongly nonlinear effects such as saturation occur for the largest modes and one observes a much larger scatter in the attempts to correlate transition with the amplitudes of the primary instabilities. Nevertheless, e^{N}-methods remain state of the art for transition prediction in the aerodynamic design of airfoils and wings, because affordable and robust alternatives are not yet available.

The subject of the present paper is to implement the e^{N}-method in the 3D Reynolds-averaged Navier-Stokes (RANS) solver TAU [1]. We choose to use the RANS boundary layer data as the mean flow for stability analysis. The 3D RANS flow field is also used to define suitable integration paths for calculating the N-factor. The instabilities in streamwise direction are analysed with a database of Stock and Degenhart [2], pure crossflow instabilities are computed with an envelope method developed by Casalis and Arnal [3].

The results of the present numerical approach are compared to experiments in 2D for a flow around a NLF(1)-0416 airfoil with corresponding experiments of Somers [5]. Further verification is presented by using the boundary layer code QICTAP [4] and the results of the panel code XFOIL [6] with boundary layer analysis. Further tests of the method are reported for the NACA 64A010 airfoil at a high angle of attack.

First results of 3D calculations for a 1:6 prolate spheroid are also shown in this paper. The ability to provide boundary layer of the spheroid in the necessary accuracy is analyzed and comparisons to experiments of Meier and Kreplin [7], [8] are presented. These experiments have been used for stability analysis calculations before by Menter and Kreplin [9].

18.2 Solution Methods

18.2.1 Solution of Reynolds-Averaged Navier-Stokes Equations

The DLR TAU Code is based on a finite volume scheme [1]. The Reynolds averaged Navier-Stokes equations are solved on hybrid grids. With prismatic layers in the near-wall region viscous dominated flows can be represented in an efficient and accurate way. The fluxes are calculated with central or various upwind schemes. Convergence acceleration is obtained with residual smoothing and multigrid methods. For turbulent flows the Spalart-Allmaras turbulence model and several two-equation turbulence models are available.

18.2.2 Transition Prediction Method

The streamwise coordinate system (x, y, z) is aligned with the local edge velocity of the undisturbed laminar mean flow such that x is streamwise. The baseflow at a given x position is represented by $(\mathrm{U}, 0, \mathrm{W})$, i.e. parallel flow. We seek the linear growth rates of perturbations

$$r' = r(y) \exp\left[i\left(\alpha x + \beta z - \omega t\right)\right]$$

The complex amplitude function r depends on the surface-normal coordinate y only. The perturbations are governed by the Orr-Sommerfeld equations. These equations provide a local relationship between the frequency ω, the streamwise wave number α, and the wave number β in z-direction. In the spatial stability theory ω is prescribed and α and β are generally complex eigenvalues. A solution of the above equations is possible by introducing assumptions about the growth direction. Here it is assumed that the growth direction is the group velocity direction [10]. The group velocity direction may be approximated by the direction of the boundary layer edge velocity as the angle between these directions is only a few degrees, according to experience from swept wings [10], [11]. That is, $\beta_i = 0$ and amplifications in the direction normal to the external streamline are neglected. For a given boundary layer the frequency and lateral wave number are fixed for any physical mode. Solution of the eigenvalue problem subject to the homogeneous boundary condition yields the complex α at each x-position. The growth rate is given by the imaginary part $-\alpha_i(\omega, \beta, x)$. The N-factor for a physical mode is determined by integrating the growth rate downstream from the neutral point. In the present work, we consider an N-factor based on the maximum growth over either the streamwise waves ($\mathrm{N_{TS}}$) or the real lateral wave numbers ($\mathrm{N_{CF}}$) as follows:

$$\mathrm{N_{TS}} = \max_{\omega}\left[\int_{x_0}^{x} -\alpha_i(\omega, 0, x)dx\right] \quad , \quad \mathrm{N_{CF}} = \max_{\omega}\left[\int_{x_C}^{x} -\alpha_i(\omega, \beta, x)dx\right]$$

Thus TS waves and travelling crossflow instabilities are treated as distinct. Transition as a result of the break down of primary unstable waves is assumed to occur whenever the N-factor exceeds an empirically defined threshold value. For the calculation of growth rates we make use of numerically efficient database methods. The flow is assumed to be incompressible, $\mathrm{M_e} \approx 0$. The stability of the 2D Tollmien-Schlichting waves is analysed with a database method from Stock and Degenhart [2]. The method is based on numerical stability data for a family of generic boundary layers that are obtained as the solutions of the Falkner-Skan equations. The stability results are parameterized as functions of the shape factor H, which is the ratio of the displacement thickness δ^* and the momentum thickness θ, and the Reynolds number based on the momentum thickness Re_θ and the prescribed real frequency ω. The results were stored in a database and this is used in the present work.

The crossflow instabilities are analysed with the database method developed by Casalis and Arnal [3]. The method seeks stability solutions for a given wave angle by projecting the local velocity profile of the boundary layer into the direction of the wave-number vector. This results in temporal amplification factors which can be transformed into spatial growth rates by using Gaster's relation. Casalis and Arnal present empirical definitions of the shape of half parabolas that relate the growth rate for a given frequency to the inflection position of the velocity profile, and to the velocity at the inflection position, referenced by the boundary layer thickness and the edge velocity. This strategy allows a rapid calculation of growth rates to be used in the N-factor integration.

18.2.3 Coupling of RANS Solver and e^{N} Method

The RANS solver and the e^{N} methods are coupled in following steps:

- The integral boundary layer quantities are calculated from the velocity information of the solver. Along distinct integration paths they are used to calculate the input parameters Re_{δ^*} and H for the Tollmien-Schlichting database.
- Velocity profiles in the boundary layer are transferred in a coordinate system in which the first axis is in direction of the group velocity, the second one normal to the surface and the third one orthogonal to them. The profiles of the velocities and their derivatives are non-dimensionalized with the displacement thickness δ^* and the velocity $\mathrm{U_e}$ at the edge of the boundary layer and then used as input for the crossflow database along the integration paths.
- With the databases the envelope functions of $\mathrm{N_{TS}}$ and $\mathrm{N_{CF}}$ for multiple frequencies ω are calculated along the integration paths. At a limiting value indicating transition from the laminar to the turbulent state a transition point is defined. Special care is taken that a unique iterative update of the transition point is provided. Beyond the line of transition points calculated for each integration path turbulence modelling is switched on within the solver.

Following these steps the edge of the boundary layer has to be detected first. Two criteria have been implemented. For the first criterion the velocity at the boundary layer edge is calculated from the wall pressure with the Bernoulli equation for compressible flows. Then the corresponding boundary layer thickness δ is found on a line normal to the surface at the intersection of the velocity profile with $0.99 \cdot \mathrm{U_e}$. If $\mathrm{U_e}$ is not reached, the point with the maximum velocity is taken instead. For the second criterion a diagnostic function derived from the Baldwin-Lomax turbulence model is used. This has been used successfully by Stock [12] and Stock and Haase [13]. At first a point is searched, where the diagnostic function

$$F(x_n, y_n) = y^a \left[\frac{du}{dy}\right]^b$$

has its maximum. Then the boundary layer thickness δ can be calculated as

$$\delta = \varepsilon \cdot y_{n,\max} \qquad \text{with} \qquad y_{n,\max} = y_n|_{F=F_{\max}} \; .$$

Once the edge is found and the corresponding velocity U_e is defined, the displacement thickness and the momentum loss thickness are computed using the trapezoid rule. The shape factor and the Reynolds length referenced with the displacement thickness follow directly

$$H = \frac{\delta^*}{\theta} \quad , \qquad \mathrm{Re}_{\delta^*} = \frac{\delta^* \mathrm{U_e}}{\nu_\mathrm{e}}$$

In 3D flows the amplitude ratios N of the disturbances have to be calculated along the direction of the group velocity vector [10]. Again, this direction is approximated by the direction of the external streamline along the boundary layer edge. Accordingly, a module to integrate streamlines along the surface based on velocities of the boundary layer edge is implemented into the Navier-Stokes solver. For the integration a 4th order Runge-Kutta scheme has been implemented. The velocity $\underline{\mathrm{U}}$ at a distinct point $\underline{\mathrm{X}}$ is calculated with inverse distance weighting, see [14]:

$$\underline{\mathrm{U}} = c_0 \underline{\mathrm{U}}_0 + c_1 \underline{\mathrm{U}}_1 + \ldots + c_n \underline{\mathrm{U}}_n$$

with the velocity $\underline{\mathrm{U}}_0$ in the centre of the cell and $\underline{\mathrm{U}}_1, \ldots, \underline{\mathrm{U}}_n$ as the velocities in the centres of the surrounding cells. The weighting coefficients c_i are calculated from the distance d_i to the point $\underline{\mathrm{X}}$

$$c_i = \frac{1/d_i^2}{\sum_{j=0}^{n} 1/d_j^2} \quad \text{with} \qquad d_i = \|\underline{\mathrm{X}}_i - \underline{\mathrm{X}}\,|$$

where $\underline{\mathrm{X}}_i$ denote the surrounding grid points. Once in a grid cell the integration paths have to be pulled to the surface, because in thicker boundary layers the velocities at the edge are only almost parallel to the wall. The size of the Runge-Kutta step depends on the mean distance of the grid cell to the neighbour cells. A resolution of five steps for this distance was found out to be sufficient. For each cell only the values of the point next to the cell centre are given to the database. For reasons of simplicity the boundary layer quantities for the point $\underline{\mathrm{X}}$ are approximated by the value of the next cell centre.

The numerical update of the transition location requires special attention. Firstly, the iterative flow solutions exhibit transients where the transition location moves either downstream or upstream, depending on the initial solution. That is, the iterative update must allow the definition of a new transition location in previously turbulent flow regions. Secondly, switching on the turbulence model at the predicted transition location will numerically affect the laminar boundary layer just upstream of that point, by the inherent discretization errors of the numerical scheme. This numerical error will in turn have adverse effects on the further transition prediction result if no adequate treatment is introduced into the transition location update. We have therefore developed a new method for extracting reliable transition location estimates from stability results based on RANS mean flow data. The method is based on streamwise linear extrapolation of N-factors from flow regions with reliable laminar mean flow into the transitional or even turbulent locations. The extent of extrapolation in streamwise direction is typically 30 times the local displacement thickness. Note that the e^N-method itself is an extrapolation technique of linear stability results to the transition point upstream of which nonlinear effects dominate in reality. Therefore, a local extrapolation of N-factors is an acceptable modification of the overall e^N method. Further, we use under-relaxation in the iterative update of the numerical transition location.

18.2.4 Code Implementation Issues

The current extension to the DLR TAU code can only be run in a sequential version. For fully parallel calculations the extension has to be adapted to the domain decomposition method. The continuation of wall normal lines and integration paths in neighbouring domains is a challenging problem. The solution is absolutely necessary in order to perform calculations for complex 3D cases with grids providing the suitable resolution at decent turn around times.

18.3 Results

18.3.1 Calculations in 2D for Airfoils

A grid with high resolution of the boundary layer has been created for the NLF(1)-0416 airfoil with the grid generator Centaur [16] in order to obtain velocity profiles from which the integral boundary layer quantities Re_{δ^*} and H can be calculated with high accuracy. The grid is shown in Fig. 18.1. The resolution in the structured part of the grid is 256 cells along the surface and 64 normal to it with a total of 24000 cells. For the current investigations the result of the Navier-Stokes calculation with with the Spalart-Allmaras turbulence model is compared to the solution of the Panel code with boundary layer calculation, XFOIL [6], at $M_\infty = 0.3$.

The pressure distribution for the case $M_\infty = 0.3$, $\mathrm{Re}_\infty = 4 \cdot 10^6$, $\alpha = 2.03°$ with fixed transition $x_{\mathrm{tr,upper}}/c = 0.35$, $x_{\mathrm{tr,lower}}/c = 0.60$ is shown in Fig. 18.2. The results from the TAU Code and XFOIL show only minor discrepancies in the front part of the profile. The resulting shape factor of the boundary layer is displayed in Fig. 18.3 as a comparison between the results of XFOIL, the boundary layer code QICTAP and the TAU code. The boundary layer quantities from XFOIL and QICTAP show a good agreement, although the pressure distribution from XFOIL is slightly different to the one from the TAU-Code which is used to calculate the boundary layer quantities with QICTAP. Detailed investigations about the effects of residual convergence have been also carried out. The shape factor at various timesteps is plotted in Fig. 18.4. It is seen that approximately 12000 iterations are necessary to calculate sufficiently accurate boundary layer data for this case. The effect of grid resolution on the shape factor is seen in Fig. 18.5. 64 cells normal to the surface seem to be sufficient. More details on verification studies are found in Ref. [15].

One sample of the iterative search for the correct transition location of the laminar research airfoil is shown in Fig. 18.6. About four steps with 5000 iterations each are necessary. Experience in finding the right relaxation parameter still has to be gathered with the method. Fig. 18.7 shows the predicted transition locations in comparison to experimental data for various α. The transition location is predicted within a range of 3% of the cord length.

A more challenging problem of 2D transition calculation is given for the flow over NACA 64A010 airfoil at a larger angle of attack. Here one encounters a short laminar separation bubble with turbulent re-attachment. The flow was initialised fully turbulent and the laminar separation bubble seen in Fig. 18.8 was obtained as a converged flow solution with transition predicted at values $N_{TS} = 9$. The flow field of the bubble is displayed in Fig. 18.9. The flow is very sensitive to the transition location, that is, a location too far downstream would result in an unsteady, oscillatory flow field. However, the iterative update of the transition location performs admirably even for this very sensitive flow as seen in Fig. 18.10.

18.3.2 Calculations in 3D for a 1:6 Prolate Spheroid

As a result from the investigations carried out by Meier and Kreplin [7] the experimentally determined transition location of the 1:6 prolate spheroid at various angles of attack for $M_\infty = 0.13$ and $\mathrm{Re}_\infty = 7.2 \cdot 10^6$ is shown in Fig. 18.11. With an increasing angle of attack the transition location on the upper side moves upstream and downstream on the lower side. At $\alpha \geq 10.0°$ a separation occurs in the turbulent part on the leeward side of the spheroid. The grid used for the 3D calculations of the 1:6 prolate spheroid was also generated with Centaur, see Fig. 18.12. It has about 40000 surface nodes with a resolution of 64 cells normal to the surface. The grid consists of 2.6 million prisms in the wall near region and 0.6 million tetrahedra between near-wall region and farfield. The surface pressure distribution and the integration paths calculated with velocities from boundary layer edge for $M_\infty = 0.30$, $\mathrm{Re}_\infty = 7.2 \cdot 10^6$ and $\alpha = 0.0°$ are shown in Fig. 18.13. The streamline tracing algorithm provides integration paths in a good quality. The transition location is defined at a point where the N-factors calculated with the TS-database become larger than 7.2. The slightly non axis-symmetric behaviour indicates the effect of discretization error that the unstructured method produces throughout the overall transition calculation process. The prescribed N-factor is an empirical value, which was determined by correlating theoretical results with experiments. The corresponding investigations were carried out by Menter and Kreplin and are described in [9].

Significant upstream movement of the transition location on the upper side of the spheroid and the downstream movement on the lower side is obtained with an increase of the angle of attack to $\alpha = 5.0°$, see Fig. 18.14. The shape factors in Fig. 18.15 on the upper side show some overshoots in front of the transition location which cause a decrease of the N_{TS}-factors in this area, see Fig. 18.16. The overshoots are probably caused by convergence issues. The surface pressure distribution, the integration paths and the transition location for $\alpha = 10.0°$ are shown in Fig. 18.17. On the lower side significant discrepancies to the experiment occur, as transition location is not found within 90% of the length of the spheroid. Note that the $\alpha = 10.0°$ case did not exhibit N_{CF} factors that came close to transition thresholds. Besides the effect of the simplifications and the convergence issues mentioned above, grid resolution may also have an influence on the result. On a way to a reliable tool more detailed investigations will have to be carried out to understand the relative effects. On the other hand we also note significant uncertainties that remain with the existing experimental data since it lacks detailed information about the dominating transition scenario.

A comparison of the pressure distribution in various sections between the Navier-Stokes calculation with the Spalart-Allmaras turbulence model and the analytical solution [17] for the case $M_\infty = 0.30$, $\mathrm{Re}_\infty = 7.2 \cdot 10^6$, $\alpha = 10.0°$ is shown in Fig. 18.18. There is a good agreement on the windward side. The discrepancies of the leeward may be caused by the turbulence model of the RANS calculation.

Distributions of displacement thickness δ^* calculated from the boundary layer velocity distributions of the Navier-Stokes computation and the experiment for the case $M_{\infty,\mathrm{calc}} = 0.3$, $M_{\infty,\mathrm{exp}} = 0.1$, $\mathrm{Re}_\infty = 7.2 \cdot 10^6$, $\alpha = 10.0°$ are shown in Fig. 18.19. The displacement thickness is in good agreement at locations where the flows in calculation and experiment are both laminar or turbulent. Major discrepancies occur, where the numerical solution is laminar and turbulent flow is reported in the experimental data.

18.4 Conclusions

The present work demonstrates that it is possible to calculate boundary layer quantities from the flow field of a 3D Navier-Stokes flow solver accurate enough that stability analysis methods can be employed for transition prediction. First applications of a Tollmien-Schlichting database method to airfoils yield predicted transition locations very close to experimental data. Results for a 3D test case given by a 1:6 prolate spheroid indicate that numerical coupling of RANS solver and 3D stability analysis is feasible. The variation of computed transition location with angle of attack is in fair agreement with existing experimental data if the uncertainties on experimental disturbance environment and corresponding transition scenarios are taken into account.

18.5 Acknowledgement

The present work was performed as part of the German research programme MEGAFLOW and funded by the German Aerospace Center DLR. Numerous and fruitful discussions with T. Gerhold, R. Heinrich, H.-P. Kreplin, N. Kroll, A. Krumbein, H.-W. Stock, M. Widhalm and J. Wild are gratefully acknowledged.

References

1. T. Gerhold, M. Galle, O. Friedrich, J. Evans: "Calculation of Complex Three-Dimensional Configurations Employing the DLR-TAU-Code", AIAA paper 97-0167 (1997).
2. H.-W. Stock, E. Degenhart: "A Simplified en Method for Transition Prediction in Two-Dimensional, Incompressible Boundary Layers", Z. Flugwiss. Weltraumforsch. 13 (1989) pp. 16-30.
3. G. Casalis, D. Arnal: "Database method - Development and validation of the simplified method for pure crossflow instability at low speed", ELFIN II Technical Report No. 145, ONERA CERT, (1996).
4. H.-W. Stock, H.P. Horton: Ein Integralverfahren zur Berechnung dreidimensionaler, laminarer, kompressibler, adiabater Grenzschichten. Z. Flugwiss. Weltraumforsch. 9, Heft 2(1985) pp. 101-110.
5. D.M. Somers: "Design and Experimental Results for a Natural-Laminar-Flow Airfoil for General Aviation Applications", NASA Technical Paper 1861 (1981).
6. M. Drela: XFOIL: "An Analysis and Design System for Low Reynolds Number Airfoils", in T.J. Mueller (Editor): Low Reynolds Number Aerodynamics, Springer Verlag (1989) pp. 1-12.
7. H.U. Meier, H.-P. Kreplin: "Experimental Investigation of the Boundary Layer Transition and Separation on a Body of Revolution", Z. Flugwiss. Weltraumforsch. 4, Heft 2 (1980) pp.65-71.
8. H.U. Meier, H.-P. Kreplin: "Boundary Layer Separation Due to "weak" and "strong" Viscous-Inviscid Interaction on an Inclined Body of Revolution", In. AFWAL Viscous and Interacting Flow Field Effects (1984) pp. 79-96.
9. F. Menter, H.-P. Kreplin: "Stability Analysis with the e^N-Method for the Prolate Spheroid at Zero Angle of Attack", IB222-88 A28, DFVLR-AVA Göttingen (1988).
10. D. Arnal: Boundary Layer Transition: "Predictions based on Linear Theory", AGARD-VKI Special Course on "Progress in Transition Modelling", AGARD Report 793 (1994).
11. H. Schlichting, K. Gersten: "Boundary Layer Theory", 8th ed. 2000, Springer, Berlin, Heidelberg, New York (2001).
12. H.-W. Stock: "Determination of Length Scales in Algebraic Turbulence Models for Navier-Stokes Methods", AIAA Journal, Vol. 27, No. 1 (1989) pp. 5-14.
13. H.-W. Stock, W. Haase: "Feasibility Study of e^N Transition Prediction in Navier-Stokes Methods for Airfoils", AIAA Journal, Vol. 37, No 10, October 1999.

14. I.A. Sadarjoen, T. van Walsum, A.J.S. Hin, F.H. Post: "Particle Tracing Algorithms for 3D Curvilinear Grids", in G. Nielson, H. Müller, H.Hagen (Editors): Scientific Visualization - Overviews, Methologies and Techniques, IEEE (1997) pp. 299-332.
15. C. Nebel, R. Radespiel, T. Wolf: "Transition prediction for 3D flows using a Reynolds-averaged Navier-Stokes code and N-factor methods", AIAA-Paper No. 2003-3593, (2003).
16. Centaursoft: Centaur^{TM}. http://www.centaursoft.com.
17. K. Maruhn: "Druckverteilungen an elliptischen Rümpfen und in ihrem Außenraum", Jahrbuch der deutschen Luftfahrtforschung (1941) pp. 135-147.

Figures

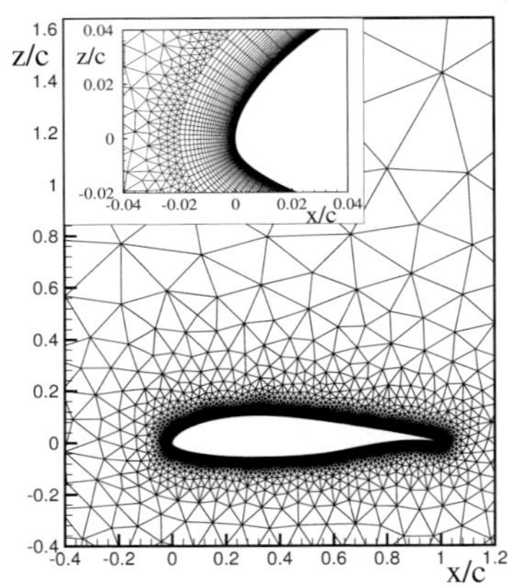

Fig. 18.1. Hybrid grid for NLF(1)-0416 profile

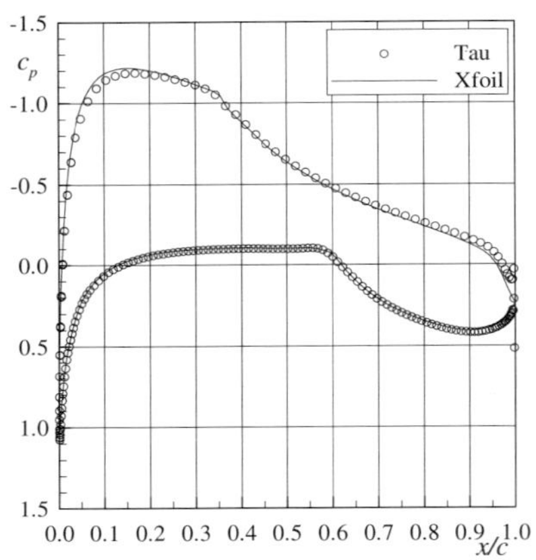

Fig. 18.2. Pressure distribution for NLF(1)-0416 airfoil, $M_\infty = 0.3$, $\mathrm{Re}_\infty = 4 \cdot 10^6$, $\alpha = 2.03°$, $x_{\mathrm{tr,upper}} = 0.35$, $x_{\mathrm{tr,lower}} = 0.60$

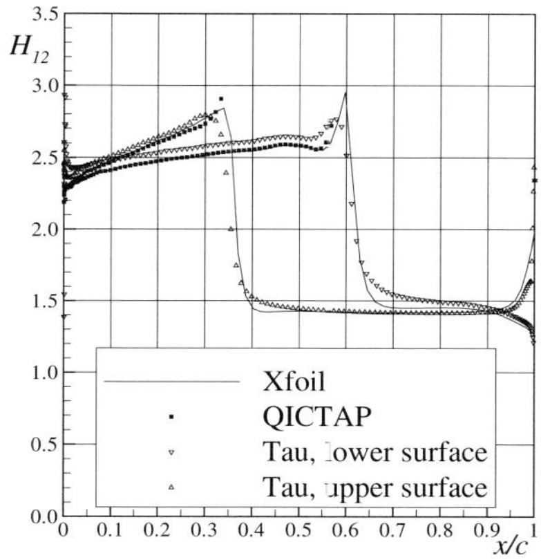

Fig. 18.3. Shape factor H for NLF(1)-0416 airfoil, comparison between XFOIL, QICTAP and TAU, $M_\infty = 0.3$, $\mathrm{Re}_\infty = 4 \cdot 10^6$, $\alpha = 2.03°$, $x_{\mathrm{tr,upper}} = 0.35$, $x_{\mathrm{tr,lower}} = 0.60$

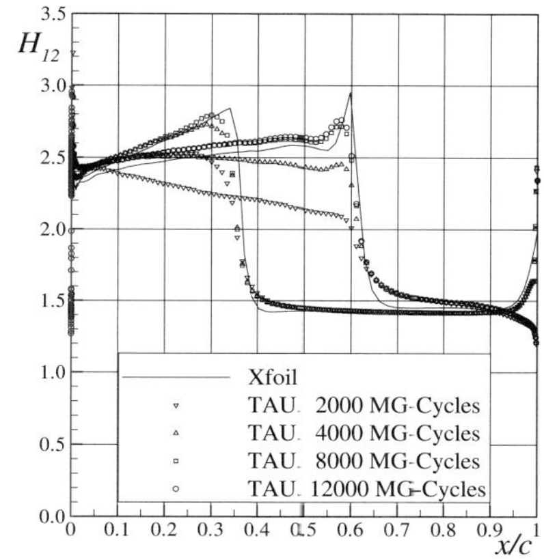

Fig. 18.4. Shape factor H, comparison between solutions at various iterations, $M_\infty = 0.3$, $\mathrm{Re}_\infty = 4 \cdot 10^6$, $\alpha = 2.03°$, $x_{\mathrm{tr,upper}} = 0.35$, $x_{\mathrm{tr,lower}} = 0.60$

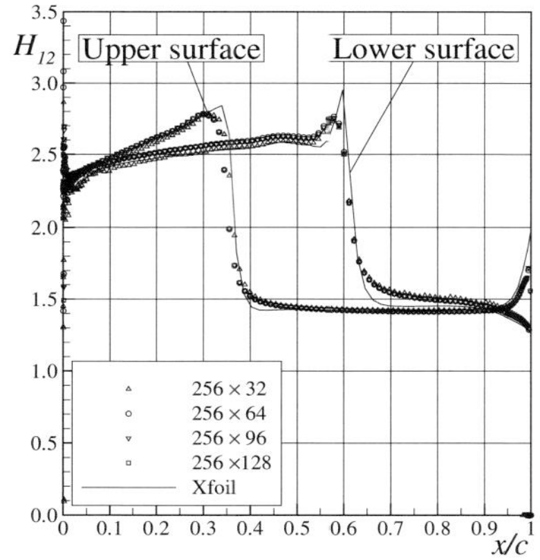

Fig. 18.5. Shape factor H, comparison between solutions with different grid density normal to surface, $M_\infty = 0.3$, $\mathrm{Re}_\infty = 4 \cdot 10^6$, $\alpha = 2.03°$, $x_{\mathrm{tr,upper}} = 0.35$, $x_{\mathrm{tr,lower}} = 0.60$

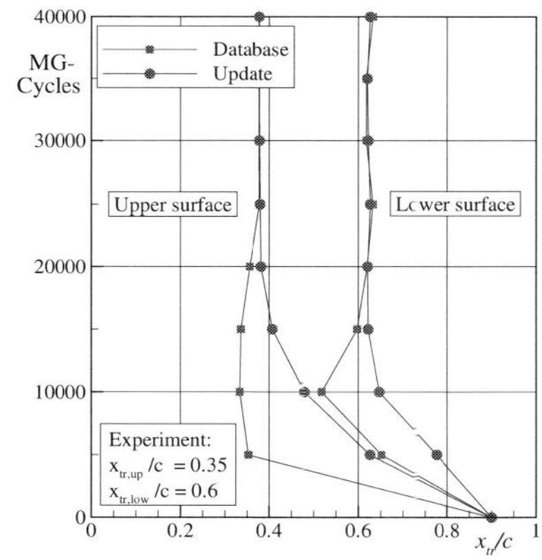

Fig. 18.6. Transition locations at various timesteps, calculation with free transition for NLF(1)-0416 airfoil, $M_\infty = 0.3$, $\mathrm{Re}_\infty = 4 \cdot 10^6$, $\alpha = 2.03°$

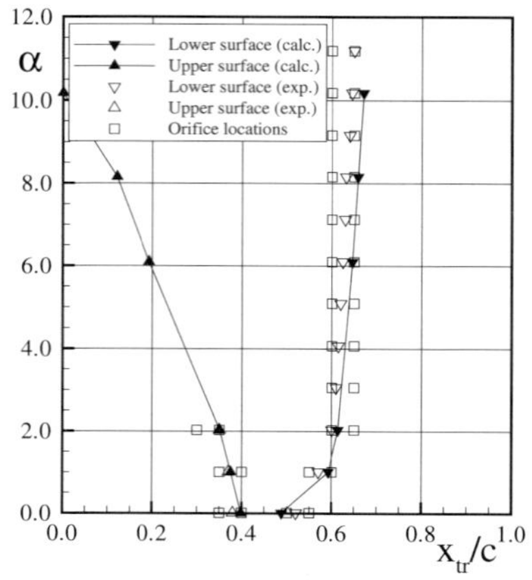

Fig. 18.7. Transition locations from calculation and experiment for NLF(1)-0416 airfoil, $Re_\infty = 4 \cdot 10^6$, $M_{\infty,calc} = 0.3$, $M_{\infty,exp} = 0.1$, $N_{TS,crit} = 11$

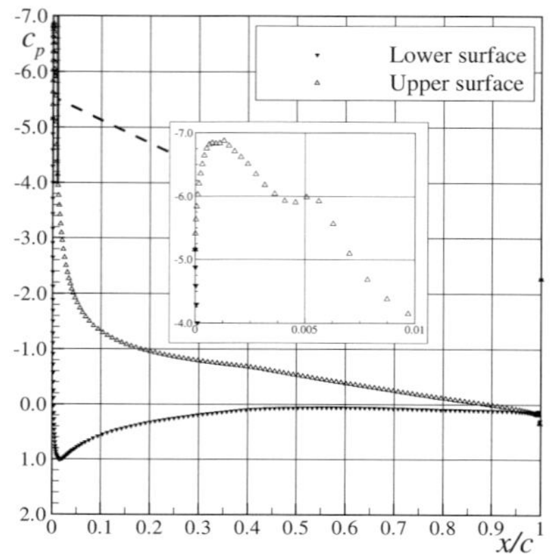

Fig. 18.8. Surface pressure distribution for NACA 64A010 profile (detail), $Re_\infty = 4.1 \cdot 10^6$, $M_\infty = 0.167$, $\alpha = 8.0°$

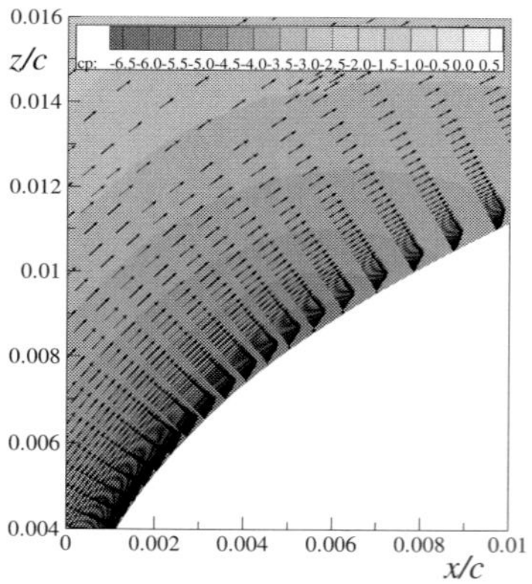

Fig. 18.9. Pressure and velocity distributions in region of the separation bubble for NACA 64A010, $Re_\infty = 4.1 \cdot 10^6$, $M_\infty = 0.167$, $\alpha = 8.0°$

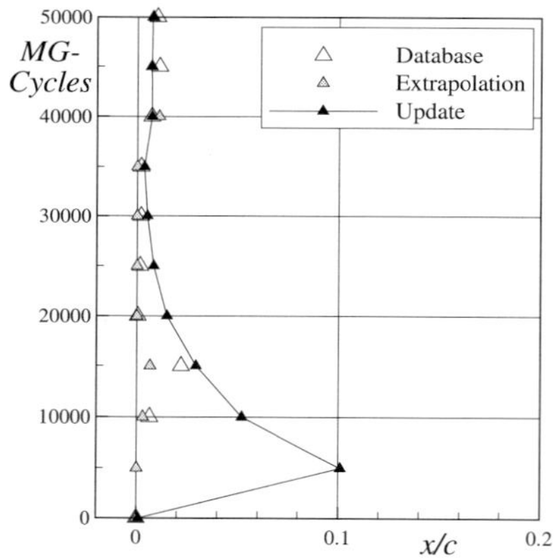

Fig. 18.10. Transition locations at various timesteps, calculation with free transition for NACA 64A010 airfoil, $Re_\infty = 4.1 \cdot 10^6$, $M_\infty = 0.167$, $\alpha = 8.0°$

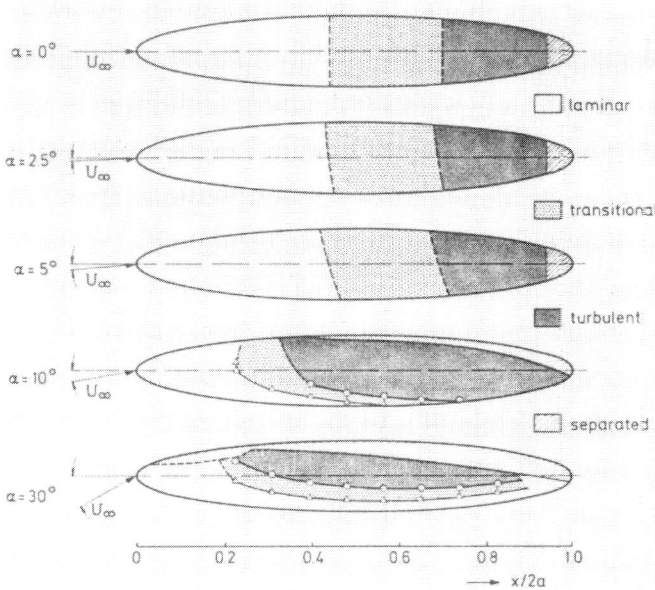

Fig. 18.11. Experimentally determined transition and separation regions on the prolate spheroid at various angles of attack, $\mathrm{Re}_\infty = 7.2 \cdot 10^6$, $M_\infty = 0.13$, from [7]

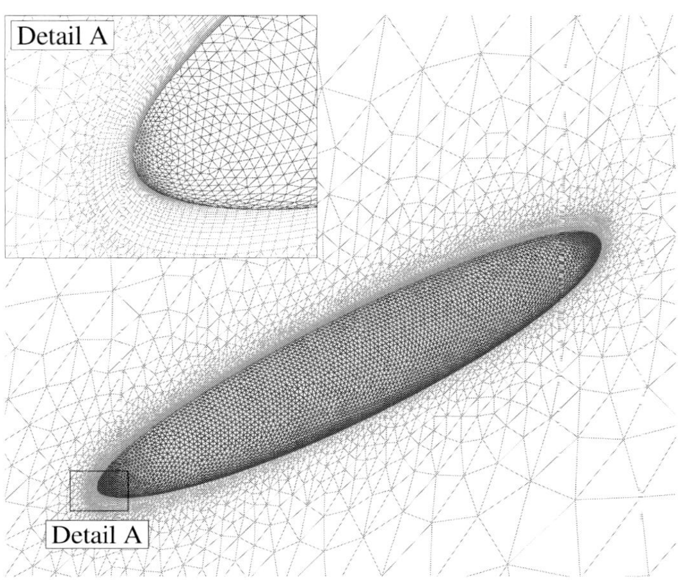

Fig. 18.12. Hybrid grid for 1:6 spheroid

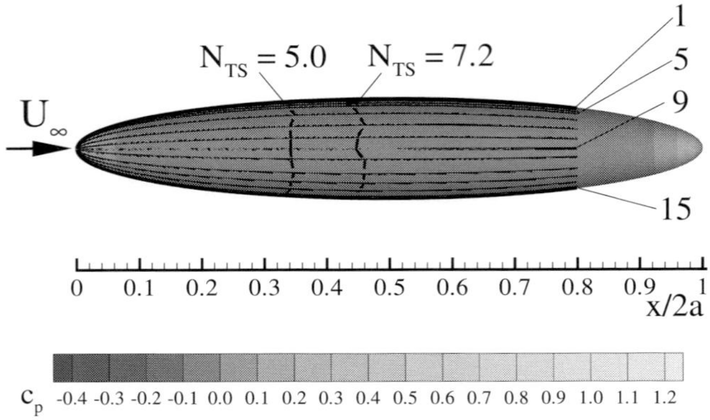

Fig. 18.13. Surface pressure distribution, integration paths and transition location, $M_{\infty,\text{calc}} = 0.3$, $\text{Re}_\infty = 7.2 \cdot 10^6$, $\alpha = 0.0°$

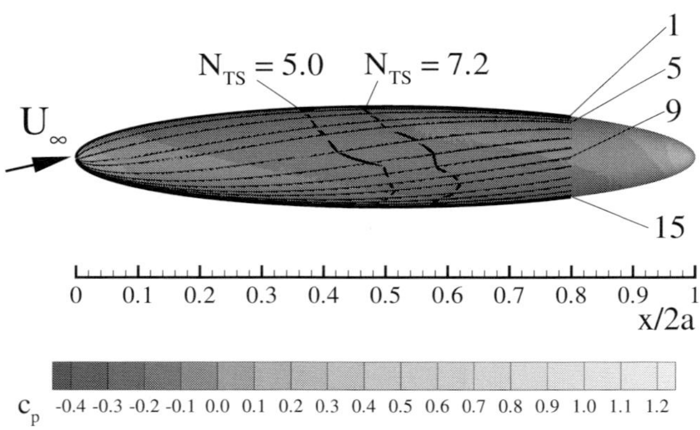

Fig. 18.14. Surface pressure distribution, integration paths and transition location, $M_{\infty,\text{calc}} = 0.3$, $\text{Re}_\infty = 7.2 \cdot 10^6$, $\alpha = 5.0°$

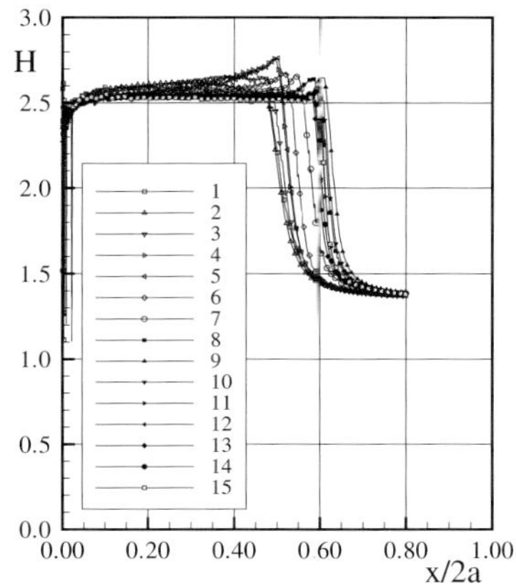

Fig. 18.15. Shape factor H from calculation for various integration paths, $M_{\infty,\text{calc}} = 0.3$, $\text{Re}_\infty = 7.2 \cdot 10^6$, $\alpha = 5.0°$, $\text{N}_{\text{TS,crit}} = 7.2$

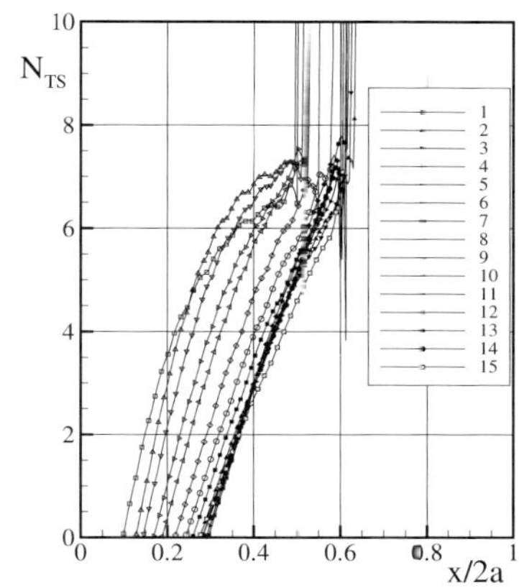

Fig. 18.16. N-factors from calculation for various integration paths, $M_{\infty,\text{calc}} = 0.3$, $\text{Re}_\infty = 7.2 \cdot 10^6$, $\alpha = 5.0°$, $\text{N}_{\text{TS,crit}} = 7.2$

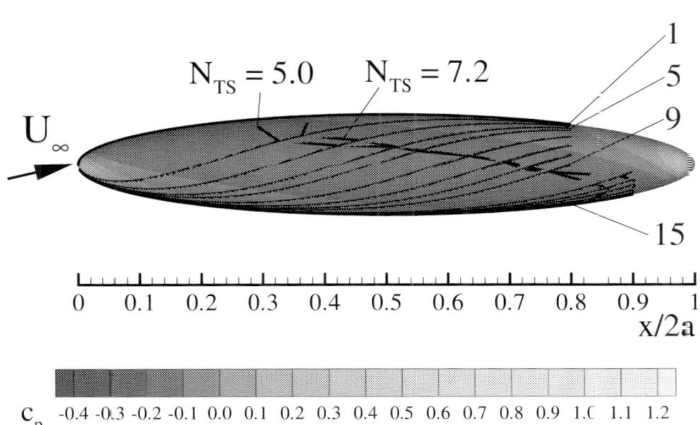

Fig. 18.17. Surface pressure distribution, integration paths and transition location, $M_{\infty,\text{calc}} = 0.3$, $\text{Re}_\infty = 7.2 \cdot 10^6$, $\alpha = 10.0°$

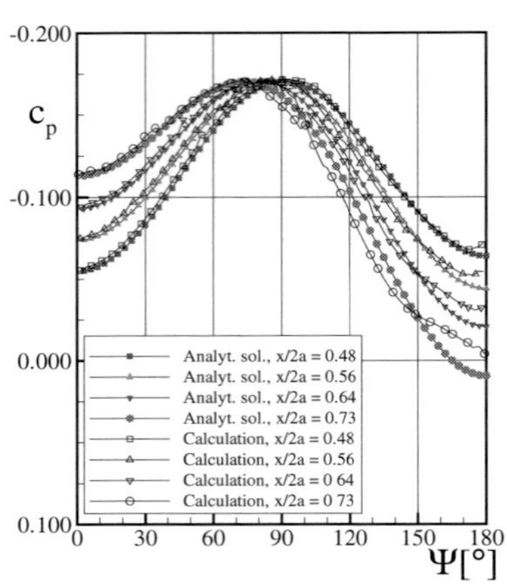

Fig. 18.18. Pressure distribution in various sections x/2a = const., $M_{\infty,\mathrm{calc}} = 0.3$, $\mathrm{Re}_\infty = 7.2 \cdot 10^6$, $\alpha = 10.0°$

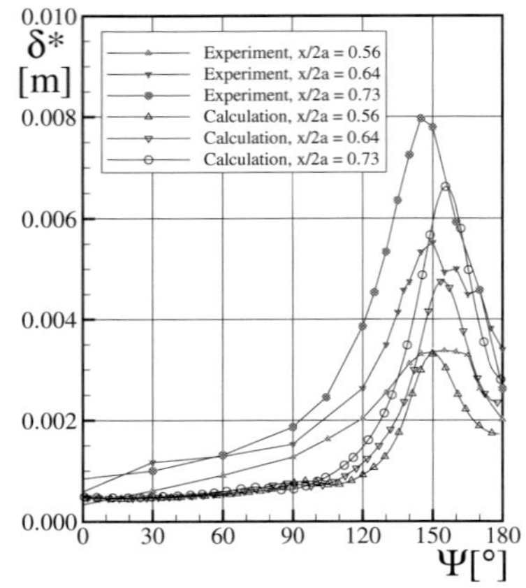

Fig. 18.19. Displacement thickness δ^* from Navier-Stokes calculation and experiment, $M_{\infty,\mathrm{calc}} = 0.3$, $M_{\infty,\mathrm{exp}} = 0.13$, $\mathrm{Re}_\infty = 7.2 \cdot 10^6$, $\alpha = 10.0°$

Part VII

Exploitation of MEGAFLOW Software

19

Application of the MEGAFLOW Software at DLR

R. Rudnik

DLR Braunschweig, Institute of Aerodynamics and Flow Technology,
Transport Aircraft, Lilienthalplatz 7, 38108 Braunschweig, Germany

Summary. The present contribution outlines several selected applications of the MEGAFLOW software at DLR, roughly according to the time-frame of the MEGAFLOW II project from 1998 – 2002. The majority of the applications is based on 3-dimensional viscous computations featuring the solution of the compressible Reynolds-averaged Navier-Stokes equation in combination with one or two transport equation turbulence models. The examples highlight a quite broad range of applications. This refers to onflow speed as well as to the mode of application, covering analysis as well as design and optimization tasks. In general a clear trend in the use of the MEGAFLOW system becomes visible. On the one hand the analyses of very specific flow and/or aircraft details on overall configurations of increasing complexity is carried out. On the other hand more design and optimization applications are requested. For both types of applications the MEGAFLOW software has become an indispensable tool for the aerodynamic and also multi-disciplinary tasks of DLR.

19.1 Introduction

The range of applications of the MEGAFLOW simulation software [1] at DLR is quite widespread according to the broad range of aerodynamic and aerothermodynamic activities. This range extends from simple 2D airfoil flows at low speeds to the computation of 3D re-entry vehicles and rockets at hypersonic speeds. In between lies a variety of applications covering transport aircraft configurations in cruise and high lift configurations, advanced supersonic transport aircraft, military fighter and transport aircraft, helicopter and rotor flows. The application of a single software to such a broad range of problems requires a high degree of flexibility and performance with respect to the simulation capabilities. Partly, this is given by the unique MEGAFLOW ability of selecting between block-structured and unstructured flow solvers with basically identical numerical algorithms. At present both methods have their pros and cons, so this flexibility is regarded as quite advantageous. On the other hand, the numerical algorithms, discretization methods, time integration schemes, and turbulence models implemented in the software have to be equally accurate and efficient for all type of specific challenges with respect to the speed range.

As it is hardly possible to give a comprehensive overview about all type of applications of the MEGAFLOW software at DLR, the present contribution concentrates on the area of commercial transport aircraft, which may be regarded as the 'core business' of the DLR applications. In the following selected examples for such type of simulations are highlighted in some detail.

19.2 Engine Airframe Integration for Cruise Configurations

The development process of the Fairchild/Dornier F/D 728 was heavily supported by DLR activities, especially with respect to engine/airframe integration using the MEGAFLOW software. While in the beginning a lot of studies for engine positioning have been carried out using the block-structured FLOWer code and the MEGACADS grid generation systems, the request turned on later towards a very detailed modeling of configurative features. This lead to a change from the block-structured to the unstructured approach with the TAU code due to its superior capability to deal with highly complex configurations. Figure 19.1 depicts a surface grid for inviscid computations around the F/D 728 configuration [2]. The engine is quite closely mounted under the wing. In addition to the flap track fairing on the wing lower surface, system fairings at the nacelle/pylon intersection, the pylon/wing intersection and a bleed-air exhaust at the pylon outboard surface underline the simulation complexity. As an example for the resulting flow field, a view from behind at the core exit jet is shown in figure 19.2. The evaluation of the total temperature distribution of the hot core jet serves to investigate, whether the chevron nozzle has an influence on the thermal loading at the pylon lower side. The analyses shows, that the thermal loads, which result basically from the core jet, are hardly affected by the chevron presence. Based on appropriate CAD geometry files a complete CFD loop from geometry change to post processing was carried out within 1 to 2 days.

19.3 Studies on Wing Tip Extensions

In addition to the engine/airframe integration studies on the F/D 728, the Megaflow software has been applied for investigations of different wing-tip extensions [3]. These extensions were of special importance for the corporate jet derivative of the F/D 728, called Envoy 7. A total of seven different wing tip extensions has been investigated based on the block-structured FLOWer code for the wing/fuselage configuration. The numerical set-up is tuned for very short turn-around times. Consequently, an inviscid fuselage representation combined with a viscous wing simulation has been selected. An C-H grid topology is applied featuring 256 x 112 x 57 cells in the computational directions. The Baldwin-Lomax turbulence model with Degani-Schiff extension is used [4]. The resulting isobars on four of the seven different wing tips are displayed in figure 19.3. The assessment of the wing tip extension was based on the drag improvements as well as on the resulting penalties in wing root bending moment. Compared to a reference configuration without wing tip extension a reduction in total drag of up to 5 % could be achieved, as indicated in the table in figure 19.3. The increase in root bending moment is of the same order. The best overall performance is obtained by the plain shark flat and the super shark configuration. The numerical results could be confirmed by corresponding wind tunnel experiments, underlining the code capabilities as a fast analysis tool for new aircraft developments.

19.4 Numerical Investigation of Spoiler Effectiveness

Another example of a challenging application is the prediction of the impact of deflected spoilers during a high speed operational descend. The investigations concentrated on studies of different spoiler concepts, evaluation of their effectiveness, and the assessment of the risk of an adverse interaction of the wake flow behind the spoilers and the empennage [5]. For this purpose RANS computations of a 4 jet commercial aircraft have been carried out using the unstructured TAU code with the Spalart-Allmaras-Edwards turbulence model [6]. The grids have been generated for wind tunnel and flight Re.-No.s with the commercial CENTAUR grid generator, containing up to 11.3 million nodes. The grids have been adapted for a non-dimensional wall spacing of $y^+ = 1$. In figure 19.4 resulting wall streamlines are depicted for a conventional spoiler configuration. The high degree of separated flow downstream of the deflected spoiler required a careful adjustment of the time integration parameters and a reduction of the CFL number. In figure 19.5 total pressure cut planes behind the wing indicate, that especially the inboard spoiler wake bears the risk of a possible interaction with the empennage. In addition to the assessment of wake-empennage interactions, these simulations provided valuable data for the estimation of the spoiler loads for the design of the corresponding devices of a wind tunnel model. Due to the small thickness of the model spoilers, these loads could not directly accessible from the wind tunnel tests.

19.5 Inverse Integrated Nacelle Design

The previous examples cover pure aerodynamic analysis. But the MEGAFLOW software has also been extended to cope with the inverse design of integrated aircraft components. A corresponding example is the design of a long-cowl nacelle mounted on the DLR F6 wind tunnel model, as depicted in figure 19.6. The design target, indicated by symbols in the nacelle pressure distribution, is to alleviate the pressure peak at 75 % nacelle chord, which is produced by the flow acceleration in the open channel between inboard pylon, nacelle, and wing surface. The low pressure areas are quite obvious in the isobars in the left part of the figure and the initial nacelle pressure distribution. The target pressure distribution has been matched well using the inverse design procedure. The benefit of the new nacelle shape affects the nacelle itself, but also the wing pressure field, demonstrated by a RANS re-computation of the complete configuration. The wing pressure distribution in figure 19.7 proves, that the strong lower wing suction peak is also alleviated, resulting in a decrease of the total wave drag of 5 drag counts. For the design method the TAU Euler analysis code is coupled to an iterative residual correction approach [7]. The grid generation and deformation is accomplished using again the CENTAUR software.

19.6 Engine Airframe Integration for High Lift Configurations

Based on thorough development and validation efforts of the hybrid unstructured approach with the TAU code and the CENTAUR grid generation software, highly complex high lift flows became more and more accessible. As an example the flow around the DLR ALVAST model in high lift configuration together with an ultra-high-bypass engine simulator CRUF has been computed [8]. Again the Spalart-Allmaras-Edwards turbulence model has been applied. In order to study engine/ airframe interference effects with the high lift system, a powered engine is incorporated in the RANS simulation. The hybrid unstructured grid consists of about 10 million grid points. The preconditioning feature of the code has been successfully used for this type of low speed solution using a code, which is based on the compressible form of the governing equations. Figure 19.8 depicts a vortex, that sheds at the nacelle trailing edge and interacts with the leading edge of the deployed slat. The impact of the mounting of two different engine concepts (Very High Bypass Ratio, VHBR — Ultra High Bypass Ratio, UHBR) on the span wise lift distribution is also shown in figure 19.8. For the VHBR concept the lift loss on the wing due to the engine mounting is roughly compensated by the lift generated by the nacelle itself. For the UHBR concept the wing lift loss is slightly stronger than for the VHBR. Nevertheless, it is overcompensation by the higher lift carried by the large nacelle. Future activities will focus on the inclusion of the slat cut-out at the pylon and nacelle strakes model all relevant high lift features of a real aircraft configuration.

19.7 High Lift Design for Future Supersonic Transport Configurations

Complementary to the classical subsonic aircraft configuration the Megaflow software has been a very valuable part of high lift design studies for a 2nd generation supersonic transport configuration. These investigations were carried out within the EC project EPISTLE, aiming at highly efficient and low noise high lift systems for supersonic aircraft. A proper high lift design is one of the most essential issues of such type of delta wing configurations to stay in compliance with the community noise regulations. The focus of these studies is laid on the design of leading edge flap systems. Corresponding design sensitivities have been examined and most promising candidates, as the double hinge line leading edge, depicted in figure 19.9, have been assessed using the FLOWer code [9]. For this purpose an 8-block structured grid with roughly 1.9 million grid points has been used. A C-O topology is applied for the grid around the double-delta wing. The RANS computations incorporate the Degani-Schiff turbulence model [4], that was found to be best applicable for thus type of flows [10]. Due to the fact, that the design point, $C_L = 0.4$; $M = 0.25$; $Re = 25 \cdot 10^6$ is characterized by a considerable amount of separated flow, see also figure 19.9, the performance verification of different leading edge concepts represents a numerical challenge. As the drag polar of the datum and the final design shows a drag reduction of 20 % is achieved for the design conditions.

19.8 Numerical High Lift Optimization for Multi-Element Airfoils

As a consequent next step after high lift analyses and design studies, the MEGAFLOW software has been extended by a setting optimization module for multi-element airfoils called CHAeOPS. The optimization method is based on a deterministic SUBPLEX strategy [11]. A thorough code adjustment study has been carried out for this purpose based on the block-structured FLOWer code [12]. This covers the determination of best suited high lift solver features, such as Spalart-Allmaras Edwards model [6], as well as the set-up of a flexible block topology for the setting variation, based on the MEGACADS grid generator. The grids used for the optimization of 3-element airfoils contain typically about 80.000 grid points in 9 blocks featuring C-block topologies around the single elements. Within this context an algorithm to determine maximum lift without the necessity to run complete polar computations improved the efficiency of the method considerably. As a demonstration results of a drag optimization for a 3-element airfoil in take-off configuration are presented in figure 19.10. A limit in pitching moment has been prescribed as a secondary constraint. In total 12 design variables are taken into account. These are slat and flap gap, overlap, and deflection. In addition the slat and flap cut-out contours are parametrized by three variables each. In the upper part of Figure 10 the initial and optimized slat and flap contours are shown, in the lower part the corresponding pressure distributions. The optimization affects the element chord, setting, and deflection angle as well as the angle of attack. The optimization results in a decrease in total drag of 21%, while the maximum lift is also slightly improved be nearly 2 %. The combination of MEGAFLOW flow solution and CHAeOPS optimization has been successfully applied to support the high lift design of the Airbus A380 and the Fairchild/Dornier F/D 728.

19.9 Conclusion

The preceding results demonstrate the capabilities of the MEGAFLOW software as used by DLR in the analysis, design, and optimization of commercial transport aircraft. Either in research activities as well as in supporting the aircraft industry demands the numerical simulation has proven to be essential for the development of current and future aircraft projects. Future requirements will be oriented towards multi-disciplinary analyses, e.g. fluid-structure coupling, or the coupling of CFD codes to computational aero-acoustic methods. For these purposes the MEGAFLOW software systems represents an efficient and validated flow solution system.

Acknowledgements

The present article is a summary of several investigations carried out at the Institute of Aerodynamics and Flow Technology. In this context the author likes to acknowledge the contributions of Heiko Frhr. v. Geyr, Dr. Ulrich Herrmann, Stefan Melber, Arno Ronzheimer, Dr. Thomas Streit, Dr. Jochen Wild, and Roland Wilhelm.

References

1. Kroll, N., Rossow, C.-C., Schwamborn, D., Becker, K., Heller, G., "MEGAFLOW — A Numerical Flow Simulation Tool for Transport Aircraft Design", 23rd ICAS Congress, Toronto, paper 2002-1.10.5, 2002
2. Ronzheimer, A., Brodersen, O., "Hybrid Grid Generation for Aircraft Configurations at DLR", Proceeding of the 7th international conference on numerical grid simulation in computational field simulation, pp.1071 - 1080, ed. B. K. Soni et.al., 25.09-28.09.2000
3. Heller, G., Dirmeier, S., Kreuzer, P., Streit, Th., "Aerodynamische Leistungsverbesserung durch Flügelspitzenmodifikation am Beispiel der Envoy 7", DGLR Jahrestagung Hamburg, 17.-20.09., DGLR Jahrbuch 2001, 2001

4. Degani, D., Schiff, L.B., and Levy, Y., "Physical Considerations Governing Computation of Turbulent Flows Over Bodies at Large Incidence," AIAA Paper 90-0096, 1990
5. Frhr. v. Geyr, H., private communications, 2002
6. Edwards, J. R., Chandra, S., "Comparison of Eddy Viscosity-Transport Turbulence Models for Three-Dimensional, Shock Separated Flowfields", AIAA Journal, No. 4, April 1996
7. Wilhelm, R., "Inverse Design Method for Designing Isolated and Wing Mounted Engine Nacelles", Journal of Aircraft, Vol. 39, No. 6, pp. 989-995, 2002
8. Melber, S., "3D RANS Simulations for High Lift Transport Aircraft Configurations with Engines", DLR IB 124-2002/27, 2002
9. Herrmann, U., "Designing High Lift Systems for Low Drag SCT", Proceedings of the lecture series of the von Karman Institute of Fluid Dynamics: CFD based Aircraft Drag Prediction and reduction, Brussels, 03.02 - 07.02.2003
10. Herrmann, U., "Validation of European CFD Codes for SCT low-speed high-lift Computations", AIAA Paper 2001/2406, 2001
11. Rowan, T., "Functionality Stability Analysis of Numerical Algorithms", PhD-Thesis, University of Texas, Austin, Texas, USA, 1990
12. Wild, J., "Direct Optimisation of Multi-Element-Airfoils for High-Lift using Navier-Stokes Equations." in Computational Fluid Dynamics 1998, pp. 383/390, Verlag John Wiley & Sons, Chichester, UK, 1998. Proceedings of the 4th European Computational Fluid Dynamics Conference (ECCOMAS) , 1998

Figures

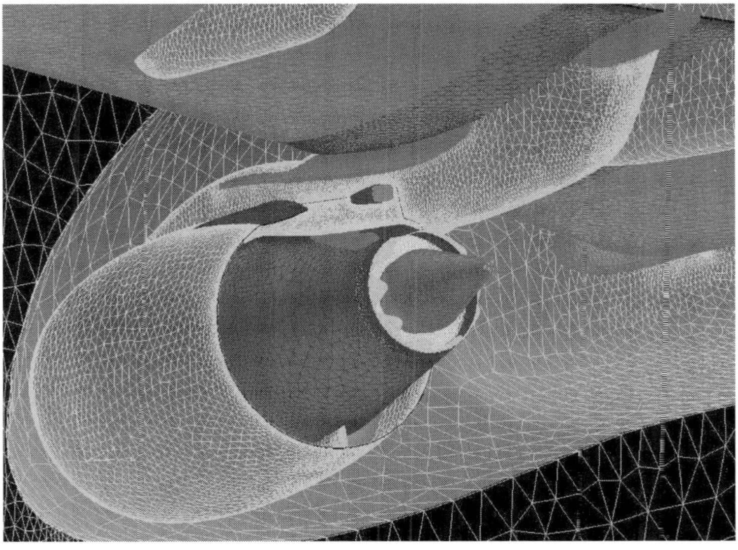

Fig. 19.1. Unstructured numerical grid of F/D 728 with GE CF34 engine

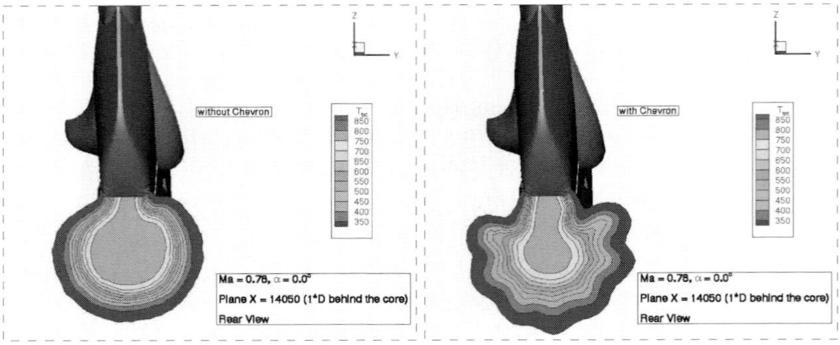

Fig. 19.2. Total temperature distribution behind core nozzle exit of the CF34 engine mounted on a F/D 728

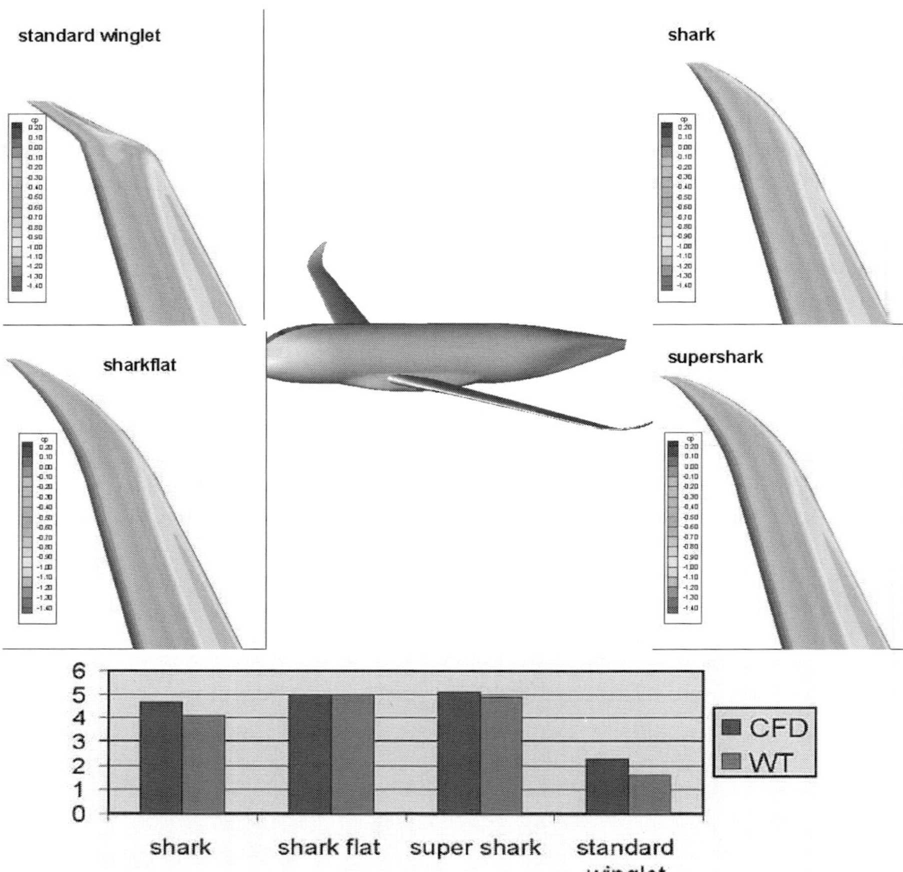

Fig. 19.3. Numerical studies for efficiency improvement by different wingtip extensions on the F/D Envoy version

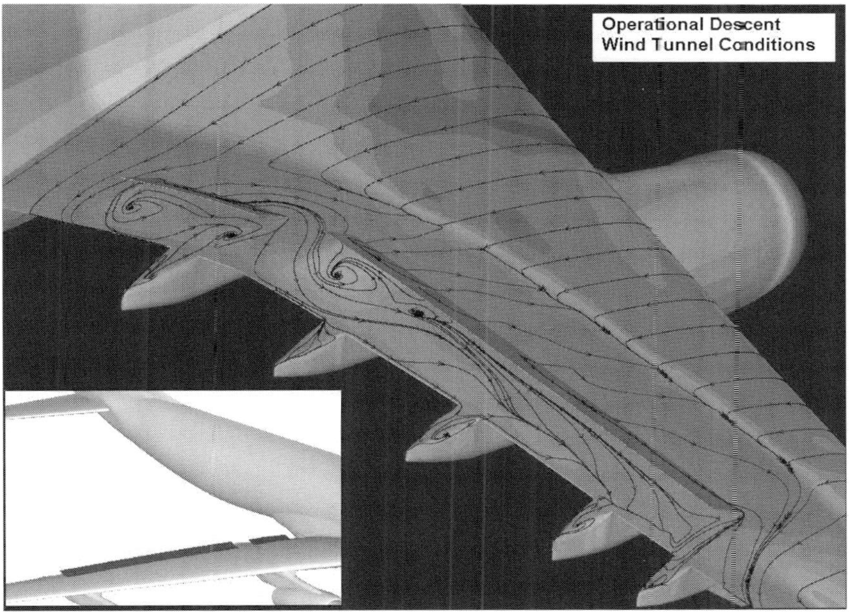

Fig. 19.4. Simulation of the effect of deployed spoilers on a commercial transport aircraft

Fig. 19.5. Cut planes behind the wake of a wing with deflected spoilers

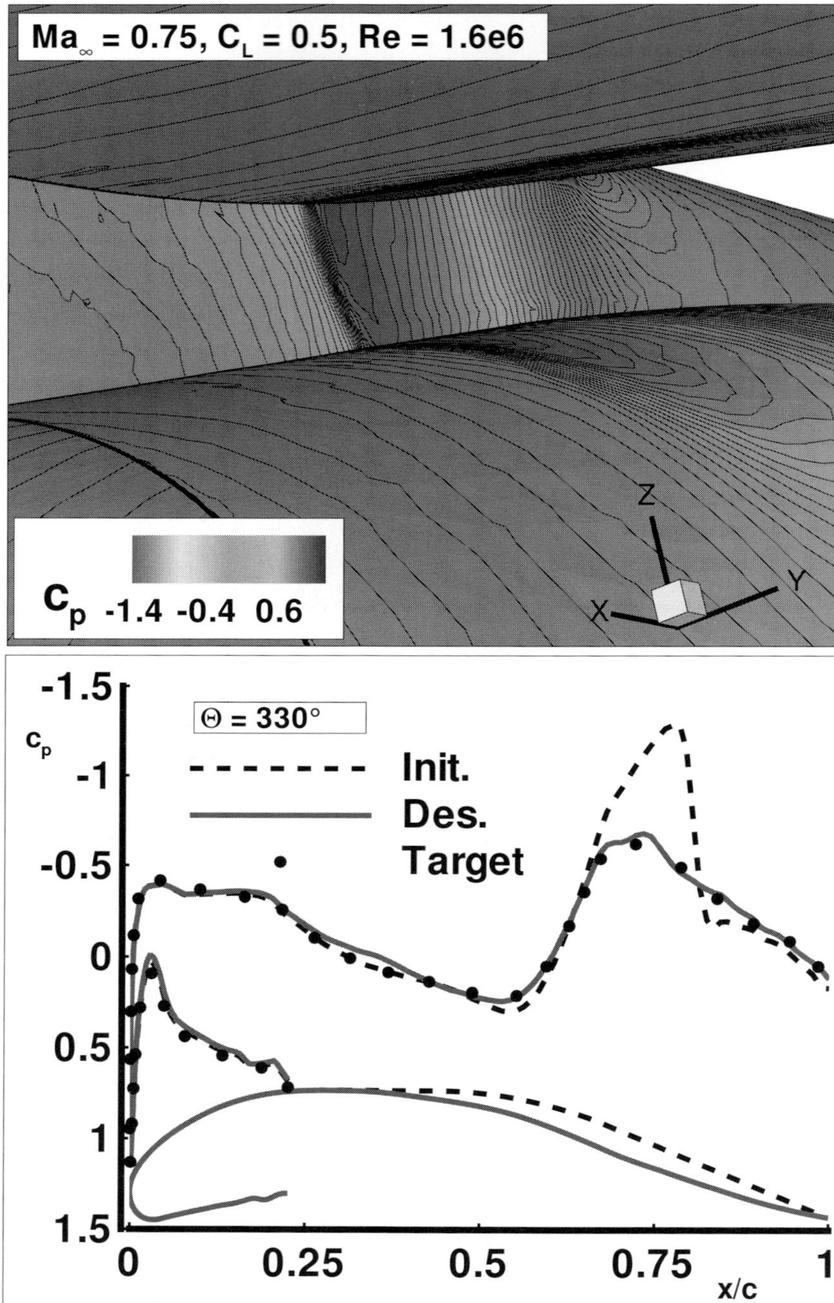

Fig. 19.6. Inverse design of an integrated nacelle mounted on the DLR F6 configuration

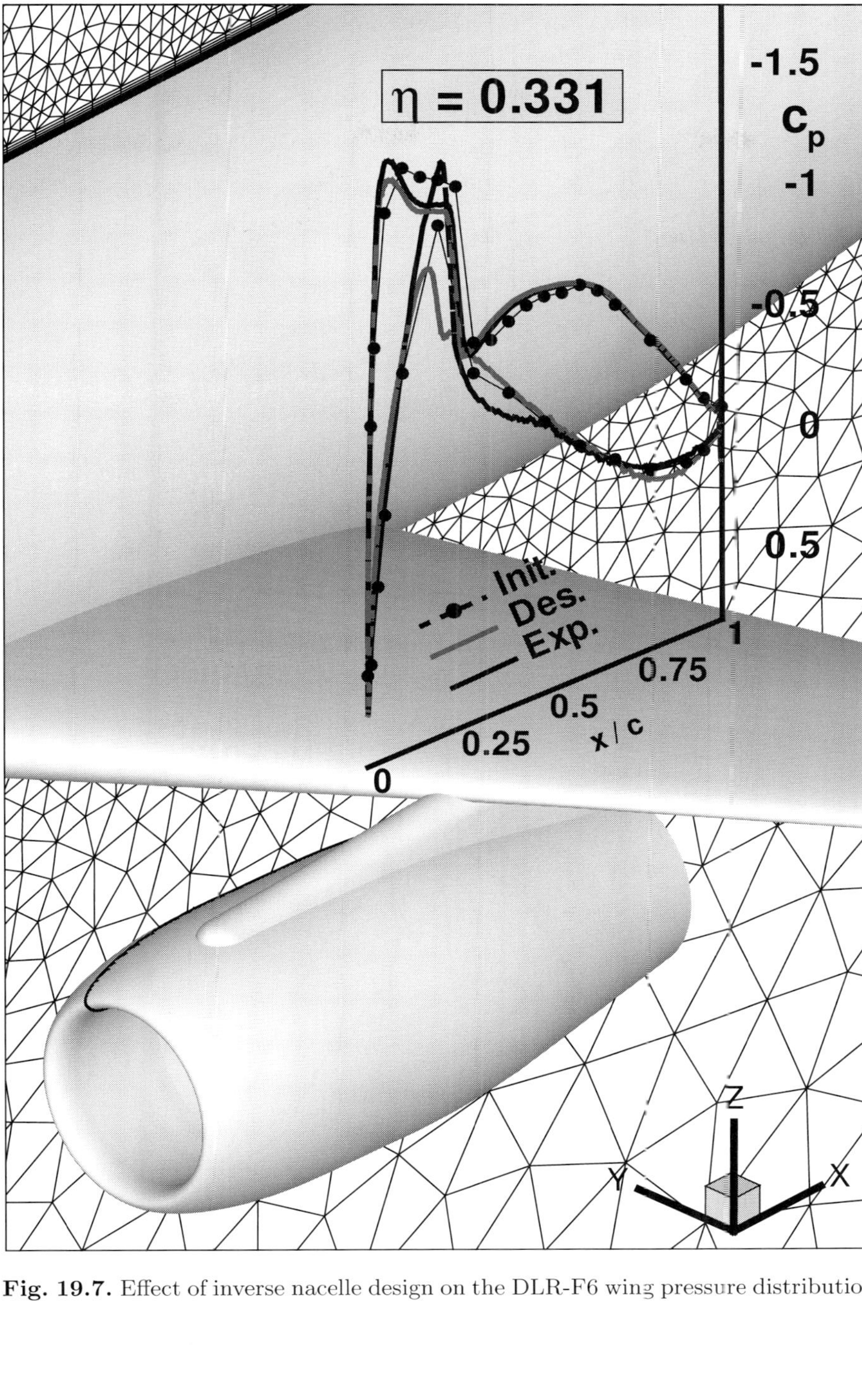

Fig. 19.7. Effect of inverse nacelle design on the DLR-F6 wing pressure distribution

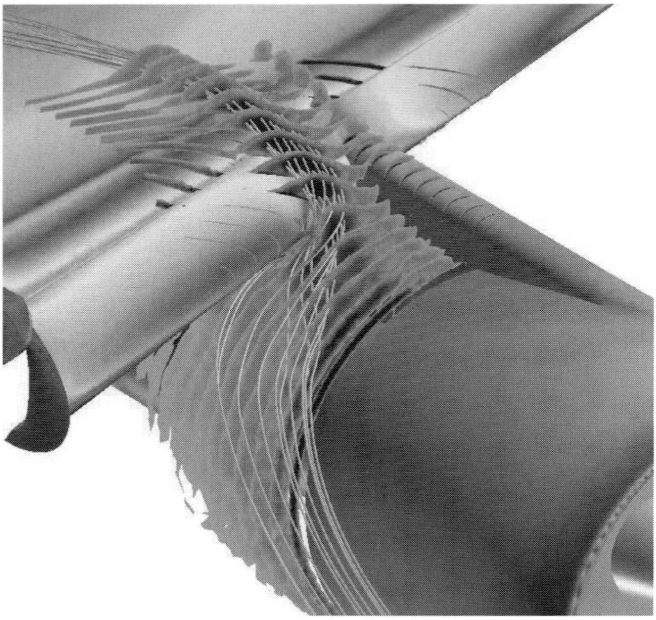

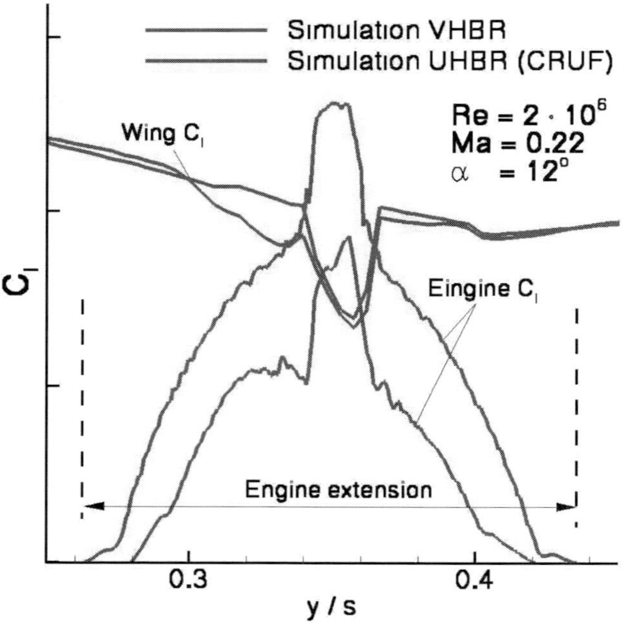

Fig. 19.8. TAU RANS simulation of the engine interference with the deployed high lift system

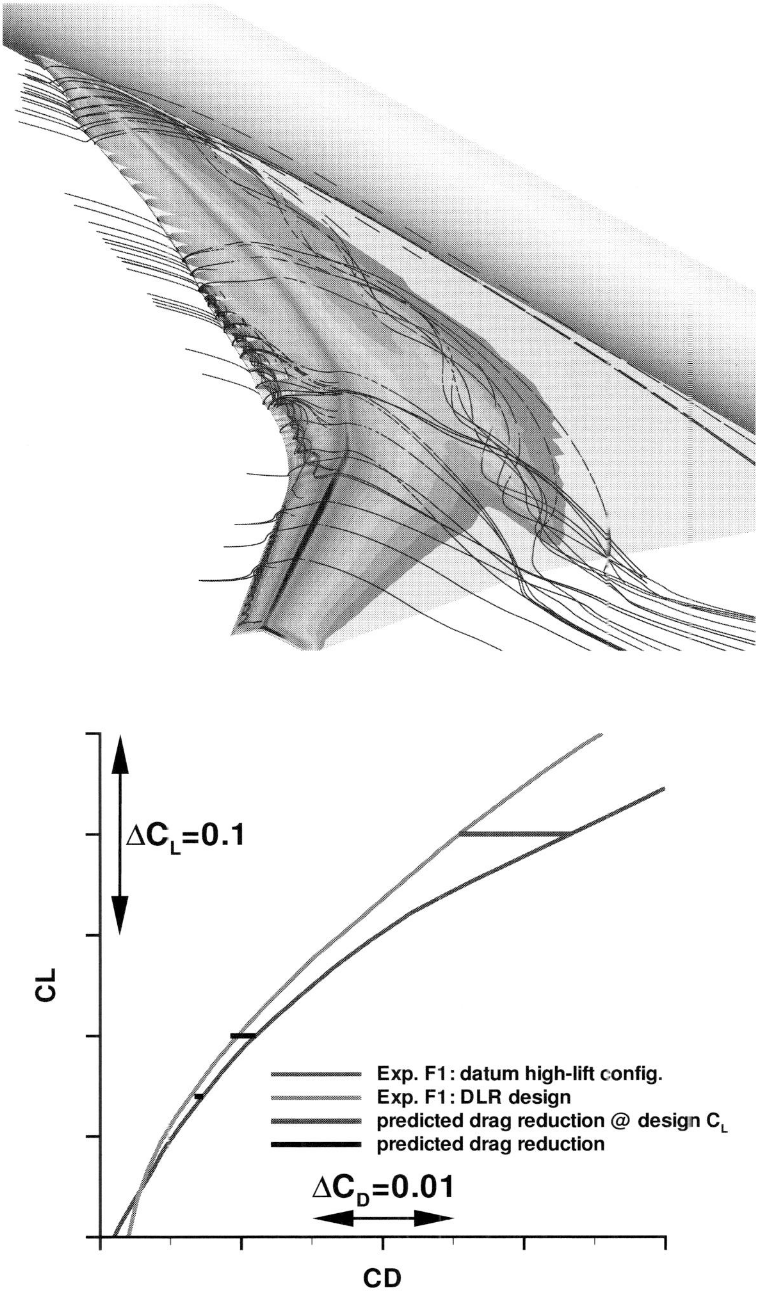

Fig. 19.9. Low noise high lift system design study for a 2nd generation SCT configuration using FLOWer computations

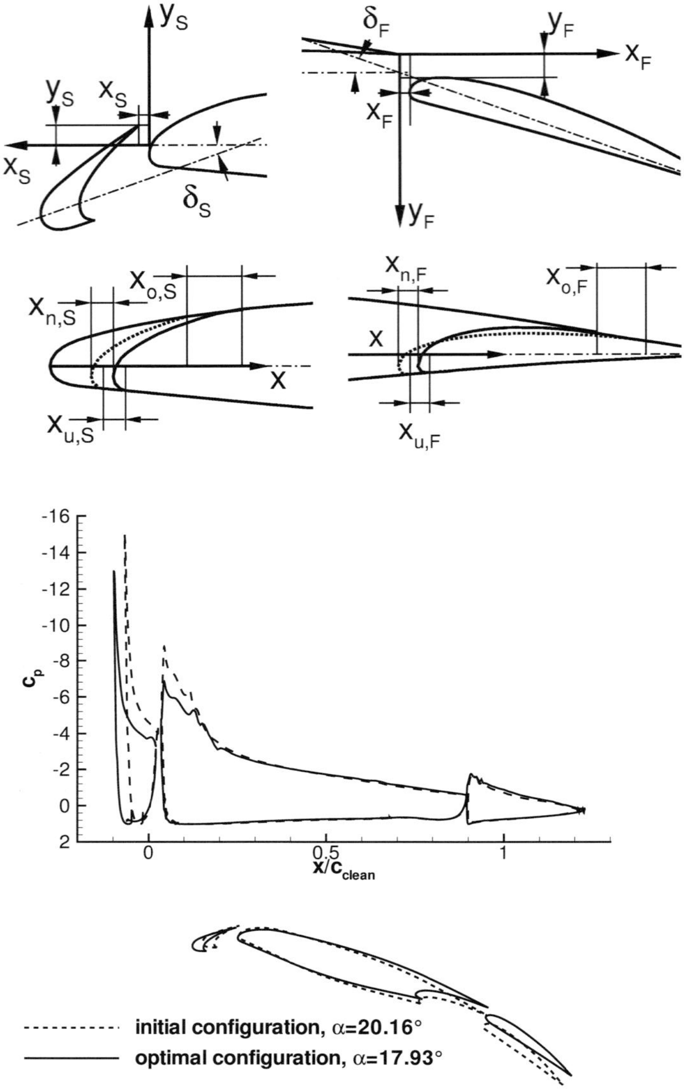

Fig. 19.10. Setting optimization of a 3-element L1T2 airfoil using the FLOWer code

20

MEGAFLOW for AIRBUS-D — Applications and Requirements

Petra Aumann and Klaus Becker

Airbus Deutschland, 28183 Bremen, Germany

Summary. The MEGAFLOW projects — I and II — are key drivers for CFD (Computational Fluid Dynamics) technology research within Germany, possibly in Europe as a whole. MEGAFLOW combines the development of most modern CFD tools and suites with the daily application of these tools for aircraft design and data processes. Due to aircraft industries being forced to optimize their designs and to give performance predictions in highest accuracy terms, while the time for design cycles are reduced, high level CFD becomes more and more a basic design tool. However, in many cases CFD reaches its limits due to the ongoing increase of complexity with respect to geometry and flow conditions. Further development of physical flow modeling and further validation of new application cases need to be continued. This paper describes AIRBUS-D requirements on methods and tools development in MEGAFLOW, validated by AIRBUS-D test cases. It outlines which requirements are fulfilled, which are still open, and which have had to be added over the years.

20.1 Introduction

In recent years, AIRBUS-D has worked together with the German Aerospace Center DLR and many German universities in several government funded projects in order to improve the capabilities provided by CFD tools in terms of efficiency, accuracy, flexibility, reliability, and robustness. The baseline was laid in the POPINDA [1] project where the decision was made to develop one common flow solver by placing the major national effort and experiences together. The resulting code FLOWer was verified and validated using industrial test cases. Within the following projects MEGAFLOW I [2], EFENDA [3] and MEGAFLOW II [4] the same way of cooperation was followed. Tools like FLOWer or later TAU were further developed, driven by increased requirements from aircraft industries. After in-line validation the outcome was easily transferred to productive application for AIRBUS-D.

20.2 Status at AIRBUS-D in 1999

By the end of the 1980's, AIRBUS-D had put much validation effort into high level CFD technology. By the end of the 1990's, CFD itself was fully accepted and used within the aerodynamic design and data processes. The structured multiblock Euler/Navier-Stokes code FLOWer, which has been developed in POPINDA and MEGAFLOW I, is the baseline CFD tool for 3D applications. FLOWer Navier-Stokes is the standard tool for simple configurations like wing/body for aerodynamic design and data productive computations FLOWer Euler is the standard tool for complex configurations, especially used for aerodynamic data for loads prediction. The hybrid code TAU, developed mainly at DLR within MEGAFLOW I, has been installed at AIRBUS-D for test and validation purposes for selected AIRBUS-configurations. This hybrid mesh based approach turned out to be so easy to handle - even for configurations of highest complexity - that TAU started to become the AIRBUS-D "way forward" tool for aero design and data production. For shape optimization, a complete automatic process was set up and made available to AIRBUS-D based on Synaps PointerPro optimization software [5] and FLOWer.
Fig. 20.1 shows a table of standard applications used in AIRBUS-D in 1999. All gray aircraft components/tool names show those configurations which are already used in production mode, while all white configurations are likely to be in used in the future for production. TAU was used for 2D high-lift configurations, and in 3D under validation for complex high-lift or engine integration configuration it was the designated future tool for complete aircraft computations. On the structured way, FLOWer was used for standard wing/body and wing/body/tails configurations. Due to the enormous effort, which has to be put into grid generation for structured/multiblock meshes, no higher complexity configurations have been considered for future applications.

From a 1999 point of view, there were some obvious deficits in the numerical Aerodynamics approach. Some examples:

- *Efficiency*: Computation of 3D high lift or cruise configurations still needed very large effort, mainly in time for preparation and geometry handling.
- *Quality*: The quality of data prediction for high Reynolds numbers was not completely validated.
- *Robustness & Flexibility*: Flow modeling for off-design flow cases was not completely validated.
- *Integration/Industrialization*: IT-Framework tools were only available for very special applications and numerical optimization methods were used for test purposes only.
- *Multidisciplinary*: No direct coupling with other disciplines like structures, cinematics, etc. was available.

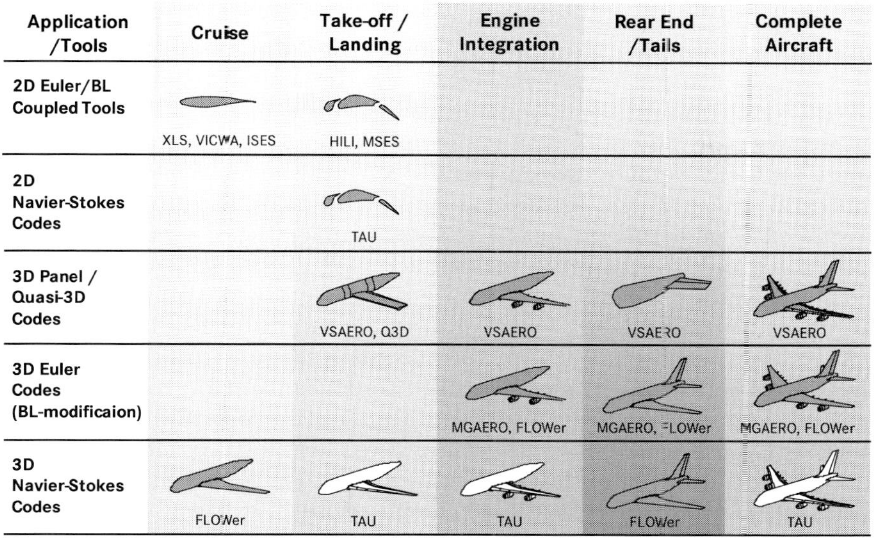

Fig. 20.1. Table of CFD applications used in AIRBUS-D in 1999

20.3 Requirements for MEGAFLOW II

As an outcome of from the MEGAFLOW I project and due to additional industrial requirements, it was clear for AIRBUS-D that a further ongoing strategic MEGAFLOW development had to be enforced.

The MEGAFLOW I project had achieved very good results from a technical point of view, but further development and investigations were needed in:

Flexibility: Further development for more flexible grid approaches were needed (unstructured, hybrid, Chimera). First investigations using the unstructured code TAU showed very promising results as well as the Chimera approach in FLOWer. Following steps were defined in order to enhance TAU for arbitrary elements and for FLOWer to improve the Chimera process chain.

Accuracy: Further improvements in accuracy for flow modeling and discretization, especially for off-design flow cases were needed. Off-design flow cases like in areas of higher Mach number or higher angles of attack have strong requirements to the physical modeling in order to capture complex flow effects like flow separation or vortices in the correct manner.

Industrialization: Further investigations in industrialization and validation process were needed (robustness, efficiency and process automation). For productive use all CFD tools starting from geometry preparation through grid generation, flow solver up to post processing needed to have a high level of robustness and efficiency. Because aerodynamic designers or data specialists are most often not CFD specialized, the complete process has to be run automatically and be as reliable as possible.

MEGAFLOW II was defined in accordance with AIRBUS-D and other German industries' requirements, although MEGAFLOW II was an internal DLR project and not overall funded by the German government. However, AIRBUS-D performed the in-house project ENUVA [6] with compatible complementary goals and work plan in order to achieve best possible progress. Apart from the detailed work plan, the following main objectives from an application point of view are defined for MEGAFLOW II:
Prediction of aerodynamic data: The aerodynamic drag of a wing/body configuration should be predicted with an accuracy of 3%. Such a computation should be finished within two days for a new geometry (without CAD preparation) and within 5 hours for a geometry or parameter modification. Reliable accuracy should be achieved for maximum lift predictions of simplified high-lift configurations. For prediction of aerodynamic data, results with dependable accuracy should be available within 4 weeks for a new configuration (without CAD preparation), or within 2 weeks for geometry modifications.
Aerodynamic shape optimization: Methods and tools for aerodynamic shape optimization should deliver reliable computations of extreme values of relevant cost functions within 2 days for 2D geometries (airfoils), and between 4 and 12 weeks for simplified 3D configurations.
Validation: The general aerodynamic design process should be accelerated by 30-40% with respect to performance assessment of main aircraft components. The prediction quality of MEGAFLOW CFD tools should be demonstrated by comparison with experimental data for industrially relevant applications (sensitivity studies, efficiency measurements)

20.4 Where are we now — in 2002!

MEGAFLOW II has been successfully completed. Many excellent tools and procedures for CFD application to industrial flow cases have been developed.

20.4.1 FLOWer 3D RANS & Euler:

In the FLOWer code improvements and modifications concerning physical modeling, discretisation approaches and turbulence modeling were made. FLOWer is still used for productive computations at AIRBUS-D for simple wing/body configurations in Navier-Stokes mode, and for more complex configurations up to complete aircraft (cruise) in Euler mode. For wing design a special validation study has shown the quality of the solution for off-design cases. Fig. 20.2 gives a comparison of local pressure coefficients of a computation with experimental data for a Mach number of 0.90 and an angle of attack of $-1°$. Globally there is a very good agreement even for this complex flow case. However, improvements are necessary near the suction peak and the shock especially at the outboard wing. Since the Reynolds-Number is not

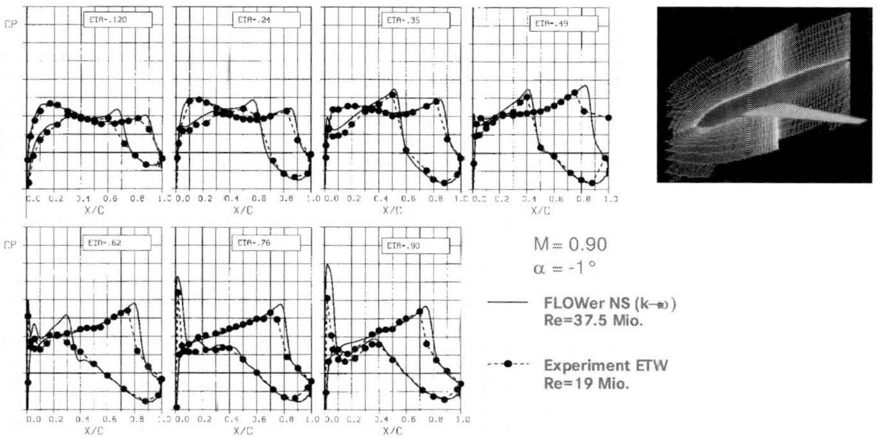

Fig. 20.2. Comparison of pressure distributions for wing/body at off-design conditions

consistent and the local wing twist is measured only for the aircraft design point, an exact comparison is difficult to judge.

For the same configuration, a study was performed in order to validate the buffet/CLmax at high Mach numbers. Fig. 20.3 shows CL-Alpha polars for 4 Mach numbers up to Mach=0.96. The general behavior is well predicted by FLOWer up to buffet/CLmax for all Mach numbers. In particular the increase of the absolute CLmax value can be found in the computations as well as in the measurements.

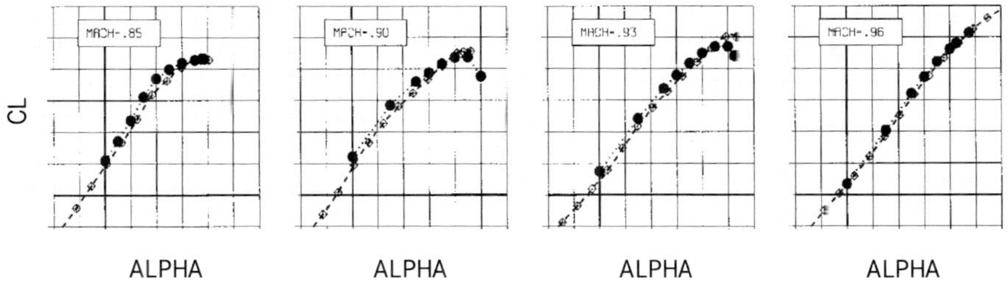

Fig. 20.3. Comparison of CL-Alpha polars for off-design Mach numbers

FLOWer in Euler and in Navier-Stokes mode is productively used for prediction of aero data for loads. Fig. 20.4 shows a comparison of a lateral aero load on the complete fuselage computed by FLOWer (in Euler mode) with experimental data. Both data show a very good agreement, especially in the front fuselage area. For these computations a wing/body/tails/engines configuration in complete model (left and right side) was used, in order to predict the beta effect in the correct way. Due to these very good results using CFD, some simple pre-Loads-Loops were already being based on aero data for loads made by CFD.

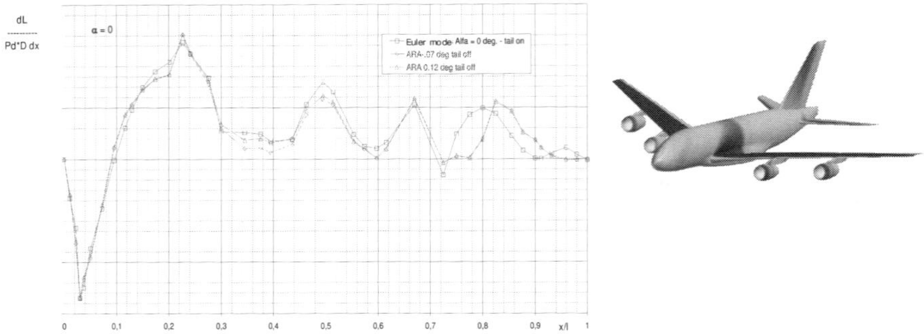

Fig. 20.4. Comparison of lateral aero load on the fuselage between CFD and experiment

In addition to the cases previously discussed, 3D FLOWer is used for various other applications at AIRBUS-D, e.g. for tail-plane design, wake flow investigations, parasitic drag support, 2D high-lift design, and several others. Finally, FLOWer is used in combination with the Synaps PointerPro optimization framework tool, which has been further developed within MEGAFLOW I and II projects. PointerPro is well used in AIRBUS-D, for 3D high-speed wing design tasks and for 2D high-lift setting optimization.

Within MEGAFLOW the complete system was validated performing an engine integration task using automatic wing shape optimization [7]. The basic wing was computed by a multi-point optimized design. For engine integration, the wing shape was optimized by minimization of wave drag. Based on the highly automatic optimization system PointerPro, in combination with BGRID/FLOWer such an optimization ran totally independent in about 47 days on an 8-CPU cluster of PCs. The results are shown in Fig. 20.5. The top image gives the wave drag distribution for the start solution; in the lower image we see the optimized solution. In total, a reduction of wave drag by 82.5% was achieved.

One of the main AIRBUS-D objectives in MEGAFLOW was to develop more flexible approaches concerning complex geometry and grid modeling. For FLOWer the Chimera approach was validated by AIRBUS-D by applying it to a high speed wing/body configuration [8], spoiler and flaps deflected. Fig. 20.4.1 shows streamlines and a cut section through a wake vortical flow field. The quality of the results is very good and promising for other future applications, but up to now fully automatic grid generation and the technology for finding overlapping areas and holes is not available for industrial applications.

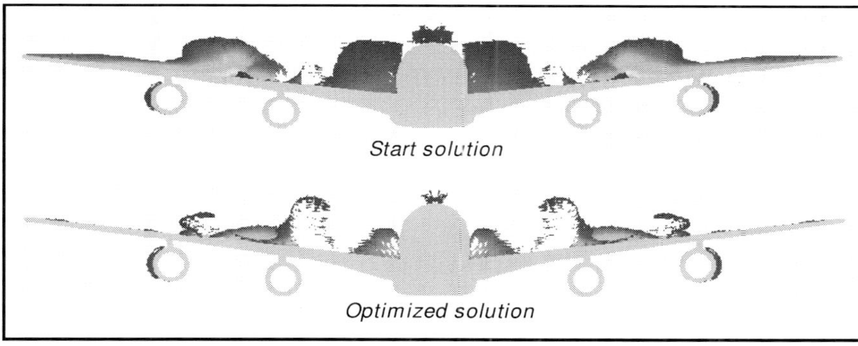

Fig. 20.5. Local wave drag resulting from wing shape optimization in presence of engines

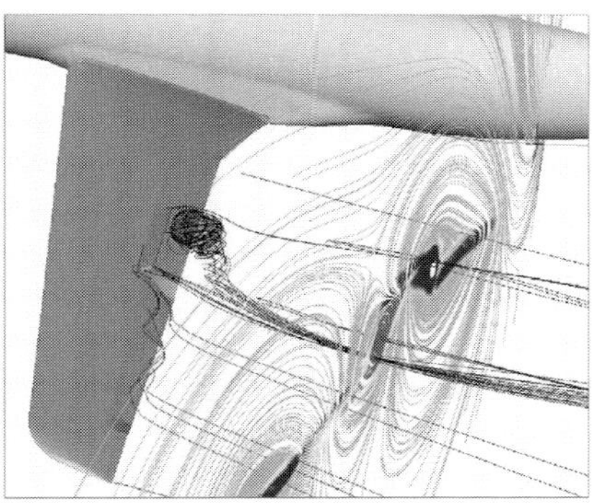

(a) High speed configuration with spoiler and flaps deflected, Mach = 0.85, Re = 8.1 Mio.

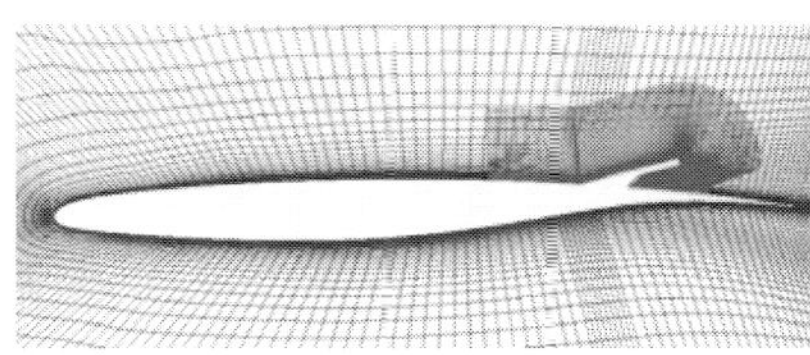

(b) Local mesh around spoiler: 3-block C-H, 1.3 Mio. nodes

Fig. 20.6. Chimera/Flower used for computation of spoiler effects

20.4.2 TAU 3D RANS & Euler:

Starting with MEGAFLOW II in 1999, TAU was already in use at AIRBUS-D for first tests and industrial validation. During the last years TAU became more and more a powerful CFD suite which is nowadays in productive use at AIRBUS-D in almost all "Aerodynamic Design & Data" processes. The significant reason for this success is the flexibility in grid generation for arbitrary geometries due to the hybrid discretization approach. Ongoing development work was made to improve the accuracy of TAU through physical and the turbulence modeling.The complete TAU process chain — including the grid generation by the CENTAUR Soft Grid Generator [9], adaptation modules, flow solver and pre-and-post-processing tools — was optimized in order to use TAU for all day industrial applications. Due to the high level of parallelization in TAU and due to increasing computational power over the recent years the turn-around-time is acceptable now for industrial applications. Some examples of validation work made in AIRBUS-D are given below.

As already stated, TAU is in productive use in almost all aerodynamic design and data processes at AIRBUS-D in Bremen. One of the processes with highest requirements on CFD tools is the high-lift design process. Fig. 20.7 shows validation results for a high-lift test case based on a simplified full span slat/flap configuration. This test case has been generated within the CEC EUROLIFT program [10]. The streamlines comparing computation with experiment show a very good general agreement; the flow separation on the outboard flap is very similar. The comparison of pressure distributions substantiates the good agreement.

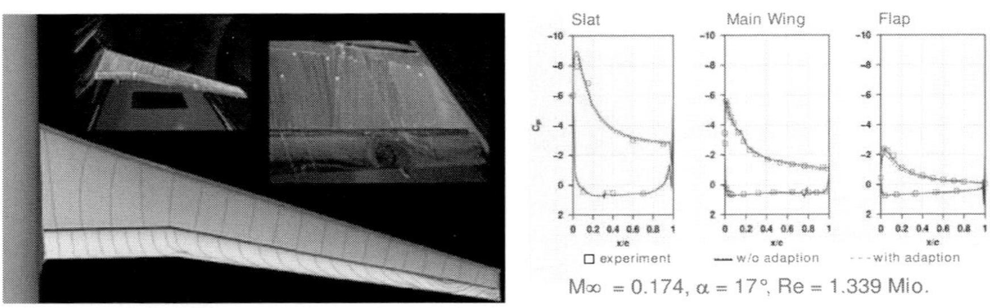

Fig. 20.7. TAU Navier-Stokes computation for simplified high-lift configuration

Based on these very good first results, AIRBUS-D continued using TAU for high-lift productive design issues in aircraft programs. For example, for A380 program, a question was how to predict and verify the interference of the hot engine jet with the PEF (PylonEndFairing) at the outboard engine position. Euler flow computations were made for the complete configuration (body/high-lift-wing/winglet/ftf/engines) in landing mode with simulated jet stream. The results are shown in Fig. 20.8. Based on these computations, the flow was analyzed in wing-tip/winglet and in slat/flap area in order to prove the general quality. Finally, an aerothermic analysis was made to check the heat transfer from the engine jet on the lower side of the pylon and on the PEF.

For the A400M, high-lift design TAU became the basic design tool. Fig. 20.9 shows the surface pressure distribution and streamlines of a simple A400M high-lift configuration. During MEGAFLOW II the modeling of propeller effects was added to TAU by implementation of an actuator disc boundary condition. In particular for the A400M configuration it is very important for AIRBUS-D to judge the power effects on the high-lift system and on the tail-plane efficiency. In Fig. 20.10 we see validation results of the power effect on a complete high-lift A400M configuration. In order to retrieve proper results the mesh has to be very fine in the area of propeller wake flow. For

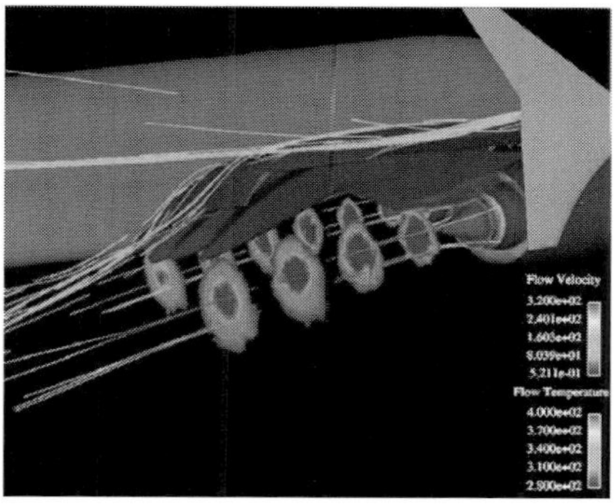

Fig. 20.8. Jet/PEF interference analysis for A380

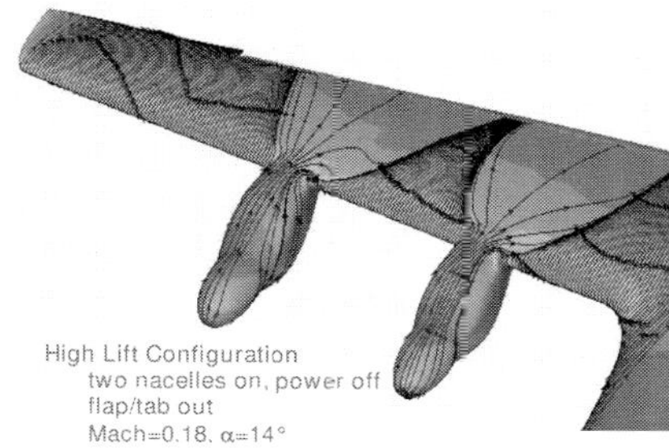

Fig. 20.9. TAU for A400M high-lift design

these computations about 13 million mesh points were used. The comparison of the pressure distribution gives very similar results between the measurements and CFD computation both for power on and off. The cutting plane is set between the first and second flap-track-fairing, close to the inboard propeller. Further investigations are being made to assess the effect of propeller modeling on the horizontal and vertical tail planes.

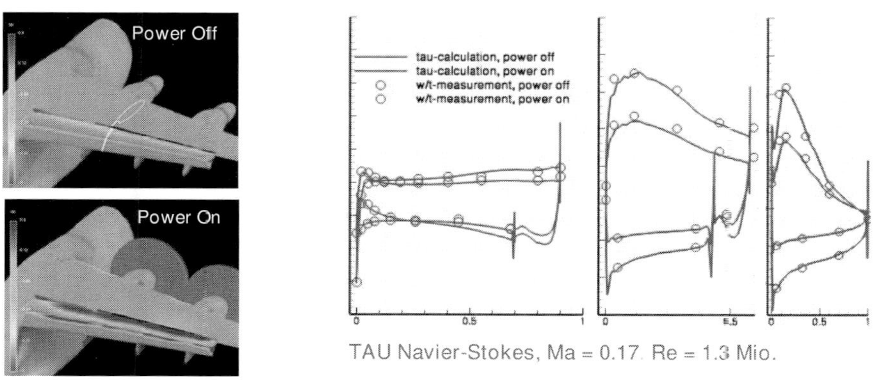

Fig. 20.10. Investigation of power effects on High-Lift design

Apart from standard 3D CFD applications, the definition of the high-lift setting design opens a large field for numerical optimization. Within MEGAFLOW II first validation test were made to run a fully automatic 2D setting optimization using Synaps PointerPro environment with TAU 2D Navier-Stokes for flow computation. Fig. 20.11 shows the results [11]. By doing this optimization of 6 parameters (setting angles and gap/overlap) at 2 multipoints, the maximum lift is increased by 14% for this high lift configuration, while the maximum angle of attack is reduced. Flow separation found in the reference solution on the slat is removed in the optimum slat/flap position.

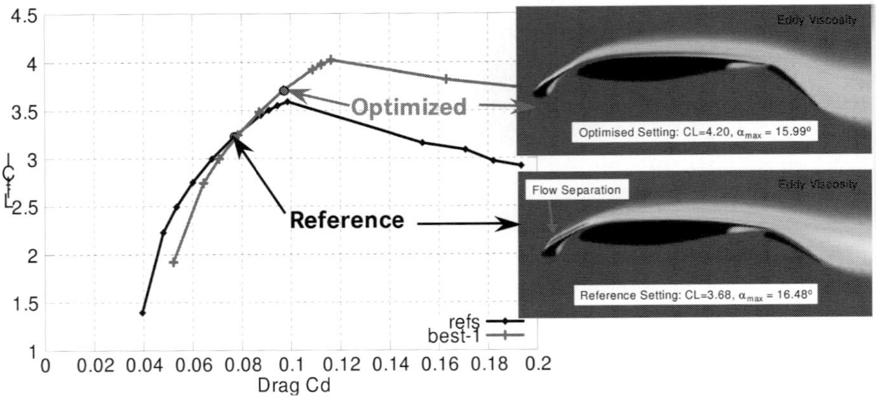

Fig. 20.11. High-Lift slat/flap setting optimization using PointerPro and 2D TAU

Another important area for AIRBUS-D is the rear-fuselage and tail-plane design process. During MEGAFLOW II TAU was validated for several tail-plane cases, and it is now in productive use for sponson, rear-fairing, horizontal and vertical tail design itself as well as for production of aerodynamic data for loads on these components. Fig. 20.12 gives an example of a computed flow for an A400M configuration in order to design and check the lateral behavior influenced by the sponson geometry. The streamlines show the complex flow situation at the rear fuselage. Flow separation induced by the sponson produces a vortex; adjacent the sponson the flow remains attached and the streamlines follow the shape of the rear fuselage geometry. These CFD computations were used to optimize the sponson geometry in order to improve the lateral stability of the complete aircraft.

As important as the lateral behavior of an aircraft is the prediction of the rudder efficiency itself. TAU was used for verification of the rudder efficiency for aerodynamic data for loads for critical off-design flow cases, results are shown in Fig. 20.13. For simplification, the gaps between rudder and vertical tail plane and between rudder and fuselage were sealed. The computations produced promising tendencies, but further work has to be done to capture the local flow separation for very high rudder deflection and side-slip angles.

All the aforementioned applications demonstrate that CFD tools developed within the MEGAFLOW projects are well integrated in AIRBUS-D aerodynamic design and data processes. There are many more examples of application, and the number increases day by day. Looking back to the requirements defined for MEGAFLOW, most of them are fulfilled whilst some are still open and need further development and improvement of CFD technol-

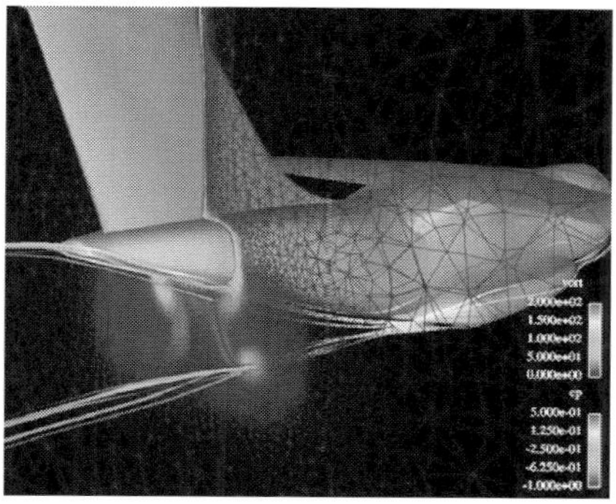

Fig. 20.12. Rear fuselage flow by TAU

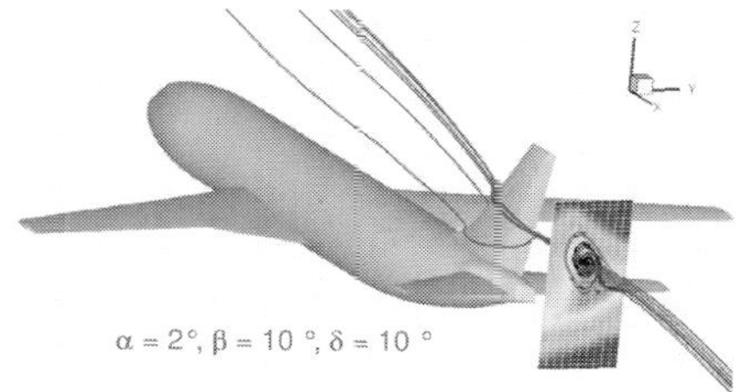

Fig. 20.13. Analysis of rudder efficiency

ogy. Fig. 20.14 gives an overview of previously defined requirements including judgment from an AIRBUS-D point of view.

Prediction of aerodynamic data

- Drag prediction of wing/body-configuration with accuracy of 3% within
  - 2 days for a new geometry (without CAD preparation)
  - 5 hours for geometry or parameter modifications
- Reliable accuracy for maximum lift predictions of simplified high-lift configurations
- Prediction of aerodynamic data with dependable accuracy within
  - 4 weeks for new configurations (without CAD preparation)
  - 2 weeks for geometry modifications

Aerodynamic shape optimisation

- Reliable computation of extreme values of relevant cost functions within
  - 2 days for 2D geometries (airfoils)
  - 4 to 12 weeks for simplified 3D configurations

Validation

- Acceleration of aerodynamic design process by 30-40 % with respect to performance assessment of main aircraft components
- Demonstration of prediction quality of software by comparison with experimental data for industrial relevant applications (sensitivity studies, efficiency measurements)

Fig. 20.14. Review of MEGAFLOW requirements

Prediction of aerodynamic data: The experiences over the last years have shown that there might be difficulties in verifying the predicted drag by experiments due to slight model modifications during the test, especially for off-design comparison. Generally, the predicted drag by CFD compares better with measurements in the design case than in any off-design case. Relative comparisons of CFD computations show very good results, but there are difficulties of CFD to agree with the absolute values resulting from the measurement. The turn-around-time of the complete CFD process is quite promising. For maximum lift predictions, there are still too large differences to measurements. The prediction of a wide range of aerodynamic data with dependable accuracy has been well achieved.

Aerodynamic shape optimization: The methods and tools for aerodynamic shape optimization have been and continue to be further developed and improved, but there is still a lack of using them by designers themselves for their daily work - further industrialization work has to be undertaken.

Validation: The general aerodynamic design process is accelerated by using CFD for daily design work, but it's difficult to measure by any accurate value. The more CFD is used within aircraft processes, the more ideas evolve and questions are newly being asked, for which CFD could provide the solutions.

20.5 Where do we have to go to?

As already mentioned, there are almost no border-lines to limit CFD applications in industrial context. Day by day we recognize an increase of various CFD applications requested by all aerodynamic design and data processes. In order to fulfill these aircraft process requirements, further on improvements in terms of efficiency, accuracy, performance, flexibility and reliability have to be made [12]. Some key requirements on future developments and improvements are listed below.

Physical Modeling: Further development effort is needed in order to improve the robustness and accuracy of turbulence models. In an industrial context the aim is to have one single turbulence model to be used with standard parameters for a large variety of applications and flow conditions, giving best possible accuracy. Automatic detection of transition lines together with improved transition modeling will help to remove uncertainties from the user and will permit the prediction of flow in off-design areas with higher accuracy. In particular for unsteady flows, further verification and validation for industrial applications has to be done, which will be in line with further improvements of models. A major issue is to provide the capability to perform these high level developments, either in research center or in the aircraft industry, with a direct coupling to industrial applications to achieve highest possible efficiency.

Numerical Algorithms: In order to improve the computational performance for highest complexity CFD applications, further development of the numerical algorithms has to be made. In terms of robustness and performance the global convergence should be accelerated whilst the convergence process itself should develop smoothly.

Error Diminishing: For an industrial application, one of the major requirements on CFD is to have a tool suite, which produces accurate and reliable results. In addition, results have to be comparably consistent with each other within an aerodynamic design cycle. Further investigations are necessary to develop automatic grid adaptation methods and to achieve more experiences appertaining mesh independent flow computations.

Operating Environments: Due to the ever-increasing number of CFD applications in AIRBUS-D, the complete CFD process has to run on a very high automatic level. This process includes geometry preparation, grid generation, flow computation and post-processing. All results used for design or data production have to be registered within a data management tool for traceability. Such a framework tool has to be easy to use, and it has to be an intelligent system in order to inform the user about any problem or difficulty which may occur in one of the many and varied process steps. Standard CFD application has to produce standard output information, and a numerical optimization system has to be easily integrated within such an environment.

Standard Data Format: In particular for AIRBUS, the question of having standard data formats, independent from flow solver and from hardware platforms, is very important. Fulfilling this requirement would allow an easy and efficient communication between different sites and tools. Such a common data format has to be de-fined having the constraint to be flexible enough for any new requirement, e.g. from multi-disciplinary simulation.

ISO-9000 Development Standards: Starting with MEGAFLOW, DLR introduced ISO-9000 development standards to their CFD development processes. Because the number of developers and specialists within research centers and AIRBUS is currently increasing, these standards have to be retained, in order to have a clear and transparent development process including a comprehensive documentation. In opposition, and by using these standards, there is a risk of losing progress of capability and competitiveness. Both issues have to be considered for future developments.

Software Architecture: A clear and open structure has to be retained for any further developed CFD tool. The main issue is to permit a high maintainability. In order to avoid large time schedules and high development cost, an effective transition has to be made from the current capabilities to any new development.

HPC Hardware: CFD suites have to be optimized, as far as possible, for both vector and parallel processors. In the AIRBUS and research center perspectives, a various number of platforms exist which all have to be used in the future in order to cater for the high demands originating from aircraft development processes. In addition to the HPC performance itself, the HPC environment (network, file server capacity) has to be optimized as well.

On top of the aforementioned technical requirements, there will be further requirements from an application point of view to use CFD for a larger field of flow problems, to continue on investigating swifter methods for numerical investigations, and to see CFD as just one part in a global multidisciplinary approach (aeroelastics, aeroacoustics, and many more).

20.6 Conclusion

This paper gives a review of MEGAFLOW requirements and applications - a 1999 and 2002 status. It summarizes requirements for future developments from an AIRBUS-D point of view. All CFD tools developed in MEGAFLOW are highly productive for AIRBUS-D. Within the past few years the role of CFD in aircraft design processes has increased very much, and it will continue to increase. All aerodynamic processes and the multi-disciplinary design have strong requirements on CFD, thus a necessity to also increase the development of CFD. All capabilities of CFD are not yet fully exploited — further industrial validation work has to be done. The MEGAFLOW cooperation network that exists between DLR research center, universities and aircraft indus-try in Germany was and continues to be very productive, due to the direct industrial involvement in definition of requirements and industrial validation. In order to fulfill future requirements the strategic development of CFD tools has to be enforced, most likely in a cooperation as MEGAFLOW.

Acknowledgements

We would like to thank all colleagues at AIRBUS involved in MEGAFLOW via the projects ENUVA, BREFLOW and Pro-HMS. In addition, parts of the results were obtained from the CEC projects EUROLIFT, HIRETT and AEROSHAPE - thanks to these project teams as well.

Finally, we must thank all colleagues from DLR and other MEGAFLOW partners for the excellent cooperation and results, in particular Prof. Dr. Norbert Kroll.

References

1. A. Schüller, (Ed.): "Portable Parallelisation of Industrial Aerodynamic Applications (POPINDA)", Notes on Numerical Fluid Mechanics, Vol. 71 (Vieweg, Braunschweig/Wiesbaden, 1999)
2. N. Kroll, C.-C. Rossow, K. Becker and F. Thiele: "MEGAFLOW - A Numerical Flow Simulation System", ICAS-Paper 98-2.7.4, September 1998
3. EFENDA: http://www.hlrs.de/organization/vis/projects/Efenda
4. N. Kroll, C.-C. Rossow, D. Schwamborn, K. Becker, G. Heller: "MEGAFLOW – A Numerical Flow Simulation Tool for Transport Aircraft Design", ICAS-Paper 2002-1.10.5, 2002
5. SynapsPointer Pro V2: "Installation and Getting Started. Design and Automation Software", Version 2.5, Synaps Ingenieur-Gesellschaft mbH, 2002

6. ENUVA: "Enhanced Numerical Methods for Aerodynamics", Airbus-D, project description, 2002
7. O. Frommann, G. Dargel: "Aerodynamische Triebwerksintegration mittels Numerischer Optimierung" DGLR-JT-2000-092, DGLR Jahrestagung, 18.-21.9.2000, Leipzig
8. R. Mertins, S. Barakat, E. Elsholz: "3D Viscous Flow Simulations on Spoiler and Flap Configurations", CEAS Aerospace Aerodynamics Research Confidence, Paper No. 65, Cambridge, UK, 2002
9. CentaurSoft, http://www.centaursoft.com
10. R. Mertins : "Dreidimensionale Navier-Stokes-Nachrechungen und Adaption für eine Hochauftriebskonfiguration unter Anwendung unstrukturierter Methoden", Paper, DGLR-Jahrestagung, 2001
11. AEROSHAPE: http://aeroshape.cira.it
12. K. Becker: "Perspectives for CFD — the Airbus view", Paper, DGLR-Jahrestagung 2002

21

Aerodynamic Analysis of Flapping Airfoil Propulsion at Low Reynolds Numbers

Jan Windte[1], Rolf Radespiel[1], and Matthias Neef[1,2]

[1] Institute of Fluid Mechanics, Technical University Braunschweig
Bienroder Weg 3, 38106 Braunschweig, Germany
Internet: www.tu-braunschweig.de/ism, E-mail: j.windte@tu-braunschweig.de
[2] now: Siemens Power Generation, Mülheim/Ruhr, Germany

Summary. The laminar flow around a NACA 4402 airfoil is investigated at a Reynolds number of Re = 6000 using the RANS solver FLOWer. Significant flow separation occurs over almost the whole regime of angles of attack for steady on-set flow conditions. Therefore, the calculated propulsive efficiencies of pure plunge motions are rather poor. This leads to the need of a combined plunge and pitch motion, which is investigated subsequently. For combined motions, cases with high propulsive efficiencies are found over a wide range of produced thrust.

21.1 Introduction

Low-Reynolds-number flows around airfoils and wings may be viewed and classified in many ways. Here we present an overview of flow regimes with low Reynolds numbers which is derived from typical flight applications.

The flow over a sailplane may be considered as an example of the upper border of low Reynolds numbers. This Reynolds number regime ranges from Re = 5 x 10^5 - 2 x 10^6. This is also the flight regime of larger drones with a mass of about 50 kg. As far as aerodynamics is considered, there is no living creature or application known that uses flapping wings as propulsive system in this regime.

The Reynolds number region of Re = 2x 10^5 - 5 x 10^5 marks the flight regime of small drones, large birds and model aircraft with a mass of about 1-5 kg. Also, some large electrically powered ornithopters like the "Kestrel", "Slowhawk" and "Cybird" can be located in this regime. While the possibility of flapping wing flight in this Reynolds number regime birds is proofed by birds as well as these ornithopters, it is still in doubt whether this is the optimum propulsion system. The flow is usually attached to the wing with partly laminar and turbulent boundary layers. Note that adverse effects of laminar separations can be largely avoided by careful aerodynamic design. This range is well suited as a starting point for numerical flow analysis of flapping wing propulsion with state of the art flow solvers including 3D-flows with moving and deforming meshes. Also, some elementary analytical solutions can be found for comparison with numerical results. A good example for recent progress in this area is reported by M.Neef and D.Hummel [1], [2]. They successfully validate the unsteady flow solver FLOWer for inviscid and viscous 2D and 3D flows and identify favourable wing motions with a balance of thrust and drag. Their analysis identifies the relative role that vortical flow structures play for flapping wings with attached flow at small to moderate frequencies of motion. **Figure 21.1** shows the pressure distribution for a flapping wing during downstroke.

Reynolds numbers from about Re = 5x 10^4 - 2 x 10^5 and a mass of around 50g is the typical regime for newly developed micro air vehicles (MAV). These, for example, are the "Black Widow" [3] with a mass of 80g using a fixed wing, the small ornithopter "Microbat" [4] with 13g and a recently developed vehicle by Jones et al. [5] at a weight of 14g with two counterphase flapping wings. Important for this flight regime is that transition usually occurs in combination with a large laminar separation bubble (LSB). In this case the flow separates from the surface in the laminar state, transitions to the turbulent state in the free shear flow and then, because of the now much stronger turbulent momentum transport, reattaches to the surface, thereby closing the LSB. The point of transition has to be predicted in this regime in order to achieve a reliable numerical flow simulation. In the case of flapping wings, accurate transition prediction is needed even for the unsteady flow. For Reynolds-averaged Navier-Stokes (RANS) simulations, transition prediction as well as accurate simulation of LSB's are not yet state of the art. Hence, numerical results in this regime should be interpreted with care. Nevertheless there exist airfoils in this regime that are optimized for flows with LSB's and therefore can be considered for applications, i.e. the airfoil SD7003 [6], see **Fig. 21.2**.

Finally, the subsequent Reynolds number regime can be identified at Re = 5x 10^3 - 5 x 10^4, with flight masses of a few grams. This is the flight regime of large insects like dragonflies up to very small birds and bats. From a technical point of view, this is the regime of small indoor ornithopters as well as upcoming "meso scale" MAV's. The aerodynamics in this region can be expected to be highly unsteady, as flow separations occur at quite low angles of attack. Furthermore, for Reynolds numbers below 1x 10^4, boundary layers are expected to be fully laminar. This gives the opportunity to achieve accurate numerical simulations in this regime, as neither transition nor turbulence needs to be modelled. As there is only molecular momentum transport in the laminar boundary layers, they cannot overcome strong adverse pressure gradients. Hence very thin airfoils yield the highest lift-over-drag values in this regime. In the middle of Fig. 21.2, an airfoil is shown that has been numerically optimized for Re = 6000 [7]. As validation of results at such low Reynolds numbers is very difficult due to the low forces, micro-gliders are used for that purpose in [7]. This way, information about a tested airfoil can be obtained by measuring the glider's mass, velocity and glide-slope. One of the airfoils used, with a chord length of about 5mm, is also shown in Fig. 21.2.

The scope of the present work is the flow analysis of flapping airfoils in this low Reynolds number regime with fully laminar flow. The objective is to understand how thrust can be obtained for flows with boundary layers that separate easily, which role the flow separation plays and what propulsive efficiencies can be obtained. The flow solutions presented are verified with respect to spatial and time accuracy. Finally we report the results of a parameter study where the plunge and pitch motion is varied and the effects of the resulting flow and forces are analyzed.

21.2 Numerical Method

The detailed features and capabilities of the FLOWer-Code used to solve the flow equations are described throughout this book and are documented elsewhere [8]. Here, to allow for arbitrary motions and deformations of the computational grid, the governing equations are formulated for a moving Cartesian coordinate system with deforming control volumes:

$$\frac{d}{dt}\int\limits_V \vec{w} dV + \int\limits_S (\vec{F^c} - \vec{F^v})\, \vec{n}\, dS - \int\limits_S (\vec{w}\vec{v}_b)\, \vec{n}\, dS + \int\limits_V \vec{g}\, dV = 0\,. \tag{21.1}$$

Herein $\vec{w}$ is the vector of conservative variables,

$$\vec{w} = (\rho, \rho\vec{v}, \rho E)^T\,, \tag{21.2}$$

while $\vec{F^c}$ and $\vec{F^v}$ represent the convective and viscous fluxes:

$$\vec{F^c} = \begin{pmatrix} \rho\,\vec{v} \\ \rho\,\vec{v}\,\vec{v}^{\,T} + p\,\vec{I} \\ \rho\,E\vec{v} \quad + p\,\vec{v} \end{pmatrix} \quad \vec{F^v} = \begin{pmatrix} \vec{0} \\ \vec{T} \\ \vec{T}\,\vec{v} - \vec{q} \end{pmatrix}\,. \tag{21.3}$$

The first two integrals in (21.1) are the same as in the Navier-Stokes equations for fixed control volumes. The third integral results from the deformation of the control volumes in the fixed Cartesian coordinate system, with $\vec{v}_b$ as the velocity vector of the moving cell boundaries. The fourth integral is generated in the transformation from the fixed to the moving coordinate system. The additional source term

$$\vec{g} = (0, \rho\,\vec{\dot{\varphi}} \times \vec{v}, 0)^T \tag{21.4}$$

contains the cross product of the angular velocity vector $\vec{\dot{\varphi}}$ and the velocity vector $\vec{v}$. Therefore, it is only active for rotations of the moving mesh. For details on the formulation using variable control volumes and moving coordinate systems see [2], [9].

21.3 Description of airfoil motion

With the z-axis perpendicular to the free stream direction, the plunge motion of the airfoil is described as:

$$z(t) = z_1 \cos(2\pi f t) \tag{21.5}$$

The angle of attack ζ from plunging follows :

$$\zeta(t) = \arctan \frac{-\dot{z}}{U_\infty} = \arctan \left(\frac{2\pi f z_1}{U_\infty} \sin(2\pi f t) \right) . \tag{21.6}$$

With arctan ≈ 1 as a simplification for small angles, ζ can be expressed as

$$\zeta(t) = \frac{2\pi f z_1}{U_\infty} \sin(2\pi f t) . \tag{21.7}$$

Introducing the reduced frequency

$$k = \frac{\pi f c}{U_\infty} , \tag{21.8}$$

the amplitude ζ_1 of the angle of attack oscillation from plunge motion becomes

$$\zeta_1 = 2k \frac{z_1}{c} . \tag{21.9}$$

The pitch motion is described as

$$\alpha(t) = \alpha_0 + \alpha_1 \cos(2\pi f t + \Phi) . \tag{21.10}$$

The angle of attack for the combined plunge and pitch motion follows to be

$$\alpha_{\text{eff}}(t) = \zeta(t) + \alpha(t) \tag{21.11}$$

$$\alpha_{\text{eff}}(t) = \arctan \left(2k \frac{z_1}{c} \sin(2\pi f t) \right) + \alpha_0 + \alpha_1 \cos(2\pi f\, t + \Phi) . \tag{21.12}$$

For small angles of attack (arctan ≈ 1), this equation can be separated into just one amplitude and one phaseshifted sinus-function. For larger angles a similar separation into an amplitude and a periodic function $X(t)$ can be carried out:

$$\alpha_{\text{eff}}(t) = \alpha_0 + \triangle\alpha_{\text{eff}}\, X(t) \tag{21.13}$$

Since unsteady flows behave quasisteady for low frequencies (up to about $k = 0.1$), it can be expected that the occurrence of flow separations is strongly dependent on this effective amplitude $\triangle\alpha_{\text{eff}}$ at low frequencies. To maintain a consistent nomenclature, for pure plunge motion the effective amplitude is set as:

$$\triangle\alpha_{\text{eff}} = \zeta_1 = \arctan \left(2k \frac{z_1}{c} \right) \tag{21.14}$$

21.4 Results

21.4.1 Airfoil geometry and flow conditions

All results are calculated for the flow around a thin NACA 4402 airfoil, using a Reynolds number of Re = 6000. The airfoil shape is chosen because its aerodynamic characteristics are fairly close to optimized airfoils for maximum lift over drag [7]. Since the compressible equations are solved, stability and convergence problems occur at very low Mach numbers. Thus, a Mach number of $Ma_\infty = 0.3$ was used, at which compressibility effects are still negligible.

Figure 21.3 shows the used computational grid consisting of 320 x 64 points in a C-Topology, with the farfield boundary 20 chord lengths away from the airfoil. For the cases with rigid body motion, all time averaged values were obtained using the third computed motion cycle, where a periodic solution was achieved.

21.4.2 Flows with steady onset conditions

Streamlines around the Naca 4402 for steady onset conditions are displayed in **Fig. 21.4**. The fully laminar flow at $Re = 6000$, even at the small angle of attack of $\alpha = 3^o$, results in a large trailing edge (TE) separation. For an angle of attack of $\alpha = 8^o$ the flow becomes strongly unsteady, thus the instantaneous flow state at the end of the time-accurate computation is shown. As can be seen, vortices are shed at the leading edge (LE) and travel over the surface, while the TE separation is completely gone. The resolution of the boundary layer is shown in detail for two sections x = const. This way it can be observed that there are at least 20 grid points in the boundary layer to resolve the velocity gradient at the wall, making sure that the wall friction is accurately computed.

Since flow separation occurs over nearly the whole regime of angles of attack, the obtained lift curve is highly nonlinear, see **Fig. 21.5**: For angles of attack up to about $\alpha = 6^o$, the TE separation grows. This results in a decrease of the effective camber of the airfoil, and therefore the slope of the lift curve is also decreased compared to the theoretical value of 2π. For angles of attack above $\alpha = 6^o$, strong vortices develop from the LE and contribute to the suction along the upper surface, thereby increasing the lift slope.

The dependency of the drag coefficient c_d on α is shown in **Fig. 21.6** for three different grid densities. As laminar boundary layers have a lower requirement of spatial resolution compared to the turbulent ones, even the coarse grid with only 160×32 points shows good results Note that all following computations were carried out using 320 × 64 points.

21.4.3 Pure Plunge Motion

Low drag coefficients are achieved in the range of angles of attack from about $\alpha = -2^o$ to 6° for steady onset conditions, because no LE separation occurs in this range. Hence, all computations with airfoil motion were performed with a mean angle of attack of $\alpha_0 = 2^o$, right in the middle of this range.

Figure 21.7 illustrates the time-accuracy of the method by comparison of the horizontal force coefficient c_z perpendicular to the free stream direction for different timesteps and a heave motion with $k = 0.1$ and $z_1/c = 0.7$ ($\triangle\alpha_{\text{eff}} = 8^o$). As a case with a very strong separation was chosen, it can be concluded that the dimensionless timestep of $t = 0.1$ is short enough to resolve the physical problem, because the result for the shorter timestep of $t = 0.05$ shows no noteworthy difference. The dimensionless time needed by the freestream to travel one chord length is calculated as

$$t_c = \frac{c}{Ma_\infty \sqrt{\kappa}}. \tag{21.15}$$

Using a chord length of $c = 1$, a Mach number of $Ma_\infty = 0.3$ and the ratio of specific heats of $\kappa = 1.4$, this results in $t_c = 2.82$. This means that the time to travel one chord length is resolved by 28 timesteps.

Figure 21.8 shows the distribution of the vertical force coefficient c_x and the horizontal force coefficient c_z over one period of a pure plunge motion with $k = 0.1$ and $z_1/c = 0.035$ ($\triangle\alpha_{\text{eff}} = 4^o$). As can be seen, there is a phase shift between force and motion, with the force leading the motion. As this is quite unusual, further explanation is needed: Because the force F_z is unsteady, and the overall circulation in the flow field has to remain constant (neglecting the dissipative losses), so called start- and stop-vortices are shed at the TE and convect in the main flow. As these vortices induce velocities on the airfoil, this leads to a phase shift, but with the motion leading the force. As this is expected from inviscid theory, it has been found for 2D cases using Euler- as well as panel-codes [10], [1]. Hence it can be speculated that the present phase shift, with the force leading the motion, is a viscous effect. The decrease of the force coefficient c_z with still increasing effective angle of attack at $t/T = 0.25$ means that a corresponding lift curve would have a negative slope in that region. This could only happen if – as a dynamic stall effect – c_l would first rise with α with about the theoretical value of 2π and then, for high angles of attack, change to the lift curve with separation as for steady onset flow of Fig. 21.5. As the phase shift is quite small for the shown case, it should be mentioned that the phase shift grows with increasing frequency k. As can also be seen in Fig. 21.8, the coefficient c_x is always positive, meaning that no net thrust is produced at any time.

The calculated propulsive efficiency,

$$\eta_P = \frac{\overline{c}_x - c_{x,\text{steady}}}{\overline{c}_{\pi,\text{in}}}, \tag{21.16}$$

calculated with the time averaged force coefficient $\overline{c}_x$ and nondimensional power coefficient $\overline{c}_\pi$, is shown for the pure plunge motion and different reduced frequencies and effective amplitudes in **Fig. 21.9**. It should be pointed out that other definitions exist for the propulsive efficiency which do not take into account the steady drag of the airfoil ($c_{x,\text{steady}} = c_d(\alpha_0)$). That way, for a steady state flight ($\overline{c}_x = 0$), the efficiency would become zero, which is an obviously meaningless result.

For the low reduced frequency of $k = 0.1$, efficiency is decreasing for high effective amplitudes due to large separations, corresponding to the LE separations discovered in the steady onset flows. At the higher frequencies $k = 0.5$ and 1.0, start- and stop vortices are shed more often, leading to a lower efficiency at low amplitudes. This basic dependency of the propulsive efficiency on the reduced frequency has already been found and described by Küssner [11] for the inviscid flow around a flat plate. Also for higher frequencies efficiency decreases with increasing amplitudes, but less rapidly because dynamic stall now occurs at higher effective angles of attack. For none of the investigated cases enough thrust has been produced to overcome the steady drag coefficient. Even if net thrust were possible for even higher amplitudes or frequencies, the efficiency would be very poor. So, considering an insect or a micro air vehicle with a finite wing and a body with additional drag, flying at $Re = 6000$ using plunge motion only is impossible. Therefore it is necessary to use a combined plunge and pitch motion.

21.4.4 Combined Plunge and Pitch Motion

The results for the combined motion and a reduced frequency of $k = 0.1$ for constant effective amplitudes between 4° and 10° are shown in **Fig. 21.10**. The phase angle between plunge and pitch motion was fixed at $\Phi = 90^o$, so that the pitch motion is used to reduce the effective angles of attack introduced by the plunge motion (see eq. (21.12)). To obtain a constant effective amplitude, both the plunge amplitude z_1/c (and with that ζ_1 see (21.9)) as well as the pitch amplitude α_1 have to grow along the graphs. As can be seen, the graphs reach nearly constant levels for high amplitudes, with the efficiency decreasing with increasing effective amplitude. It can be concluded that, at the low reduced frequency of $k = 0.1$, the efficiency depends mainly on the amount of flow separation. It can also be concluded that the effective amplitude is a well suited parameter for the investigation of flows with low reduced frequencies.

For higher frequencies, at which the flow is no more quasisteady, the effective amplitude looses its meaning and the results are not as clear as in Fig. 21.10 anymore. Here, the variation of other combined parameters like the amplitude ratio

$$\lambda = \frac{\alpha_1}{2k\frac{z_1}{c}} \approx \frac{\alpha_1}{\zeta_1} , \tag{21.17}$$

seems to be more appropriate for finding cases with high propulsive efficiency.

Figure 21.11 shows the envelopes over the cases with optimal combinations of heave and pitch motion. Note that only cases with a phase shift of $\Phi = 90^o$ were used for the envelopes. While this is the optimal phase shift for low reduced frequencies up to about $k = 0.1$, there is still a small potential to increase the efficiency for cases with high reduced frequencies by varying Φ. The propulsive efficiency, as the envelopes indicate, is nearly independent of the produced thrust, and thrust coefficients of up to about four times the steady drag coefficient are possible for the investigated case. Even higher thrust coefficients may be possible, but as pitching angles of up to 33° have already been used , this would correspondingly lead to high twisting angles for a three-dimensional case and may be difficult to achieve in practice.

Also at low reduced frequencies there exists a kinematic boundary for a finite wing: As the angle of attack from plunge motion ζ is dependent on the product $k \cdot z_1/c$ (see eq. 21.9), a given angle ζ translates into a high amplitude z_1/c for a low frequency k. For the typical planforms of insect-wings, the plunge amplitude at which the wing tips would touch is reached soon. This happens, as indicated in Fig. 21.11, at quite low thrust coefficients for $k = 0.1$. This might explain why large insects like dragonflies typically fly with reduced frequencies above $k = 0.1$.

21.5 Conclusions

The flow around the NACA 4402 airfoil has been simulated using the Navier-Stokes solver FLOWer for steady onset conditions as well as for a pure plunge and a combined plunge and pitch motion. It is shown that the combined motion is essential to obtain high propulsive efficiencies. At low reduced frequencies k, cases with high efficiencies were found by varying the effective amplitude $\triangle\alpha_{\text{eff}}$.

As propulsive efficiency decreases with increasing reduced frequencies k, it might be interesting to further increase the thrust for a given frequency. Following the suggestions by Send [12], this might be achieved by introducing an additional sweep motion in free stream direction.

References

1. M.F.Neef, D.Hummel: “Euler Solutions for a Finite-Span Flapping Wing”. Proceedings of the Conference: Fixed, Flapping and Rotatory Wing Vehicles at Very Low Reynolds Numbers. Notre Dame, 2000, pp. 75-99.
2. M.F.Neef: “Analyse des Schlagflugs durch numerische Strömungsberechnung”. Dissertation TU Braunschweig 2002, ZLR Forschungsbericht 2002-02. www.biblio.tu-bs.de/ediss/data/20021021a/20021021a.html.
3. J.M.Grasmeyer, M.T.Keenon: “Development of the Black Widow Micro Air Vehicle”. AIAA-2001-0127.
4. T. Pornsin-Sisirak, Y.C. Tai, C.M. Ho, M. Keennon: “ Microbat- A Palm-Sized Electrically Powered Ornithopter”. 2001 NASA/JPL Workshop on Biomorphic Robotics, Pasadena, CA, USA, Aug. 14-16(2000).

5. K.D.Jones, M.F.Platzer: "Experimental Investigation of the Aerodynamic Characteristics of Flapping-Wing Micro Air Vehicles". 41st Aerospace Science Meeting & Exhibit, 6-9 January 2003 / Reno, NV, AIAA-2003-0418
6. NASG (Nihon univ. Aero Student Group)'s HomePage. http://www.nasg.com/afdb/show-airfoil-e.phtml?id=1075
7. J.Kunz, I. Kroo: "Analysis, Design and Testing of Airfoils for Use at Ultra-Low Reynolds Numbers". Proceedings of the Conference: Fixed, Flapping and Rotatory Wing Vehicles at Very Low Reynolds Numbers, Notre Dame, 2000, pp. 349-372.
8. P.Aumann, W.Bartelheimer, H.Bleecke, M.Kuntz, J.Lieser, E.Monsen, B.Eisfeld, J.Fassbender, R.Heinrich, N.Kroll, M.Mauss, J.Raddatz, U.Reisch, B.Roll, T.Schwarz: "FLOWer Installation and User Handbook, Release 116". DLR Braunschweig, Doc.- Nr. MEGAFLOW-1001, Apr.2000.
9. K.Pahlke: "Berechnung von Strömungsfeldern um Hubschrauberrotoren im Vorwärtsflug durch die Lösung der Euler-Gleichungen". Dissertation TU Braunschweig 1999, DLR Forschungsbericht FB 99-22.
10. K.D.Jones, B.M.Castro, O.Mahmoud, S.J.Pollard, M.F.Platzer, M.F.Neef, K.Gonet, D.Hummel: "A Collaborative Numerical and Experimental Investigation of Flapping-Wing Propulsion". AIAA Paper 2002-0706 (2002).
11. H.G.Küssner: "Zusammenfassender Bericht über den instationären Auftrieb von Flügeln". Luftfahrtforschung 13, 1936, S. 410-424.
12. W.Send: "Otto Lilienthal und der Mechanismus des Schwingenfluges". Jahrbuch 1996 I der deutschen Gesellschaft für Luft und Raumfahrt – Lilienthal-Oberth e.V., 1996, S. 161–172.

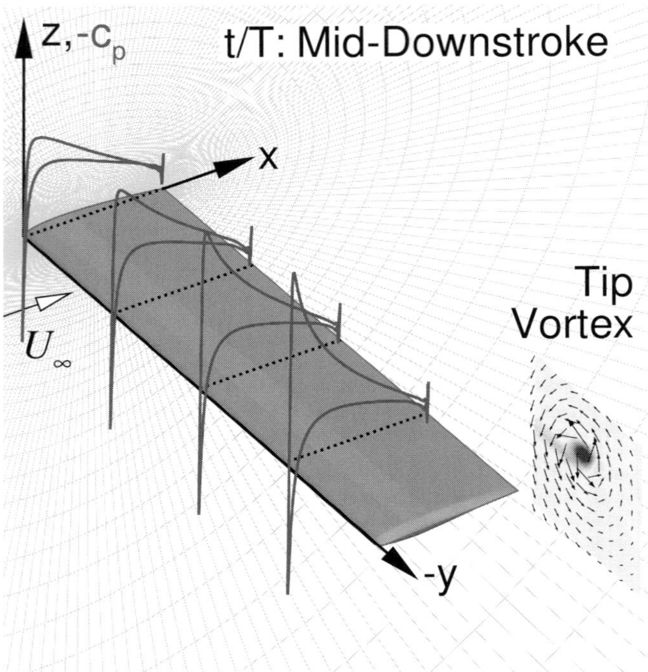

Fig. 21.1. Pressure distribution for a flapping wing during downstroke after M. Neef and D. Hummel [1]

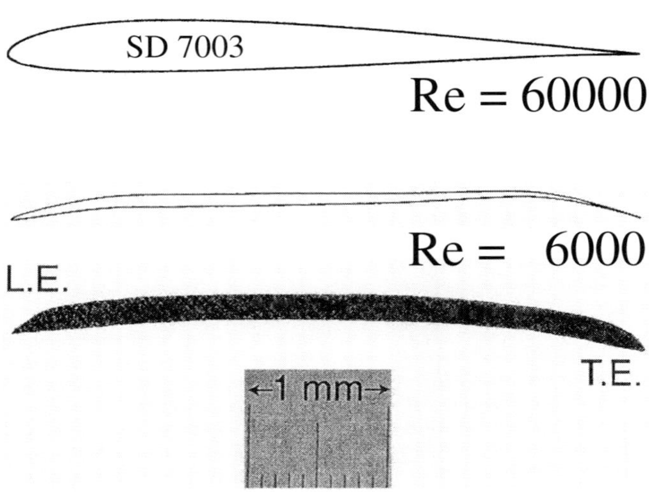

Fig. 21.2. Airfoils for low Reynolds numbers

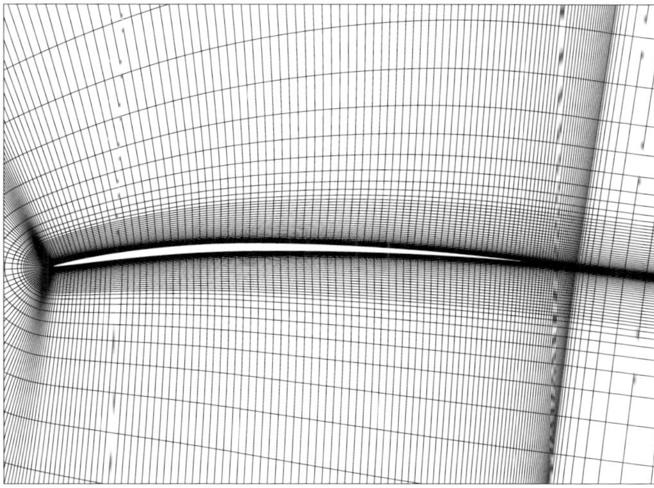

Fig. 21.3. Grid around the NACA 4402 airfoil

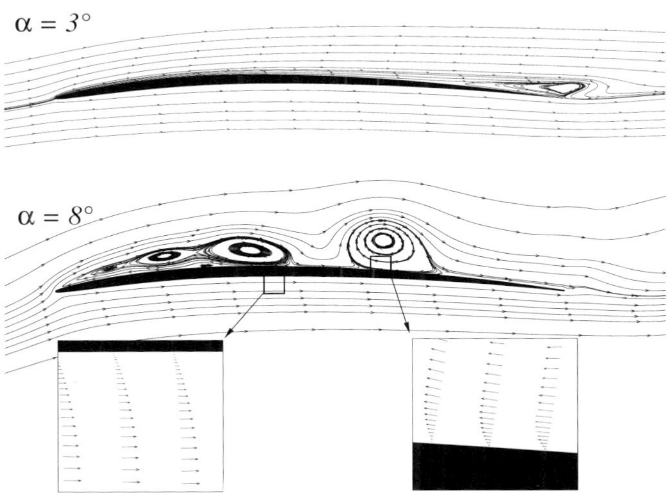

Fig. 21.4. Streamlines for steady onset flow conditions, Re = 6000

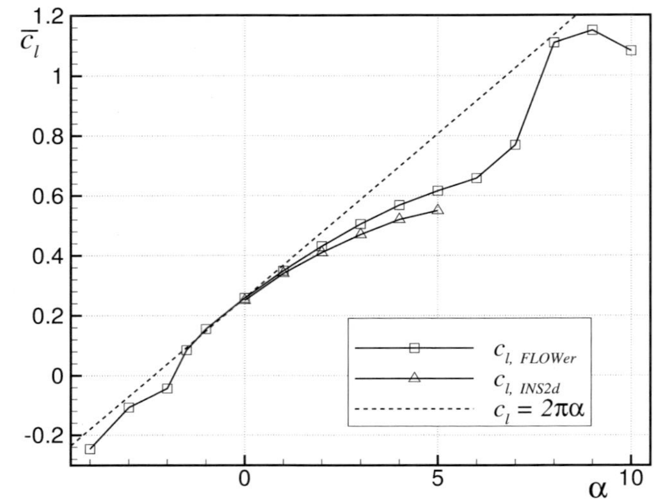

Fig. 21.5. Lift curves for the NACA 4402 obtained by FLOWer and INS2d [7]

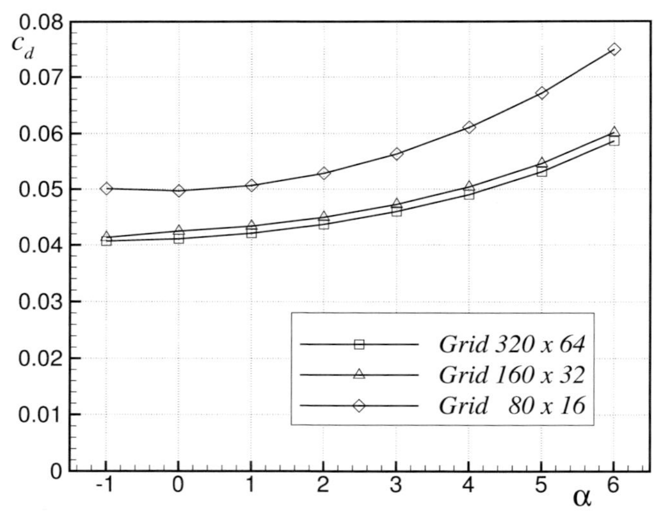

Fig. 21.6. Drag coefficient c_d for different grid densities

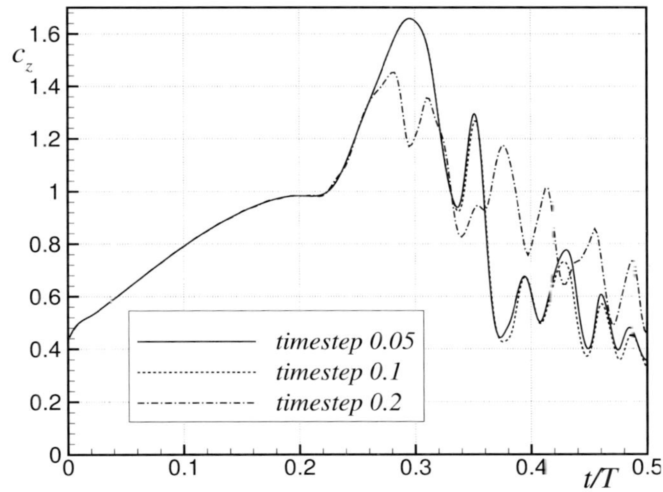

Fig. 21.7. Force coefficient c_z for different timesteps and a pure plunge motion ($k = 0.1,\ z_1/c = 0.7\ (\triangle\alpha_{\mathrm{eff}} = 8^o)$)

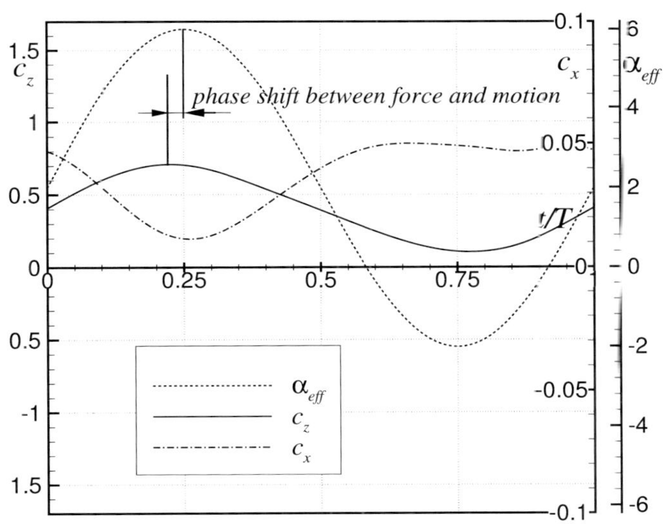

Fig. 21.8. Force coefficients and α_{eff} for one period of pure plunge motion ($k = 0.1,\ z_1/c = 0.35\ (\triangle\alpha_{\mathrm{eff}} = 4^o)$)

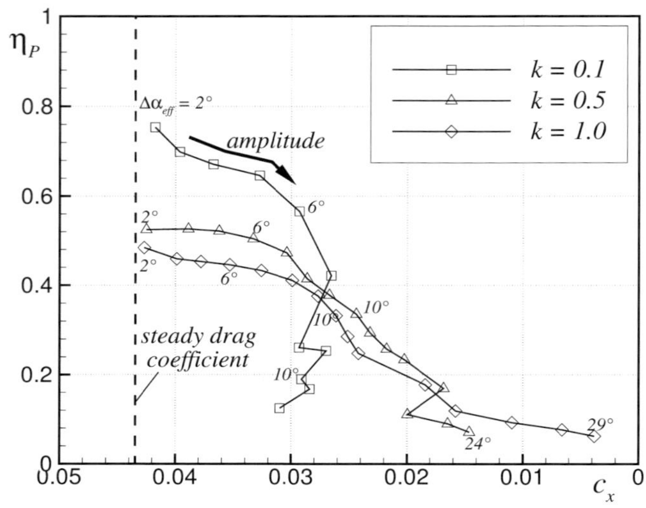

Fig. 21.9. Propulsive efficiency for pure plunge motion at different reduced frequencies k

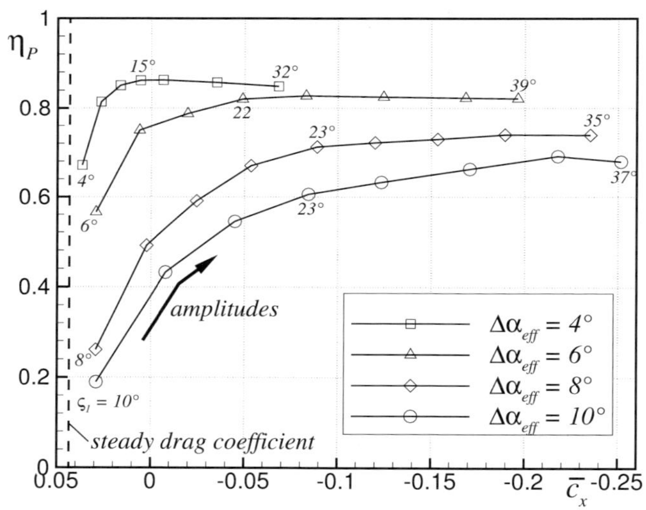

Fig. 21.10. Propulsive efficiency for combined plunge and pitch motions at $k = 0.1$ and constant eff. amplitudes $\triangle\alpha_{\text{eff}}$

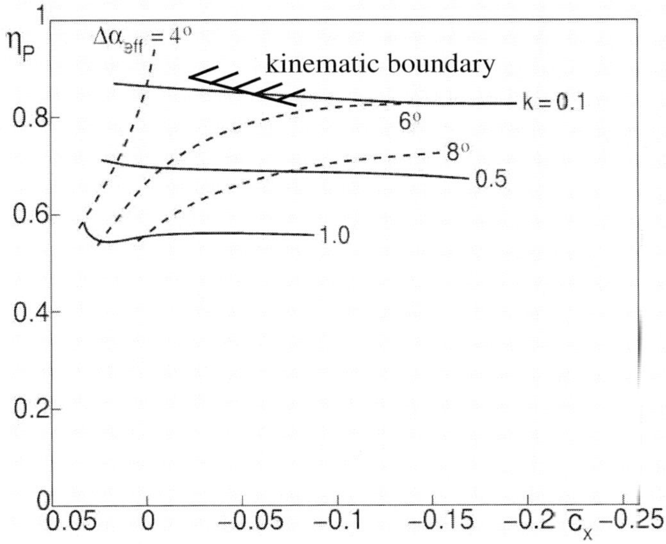

Fig. 21.11. Propulsive efficiency for optimal combinations of plunge and pitch motion at different reduced frequencies k

Notes on Numerical Fluid Mechanics and Multidisciplinary Design

Available Volumes

Volume 87: Ch. Breitsamter, B. Laschka, H.-J. Heinemann, R. Hilbig (eds.): New Results in Numerical and Experimental Fluid Mechanics IV. ISBN 3-540-20258-7

Volume 86: S. Wagner, M. Kloker, U. Rist (eds.): Recent Results in Laminar-Turbulent Transition - Selected numerical and experimental contributions from the DFG priority programme 'Transition' in Germany. ISBN 3-540-40490-2

Volume 85: N. G. Barton, J. Periaux (eds.): Coupling of Fluids, Structures and Waves in Aeronautics - Proceedings of a French-Australian Workshop in Melbourne, Australia 3-6 December 2001. ISBN 3-540-40222-5

Volume 83: L. Davidson, D. Cokljat, J. Fröhlich, M. A. Leschziner, C. Mellen, W. Rodi (eds.): LESFOIL: Large Eddy Simulation of Flow Around a High Lift Airfoil - Results of the Project LESFOIL supported by the European Union 1998 - 2001. ISBN 3-540-00533-1

Volume 82: E. H. Hirschel (ed.): Numerical Flow Simulation III - CNRS-DFG Collaborative Research Programme, Results 2000-2002. ISBN 3-540-44130-1

Volume 81: W. Haase, V. Selmin, B. Winzell (eds.): Progress in Computational Flow-Structure Interaction - Results of the Project UNSI, supported by the European Union 1998-2000. ISBN 3-540-43902-1

Volume 80: E. Stanewsky, J. Délery, J. Fulker, P. de Matteis (eds.): Drag Reduction by Shock and Boundary Layer Control - Results of the Project EUROSHOCK II, supported by the European Union 1996-1999. ISBN 3-540-43317-1

Volume 79: B. Schulte-Werning, R. Grégoire, A. Malfatti, G. Matschke (eds.): TRANSAERO - A European Initiative on Transient Aerodynamics for Railway System Optimisation. ISBN 3-540-43316-3

Volume 77: S. Wagner, U. Rist, H.-J. Heinemann, R. Hilbig (eds.): New Results in Numerical and Experimental Fluid Mechanics III. Contributions to the 12th STAB/DGLR Symposium Stuttgart, Germany 2000. ISBN 3-540-42696-5

Volume 75: E.H. Hirschel (ed.): Numerical Flow Simulation II. CNRS-DFG Collaborative Research Programme, Results 1998-2000. ISBN 3-540-41608-0

Volume 66: E. H. Hirschel (ed.): Numerical Flow Simulation I. CNRS-DFG Collaborative Research Programme. Results 1996-1998. ISBN 3-540-41540-8

Printing: Strauss GmbH, Mörlenbach
Binding: Schäffer, Grünstadt